Statistics and Experimental Design for Behavioral and Biological Researchers

An Introduction

Victor H. Denenberg
Department of Biobehavioral Sciences
The University of Connecticut

HEMISPHERE PUBLISHING
CORPORATION

Washington   London

A HALSTED PRESS BOOK

JOHN WILEY & SONS

New York   London   Sydney   Toronto

Hemisphere Publishing Corporation
1025 Vermont Ave., N.W., Washington, D.C. 20005

Distributed solely by Halsted Press, a Division of John Wiley & Sons, Inc., New York.

1 2 3 4 5 6 7 8 9 0   D O D O   7 8 3 2 1 0 9 8 7 6

Library of Congress Cataloging in Publication Data

Denenberg, Victor H   1925–
    Statistics and experimental design for behavioral
and biological researchers.

    Bibliography: p.
    Includes index.
    1. Psychometrics. 2. Biometry. 3. Laboratory
animals. I. Title
BF39.D44   150′.1′8   76-21841
ISBN 0-470-15202-8

Printed in the United States of America

Contents

vii   *Preface*

Chapter 1

1   **WHAT STATISTICS IS ALL ABOUT**

2   Descriptive and Inferential Statistics
3   Statistics and Microscopes: An Analogy
3   The Approach of this Book
4   Summary of Definitions

Chapter 2

5   **MEASUREMENT: THE CHOICE OF AN ENDPOINT**

5   An Experimental Example: Measuring Maternal Behavior
8   Some Formal Considerations of Measurement
12   Parametric and Nonparametric Statistics and Scales of Measurement
14   The Plan of the Rest of the Book
14   Summary of Definitions

Chapter 3

15   **DESCRIPTIVE STATISTICS: FREQUENCY CURVES, AVERAGES, AND VARIABILITY**

15   The Nature of Our Data
15   Tabular Description: The Frequency Distribution
19   Graphic Description: The Histogram
19   Graphic Description: The Frequency Polygon
21   Types of Frequency Polygons
25   Determining an Average Value
30   A Measure of Variability
36   Computational Formulas for Sums of Squares, Variance, and Standard Deviation
37   Summary of Definitions
37   Problems

Chapter 4

41   **MAKING INFERENCES ABOUT POPULATION PARAMETERS: GENERAL PRINCIPLES**

41   Populations and Samples
47   Population Parameters and Sample Statistics
52   Summary of Definitions

Chapter 5

53   **THE NORMAL CURVE**

56   Table of the Normal Curve
58   Probability and the Normal Curve
59   Application of the Normal Curve: Making Inferences about the
     Population Frequency Distribution
60   Application of the Normal Curve: Making Inferences about the
     Population Mean
66   Testing Hypotheses about the Population Mean
67   Assumptions Underlying the Use of the Normal Curve Tables
68   The Central Limit Theorem
69   Distinction between a Distribution of Raw Scores and a Distribution
     of Means
70   Summary of Definitions
71   Problems

Chapter 6

73   **THE *t* DISTRIBUTION**

74   Degrees of Freedom
76   Characteristics of the *t* Distribution
77   Table of the *t* Distribution
79   When to Use the *t* Distribution
80   Application of the *t* Statistic: Standard Errors and Confidence Limits
82   A General Formula for Estimating Confidence Limits
83   Summary of Definitions
83   Problems

Chapter 7

85   **COMPARING THE MEANS OF AN EXPERIMENTAL
     AND A CONTROL GROUP**

85   Research and Randomization
89   Experimental and Statistical Hypotheses
94   Levels of Significance
97   Testing the Null Hypothesis
101 The *t* Test for a Difference between Means
102 Summarizing the Steps Involved in Testing for a Difference between
     Means
105 The *t* Test for Matched Pairs
109 Interpretation of Standard Errors
110 Determining the Power of an Experiment
120 Summary of Definitions
121 Problems

Chapter 8

125   **THE ANALYSIS OF VARIANCE: SINGLE
       CLASSIFICATION**

126   The t Test, t^2, and Variances
128   The F Test for Two or More Groups
129   The Null Hypothesis
130   Evaluating the F Statistic: The F Table
131   Fixed Effects and Random Effects Models
134   Single-Classification Analysis of Variance with Independent Observations
150   Evaluating Qualitative Hypotheses
159   Evaluating Regression Experiments
178   Power and Sample Size for the Single-Classification Analysis of Variance
       Experiment
180   Missing Values
182   Summary of Definitions
183   Problems

Chapter 9

187   **THE ANALYSIS OF VARIANCE: NESTED DESIGNS,
       RANDOMIZED BLOCKS, AND FACTORIAL EXPERIMENTS**

187   Nested Designs
201   Randomized Blocks Designs
210   Factorial Experiments
218   Summary of Definitions
218   Problems

Chapter 10

223   **DETERMINING THE LINEAR RELATIONSHIP BETWEEN
       VARIABLES: THE CORRELATION COEFFICIENT**

223   Examples of Correlational Problems for Experimentalists
225   A Measure of Linear Relationship: The Correlation Coefficient
225   Measuring the Covariance of X and Y
226   Scatterplots of Correlation Coefficients
228   Another Definition of the Correlation Coefficient
230   The Line of Best Fit and the Principle of Least Squares
235   The Standard Error of Estimate
238   The Correlation Coefficient and Degree of Relationship
239   Determining the Significance of the Correlation Coefficient
240   A Computational Example
240   Uses of Correlation in Experimental Research
245   Correlation and Regression

250   Comparing Correlation Studies and Regression Experiments
251   Final Comment
251   Summary of Definitions
251   Problems

Chapter 11

255   **CHI-SQUARE TESTS OF NOMINAL DATA**

255   The χ^2 Statistic
256   Comparing an Observed Frequency Distribution with a Theoretical
       Frequency Distribution: The Goodness-of-Fit Test
261   Comparing Two Observed Frequency Distributions: Testing for
       Association
266   Unplanned Uses of χ^2
268   Assumptions Underlying the Use of the χ^2 Test
269   The Fisher Exact Probability Test
274   Correlated Proportions in a 2 × 2 Table
276   Measures of Strength of Association for Two Classifications
277   Summary of Definitions
277   Problems

Chapter 12

281   **ORDER STATISTICS FOR RANKED DATA**

283   Comparing Two Independent Groups: The Mann-Whitney Test
286   Comparing Two Matched Groups: The Wilcoxen Test
288   Comparing Three or More Independent Groups: The Kruskal-Wallis
       Single-Classification Analysis of Variance by Ranks
289   Comparing Three or More Matched Groups: The Friedman Test
291   A Measure of Relationship: The Spearmen Rank Correlation Co-
       efficient
293   A Final Comment
293   Problems

297   *Appendix*
325   *Answers to Problems*
335   *Glossary*
341   *References*
342   *Index*

Preface

This book is written for those who do experimental research with animals in the fields of psychology, behavior, or biology. For those of you who are students I am assuming that you have had at least one laboratory course in biology or psychology in which you have used animal preparations in experiments. The only assumption I make about your mathematical background is that you understand algebra.

For a long time I have been convinced that the only way to make statistics and experimental design truly comprehensible to students is to discuss these topics within the broader context of their research experiences. If a student has had research training through a course or through working in someone's research laboratory, I believe it possible to carry out a meaningful dialogue about the place of statistics in research. I view statistics and experimental design as an intimate part of the organizational and analytical processes involved in experimentation, and it is my conviction that students who enjoy doing experimental research will also enjoy statistics and experimental design once they understand the nature of this intimate relationship.

My objection to most introductory books in statistics and experimental design is that these books are written too broadly. The attempt is to appeal to students from diverse disciplines within psychology or biology, and, as a consequence, much material is included that is never used by those in various specialties. I consider this to be an inefficient way to teach students. As an example, virtually all introductory textbooks in psychological statistics have a section or a chapter on percentiles. Yet in my more than 20 years of research experience as an experimentalist I have never used any percentile statistics. Therefore, you will not find this topic discussed in this book. There are a number of other statistical concepts which I have also excluded because they are not relevant to experimenters.

What I have done in preparing this book is think back over my two decades of research experience and look over my old notes on statistics and design with the objective of presenting only those statistics that I know will be used by experimental researchers in biology and behavior. For this reason the book is clearly not an attempt to give a broad coverage of the field. However, I assure you that the topics I have covered are ones that you will have to know if you plan to carry out experimental research using animal preparations in one of the disciplines within the life sciences.

ON CHOOSING A POCKET CALCULATOR

A very wise investment for a student taking a course in statistics is to purchase a small pocket calculator. Because different calculators have different functions, it is necessary to know what features a calculator must have to be useful in statistical work. I have listed below, in rank order, the features to look for when shopping for a calculator.

1. The most frequent operation in statistics is to take a series of observations or numbers (which we shall call X) and do two things with those numbers: (1) Add them together to get the total; and (2) square each number and then add the squares together. That is, we are to get the summation of X, and the summation of X^2. The *minimum* requirement for a pocket calculator is that the machine have sufficient memory to allow you to store and accumulate X and X^2 *at the same time*. In some machines you can add the X values to get the total, and then you have to start all over and square each X value and add the squares. This doubles the amount of time needed to get the most basic information and more than doubles the chances of making errors. If at all possible, do not purchase such a machine.
2. Once you have obtained the summation of X and X^2, the most common use you will make of those numbers is to compute the mean and standard deviation of your data. Some calculators have keys that will compute these values for you directly. This is exceedingly useful.
3. Another useful feature that may be found on more expensive machines is to be able to compute a linear regression equation of the form

$$Y = a + bX$$

4. A fourth statistical operation that can be found on some calculators is to be able to compute the Pearson product-moment correlation coefficient.
5. Fincally, a rather esoteric feature, but one available on several types of pocket calculators, is a key that will compute the factorial value of a number (these keys are usually symbolized as $x!$)

As you would expect, the price of the calculator goes up with the number of features desired. You can probably purchase a machine that will satisfy the first specification listed above for around $10 to $20. The machine I have performs all the functions listed except item 4, and that machine is now selling for around $80. However, the prices of calculators keep dropping (e.g., my machine cost $225 when first introduced), and good statistical calculators should be available at very reasonable prices.

CHAPTER 1 What Statistics Is All About

Anyone who enjoys doing experimental research should enjoy statistics, for statistics is concerned with the logical basis on which an experimenter can draw reasonable conclusions about experimental data. In other words, statistics deals with the logic of experimental inference. It just so happens that numbers and mathematical formulas are involved in this logical process as intermediary steps. Many students become discouraged by the numbers and the formulas and think that these things are the important elements of statistics. Putting it simply, the student cannot see the forest (the logic of the experimental inference) for the trees (the numbers and mathematical formulas).

A further factor that students find discouraging is the considerable amount of hack work and drudgery involved in performing statistical operations. There are columns of numbers to add, cumbersome formulas into which one must plug various figures, and always the constant necessity of checking and rechecking in order to make sure that there have been no errors. These factors interfere with the true educational objective, which is to impart to the student an understanding of the role of statistics as an integral part of the intellectual process by which an experimenter attempts to draw proper conclusions from and about the research data.

Fortunately, the advent of the computer has eliminated much of the drudgery of mathematical manipulations involved in statistics, as well as the uncertainty about having made a computational error. These days it is expected that a student has, as a minimum, access to an electric desk calculator which allows him or her to perform a variety of statistical operations with ease. Thus, it is no longer necessary to present various procedures for the manual computation of statistical values, and none of the formulas necessary to such calculations will be presented or discussed in this book.

The purpose of this book is to introduce the use of statistics in the context of experimental research. Thus this is an applied statistics book rather than a theoretical one. However, the emphasis will not be on cookbook procedures, but on the logic of statistics in experimentation. Formulas will be presented, but these will be the minimum necessary to comprehend the logic involved as well as the necessary computational procedures. Two kinds of formulas will be given. The first will be *operational* (*definitional*) formulas. These are the formulas that statisticians and researchers have tacitly or explicitly agreed to use to define a particular term (in the same way as there is a particular dictionary definition of any word). At times the definitional formula can also be used as a computational formula. However, this is generally not true. Therefore, when this is not so, the *computational* formula will also be given. The computational formula is always algebraically identical with the definitional one, but it will be in a form that is convenient to use with an electric calculator or a computer.

DESCRIPTIVE AND INFERENTIAL STATISTICS

The popular definition of a statistic is a number or set of numbers that describes some measured characteristic of an individual or of a group. For example, in the Gallup Poll approximately 1000 people may be asked about their voting preference for the next Presidential election, and the statistics reported in the newspapers are the percent of those interviewed who state that they would vote for the Republican candidate, the Democratic candidate, a third or fourth party candidate, and those who are undecided. The summation of all these percentages would total 100%. Other examples included the average *per capita* income in the United States, the mean number of trials rats take to learn a maze, the amount of time for an electrical impulse to travel a certain distance down a nerve, and so on. Whenever the number or numbers we report (which is usually an average value, although this need not always be the case) refer only to the individual or group on whom the measurement has been taken, we are in the realm of *descriptive statistics*. Literally, all we are doing is describing how this particular group of people, or these rats, or this particular segment of nerve, responded at the time we made our measurements.

However, this is only a preliminary step to our real interest. Very few researchers are interested in the 1000 people who were interviewed, the rats that were tested, or the particular nerve segment under examination, *except insofar as these tell us something about the larger group of which they are a part.* The response of the 1000 people interviewed should give us an insight into how the whole country will vote, the rats' maze performance should help us in predicting how other rats will perform in the maze, and the speed of transmission of an impulse down one nerve should give us an idea about the speed of transmission of other impulses down other nerves. In this context the subjects or elements that we actually measure are called a *sample*, whereas our real interest is with respect to the larger group of which this sample is a part. This larger group is called a *population*. When we collect statistical data obtained from a sample and attempt to draw conclusions as to how well these data characterize the population, we are in the realm of *inferential statistics*. The question of how we may draw logical inferences about a population from a sample is of major concern to us. Indeed, it is the central problem facing all experimental researchers, and one with which most of this book will be concerned.

STATISTICS AND MICROSCOPES: AN ANALOGY

Though worlds apart with respect to physical features, there are some interesting parallels between statistics as used by an experimentalist and the microscope as used by an experimentalist. The function of the microscope, of course, is to enable the researcher to see certain minute but critical details that are not apparent to the naked eye. Likewise, given a massive amount of data (e.g., information on 100 or 1000 cases), it is impossible for the experimenter to scan this information, integrate it, and come up with a conclusion concerning its meaning. Here is where statistics helps.

Now, it is not necessary for a good experimentalist to understand the theories of optics and light in order to use the microscope. One might find that an understanding of these theories is an interesting intellectual exercise and enjoy learning about for its own intrinsic value, but one certainly does not need this information to examine a slide. What the experimenter does need to know, of course, are some of the mechanics of a microscope so that he or she can properly adjust it within the limitations of the instrument (e.g., you don't use a student laboratory microscope if your objective is to study intracellular space). Similarly, it is not necessary for the experimenter to have an understanding of the mathematical theory underlying statistics. If he or she wishes to learn this, all well and good, but it is not a necessity in order to use statistics in experimental research. As with the microscope, the experimenter has to learn about mechanical operations and about limitations. The mechanical operations of statistics refer to the formulas that can make certain features of the data stand out clearly. The limitations refer to the conditions under which it is appropriate to use a particular formula as well as the breadth of generalizations and conclusions. I have yet to hear of a student who has been discouraged from experimental research because it was necessary to learn how to use a microscope. If you approach the study of statistics in this same frame of mind, you will find it easy to learn.

THE APPROACH OF THIS BOOK

The approach we shall follow in developing statistical concepts and procedures will be by means of illustrations drawn from experimental research. A research problem will be presented, together with a brief discussion of how we could attack the problem experimentally. The statistical principles and procedures will follow in a logical sequence as we go about designing an experiment to answer the question we posed. As we develop more statistical principles and procedures, our experimental problems will be logically expanded, thus allowing us to incorporate additional statistical principles.

So those of you who think experimental research is fun and exciting, come along and find out how a basic understanding of statistical procedures can make it even more exciting.

SUMMARY OF DEFINITIONS

Because the definitions are so important in understanding the concepts underlying statistics, all new terms used in a chapter will be summarized at the end of the chapter. In addition, all the terms from all chapters are listed in the Appendix.

Descriptive statistics The use of numbers (statistics) to describe, characterize, and summarize data obtained from the sample in one's study.

Inferential statistics The logical procedures whereby one makes inferences about a larger group (the population) from information obtained from a sample.

Population A hypothetical unit of persons, organisms, or elements, quite large in size (often infinite) which we, as researchers, wish to characterize and draw inferences about.

CHAPTER 2 Measurement: The Choice of an Endpoint

The usual way that research starts is with an individual interested in a particular phenomenon or process who wants to find out more about it. From this rather vague and general beginning we proceed by a number of steps—and a considerable number of assumptions—to a set of experimental procedures that, presumably, provide a better understanding of the phenomenon in which we are interested.

The first problem to face in going from a vague, general idea to specific experimental operations is that of defining and measuring the phenomenon. This involves the choice of an endpoint or a dependent variable. It is not necessary to restrict ourselves to a single endpoint or criterion measure, and often we measure several variables at the same time or in close temporal proximity to each other. Regardless of the number of endpoints measured, the same logical considerations apply. For convenience, during this section of the text we shall consider that the experimenter is measuring a single variable.

AN EXPERIMENTAL EXAMPLE: MEASURING MATERNAL BEHAVIOR

To a considerable extent, the endpoint we choose to measure defines the statistical procedures that will be used to evaluate our data. This can be demonstrated by the following example. Assume that we are interested in studying hormonal factors influencing maternal behavior. The species with which we choose to work is the rat. How do we measure "maternal behavior"? Research has shown that maternal behavior is not a single unitary dimension, but, instead, involves a variety of different behavioral activities, several of which are more or less independent of each other. For example, when pregnant the rat builds a nest. When the young are born they are nursed, retrieved, cleaned, and cared for by the mother. Which of these behaviors should we choose to measure as "maternal behavior"? A reasonable approach is to assume that no one of these can be used to measure the whole complex of maternal behavior. This assumption applies unless we establish that these various activities are so highly correlated that it is possible to predict all the other behaviors from the measurement of any single one of them. This is not the case for maternal behavior.

(The matter of correlation, which is concerned with determining the degree of relationship between two variables, will be taken up in Chapter 10.)

In actual practice, the selection of the criterion measure depends on a number of factors, including statistical considerations, convenience of obtaining measurement data, and an evaluation of the meaningfulness of the measurement with respect to the phenomenon being studied. In this instance, the choice of the endpoint for maternal behavior should have some meaningful relationship to the biological phenomenon of maternal care.

Measuring Survival

For example, we could argue that the function of maternal behavior is to raise the young pups to a point where they are self-sufficient. With such a rationale, the obvious endpoint to use is the success of the mother in keeping the young alive until weaning. Therefore the simple measure of the percent of young surviving from birth until weaning would be an appropriate statistic. But maternal care actually implies more than just keeping the pups alive until weaning. In an evolutionary sense it is important that the young stay alive long enough to reproduce and to maintain their own young until weaning. So it may be necessary to keep the offspring under observation until they are sexually mature and have reared one litter successfully.

Even though it appears reasonable to use as an endpoint the percent of young reproducing successfully, upon further consideration it may not be appropriate. Remember that the initial question we posed was to investigate hormonal determinants of maternal behavior. We may ask whether hormonal manipulations are going to have a major effect on survival of the young and their later reproductive efficiency. If so, this may be because the hormonal manipulations interfere with the nursing of the young, the care and cleaning of the young, the type of nest that the female builds, the mother's behavior toward her young, the biochemistry of the mother's milk, and so on. If all that we measure is the percent that survive and have one successful litter, we would not know, given that a certain percent did die or could not reproduce, which of the conditions listed above (or some other condition) was the causative agent.

Measuring Nursing

Thus we might start by considering another criterion. The phenomenon of nursing is common to all mammals, and it is obviously relevant to the survival of the young. But how do we measure nursing behavior? Observations of the mother–young interaction reveal that some or all of the young are nursing at a particular time, but this says nothing about the amount of milk these young are receiving. Thus, there are at least three major ways by which we could "score" nursing behavior. The first is in an all-or-none fashion, in which we would record for any particular observation whether or not there is any evidence of nursing. A score of 1 would be given if one or more of the pups were suckling the mother; a score of 0 would be given if no pups were suckling. A second approach would be to determine the percent of pups nursing at any particular time. The third approach, and one that has a number of experimental as well as statistical advantages, is to find a way to quantify the amount of nursing that each of the young receives. Because the consequence of nursing is to cause the infant to grow, an indirect measure of the mother's nursing capability is to obtain the body weight of the young and determine the increment of growth each day. Thus, we could weigh the young at the same time each day and determine the weight increase over the previous day. That score (say, to the nearest 10th of a gram) would be used as an index of the mother's nursing capability.

We have seen through the above discussion that there are two general forms of measurement. One is an *all-or-none or percentage recording,* and the other is a *quantitative numerical score.* The first is reflected by the occurrence or nonoccurrence of nursing or the percent of animals that nurse, whereas the other is reflected by the weight of the pups.

Measuring Nest Building

There is a third form of measurement, which may be readily illustrated by considering another endpoint in the measurement of maternal behavior, namely, nest building. The rat builds a crude nest within which she places her young. The easiest thing to record here is the presence or absence of such a nest. Because virtually all parturient rats build a nest, it is unlikely that there would be any major variation among animals on this measure. And it is necessary to have variation in order to obtain appropriate measurements. How could we measure the variation in nest building? Suppose that we have a dozen animals, all of whom gave birth on the same day. We could inspect their nests, decide which one was the best, the next best, next best, and so on. We could, in other words *rank* the nests from the best to the worst. The "ranked scores" would range from 1 to 12, and we would then be able to relate these ranked scores to other characteristics of the experimental procedure.

There are difficulties with such a score and one recognizes, almost intuitively, the problems involved, including the fact that the score of a particular animal is dependent on the decision we make about the other animals. That is, an animal's place in the rank order is determined by the nature of the other animals in that group. Change the characteristics of the other animals, and you are likely to change the rank order of this particular animal. One way to avoid this problem is to find a method to measure directly a quantitative characteristic of maternal nest building that is independent of the behavior of the other animals. This makes for a better type of measurement system. One way to quantify nest building is to weigh the material used in the nest each day. The assumption here is that the greater the weight of the material, the better is the quality of the nest. Although this makes for a good quantitative score, it has one major drawback:

The measurement procedure itself—removing the nest each day and weighing it—*interferes with the phenomenon under investigation.* This interference might destroy the usefulness of the measurement technique. This, however, is not a statistical question but an experimental one. It requires that we as the experimenters have enough knowledge about the phenomenon being investigated to decide whether we can afford to interfere in such a manner without upsetting the phenomenon.

SOME FORMAL CONSIDERATIONS OF MEASUREMENT

Now that we have looked at some of the problems of measurement within the context of an experimenter trying to make decisions about a phenomenon under investigation, we can examine measurement in a more formal and abstract fashion. In general there are four different measurement scales available to a researcher. These are called *nominal, ordinal, interval,* and *ratio.* These scales differ in the assumptions underlying their use, the amount of information yielded, and the uses to which the numerical data can be put.

Nominal Scale

The nominal scale is the simplest of the scales and, as the name implies, really indicates the "naming" or categorization of an object. Objects are classified into mutually exclusive and totally inclusive classes. Thus, this type of scale is formally the same as the taxonomic scale used in biology to classify living organisms. The minimum number of scales is two and may extend to as large a number as we wish, as long as they remain mutually exclusive and totally inclusive. This is a "qualitative" type of classification, in that there is no implication of one class being "better" or "poorer" than another class. As experimenters we may, if we wish, interpret our data in a framework of better or worse, but this is an interpretation that we make as researchers who have intimate knowledge of our subject matter, and not because of any statistical or measurement considerations.

An example of a nominal scale was provided in the preceding discussion where an all-or-none score was given on the basis of classifying rats according to whether they did or did not build a maternal nest. Another example is coat color of a mouse or a guinea pig. Coat color has been commonly used in genetic research, not because of any intrinsic interest in coat color itself, but because it has been an easily observed and convenient marker by which one could understand gene action. The geneticist examining coat colors does not make any value judgment with respect to whether one color is better than another.

Because the data are scored with respect to the occurrence or nonoccurrence of an event, the only type of measurement obtained is a binary 1 or 0 for the individual, which, when summed over a number of individuals, allows calculation of a percentage figure. In terms of formal mathematics, the major characteristics of the nominal scale are symbolized by the expressions "equal to" and "not equal to." Their symbols are, respectively, $=$ and $\neq$. The statistics commonly used to evaluate these data include the chi-square test and the binomial distribution. These will be discussed in Chapter 11.

Ordinal Scale

The nominal scale simply classifies without any ordering or arranging of data along a presumed dimension (again, we are speaking of measurement in a mathematical sense, not with how experimenters may wish to interpret their data). The next logical step after categorizing things into separate classes is to arrange these classes such that one may be though of as having "more than" the other, being "better than" another, or being "stronger than" another. This assumes that there is a common property present among the various classes and that it is possible to make judgments or to measure this common property at least to the extent that the classes can be *ordered* with respect to that property. It is at the level of the ordinal scale that we first have measurement in the sense of quantification, because it is here that we first find the assumption of a common factor present among all classes, with the factor varying in degree. Think back to our example of maternal behavior. When we ranked our mothers as to who built the best nest down to the one who built the poorest nest, we were employing ordinal measurement with the assumption that there was a common factor or attribute called "nest building" which varied in degree, and that by careful visual examination we would be able to order these nest appropriately. It must be emphasized that ordering objects from greatest to poorest says nothing about the amount of the attribute that is possessed. The *distance* between the first rank and the second rank may, or may not, be equal to the distance between the second rank and the third rank. All that we assume with the ordinal scale is that it is possible to state that one object has more of a particular property than another object, but in no way does this allow us to state *how much* more or how much less of this attribute is present. In terms of formal mathematics, the major characteristics of the ordinal scale are symbolized by the expressions "greater than" and "less than." Their symbols are, respectively, $>$ and $<$.

An example of an ordinal scale in the area of animal behavior is the pecking order of birds. When observing a flock of birds one notices that there is one animal that is dominant over all the others, the *alpha* animal. There is then, in succession, a *beta* animal, a *gamma* animal, all the way down to the one that is the lowest in the pecking order.

The major difficulty with the ordinal scale of measurement is that we are limited to mathematical operations involving "greater than" and "less than" characteristics. This means that the operations of addition and subtraction cannot be employed with such scales, and we are restricted to statements concerning the inequalities. This leads to a discussion of the interval scale.

Interval Scale

The nominal scale allows us to place our phenomenon into any of several different categories or classes that are considered to be qualitatively different. When we progress to the ordinal scale we are able to arrange those categories along an assumed continuum or dimension with an inequality assumption so that one is ranked greater than another, a third ranked greater than a second, and so on. Until now we have not used any numbers in our measurements. This occurs for the first time with the interval scale. The importance of being able to assign numbers to our observations is that we can now specify quantitatively the distance between our observations. Consider our hypothetical experiment in which we weighed the nesting material of our experimental animals and assume that we found that one nest weighs 15 grams and the second nest weighs 13 grams. We can not state that the first nest weights 2 grams more than the second nest. Suppose that a third animal built a nest that weighs only 7 grams. This animal's nest score is 6 grams less than the second animal, and we can conclude that the first two animals are rather similar in their nest building and are quite different from the third animal. It is this type of statement, expressed in more quantitative form, that is permissible with the interval scale. The reason for calling this an "interval" scale is that the interval or distance between observations can be measured quantitatively. Given the interval scale, the mathematical properties of addition, subtraction, multiplication, and division are permissible for the purposes of obtaining measurement data. This is the scale that will suffice for virtually all biological and behavioral research and is the one toward which a researcher should strive when designing experiments. With this scale all the powerful statistical procedures, including t test, analysis of variance, and correlational procedures, are permissible.

Underlying the interval scale is the assumption of "additivity." This assumption is that the distances between observed points may be added together to give the total distance. In terms of formal mathematics, the major characteristics of the interval scale are symbolized by the expressions "addition to" and "subtraction from." Their symbols are, respectively, $+$, and $-$.

Consider a simple example from the physical world. If there are three posts in a line, we can take the distance from post 1 to post 2 and the distance from post 2 to post 3, add these together and get the total distance from post 1 to post 3. In a physical measurement problem this can be verified empirically to be sure that the assumption of additivity holds. The same logic applies to a behavioral or biological interval measurement scale. However, the additivity assumption cannot be tested directly as was done with the post example. We shall have more to say on this matter at the end of the chapter.

As far as statistics go, the three scales of nominal, ordinal, and interval are all that are relevant to us now. However, there is a fourth scale that comes into play in the area of measurement, which is not the same as the area of statistics. Measurement is concerned with the logic and the procedures involved in assigning a score to each element of the subject matter under study. Statistics, on the other hand, assumes that the measurement score is valid and is concerned with assumptions about the nature of the population (e.g., the assumption that the population has a certain form of distribution), and the logic of drawing inferences about the population from one's sample data. Once we have the additivity assumption in the interval scale, we have all that is needed to go ahead and use all the usual formulas that are common to statistical procedures. But there is a type of measurement statement that cannot be made from the interval scale alone and that requires the use of a *ratio scale*, which has an additional assumption built into it.

Ratio Scale

The interval scale tells the distance from one observation to another. However, it is not possible from interval scale data to make the following type of statements: Rat A has built a nest that is four times as good as the nest built by Rat B: the experimental mouse took twice as long to learn the problem as the control mouse; after having treated the animal with drug X, its aggressive behavior was reduced by 75%. Such statements are *ratio* statements and involve comparing one score to another in the manner of a ratio (i.e., dividing one score by another.) In order to do this type of division, it is necessary that an *absolute zero* be present.

A common example will help illustrate this. Consider two objects, one of which has a temperature of 40°C, and the other with a temperature of 10°C. We know that the first is warmer than the second (ordinal scale measurement), and we also know that the first is 30° warmer than the second (interval scale measurement). However, it is *not* correct to state that the first is four times as warm as the second, because the centigrade scale has an *arbitrary* zero set at the point where water freezes. Although this is a reasonable place to put a zero point (certainly far more reasonable than the zero of the Fahrenheit scale), this zero has nothing to do with the intrinsic nature of temperature itself. When we shift to the Kelvin scale, where an *absolute* zero (i.e., the lack of molecular motion) is present, and convert from centigrade to Kelvin by adding 273°C, we can see that 40°C is actually 1.106 times as warm as 10°C $[(273 + 40/(273 + 10)]$.

It is very uncommon to have an absolute zero in psychological measurement, although some examples exist, primarily in the areas of psychophysical research. The difficulty, of course, is that of trying to establish an absolute zero when one works with psychological data. Consider, for example, the intelligence test. The IQ is assumed to have ratio scale properties, but this is more of an assumption than a demonstrated fact. There are various test items that are given to children when measuring their intelligence. Suppose that a child fails to pass any item. Does this mean that he has zero intelligence? Of course not, because we can always construct an item to be put into the test which the child could pass. Until we have developed an adequate theory of intelligence that allows us to escape from this bind, we shall not be able to talk in terms of an absolute zero. Until then, IQ scores will have only the properties of an interval scale.

In actual practice, researchers treat their data as though they had ratio scale properties—that is, as though there were an absolute zero present. Again, this way of treating the data has nothing to do with its statistical or measurement properties, but with the interpretation one makes of data in terms of ratio statements. Even though we know that such statements are inaccurate because we do not have, in general, an absolute zero, to a crude first approximation this is a useful way of describing the findings.

Table 2.1 summarizes the characteristics of the four measurement scales.

PARAMETRIC AND NONPARAMETRIC STATISTICS AND SCALES OF MEASUREMENT

Certain statistical procedures require that assumptions be made about the nature of the population to which one wishes to generalize (e.g., a common assumption is that the population has a normal distribution). Any statistic that makes one or more assumptions about characteristics of the population is called a *parametric statistic.* Included in this group are such commonly used statistical procedures as the *t* test, the analysis of variance, and correlations. Statistical procedures that do not make assumptions about the nature of the population distribution are called distribution-free or *nonparametric statistics,* and include the various procedures to analyze rank-order and nominal data.

TABLE 2.1 Summary of the characteristics of the four scales of measurement

Scale of measurement	Its logical properties	Its mathematical properties	Some examples	Usual statistics
Nominal	Classifies into mutually exclusive and totally inclusive groups	$=, \neq$	Coat color, built nest or did not build nest	chi-square, binomial distribution
Ordinal	Ranks in order of amount of attribute possessed	All above plus $>, <$	Pecking order of animals, hardness of rock	rank-order statistics
Interval	Assigns numbers to events so that the distance (interval) between events is additive	All above plus $+, -$	Temperature (C or F), IQ scores, response latency	*t* test, analysis of variance, correlation
Ratio	Allows one number to be divided into another number to determine difference in magnitude	All above plus absolute zero	Temperature (Kelvin) body weight, number of pups born	*t* test, analysis of variance, correlation

In addition to this distinction, some statistical authorities also claim that parametric statistics can be used only when the data have been measured on an interval or ratio scale. If the assumptions underlying the interval scale are not fulfilled, these people claim that it is necessary to use nonparametric statistics to evaluate the data. The rationale behind this statement is that the assumption of additivity (the key assumption of the interval scale) is necessary in order to talk about population distributions. Not every statistician agress that is is necessary to have an interval scale of measurement in order to use parametric statistics, and this matter is still being argued in the statistical literature.

However, the experimental researcher cannot wait for these and other theoretical issues to be settled. In a workaday world the researcher has questions to answer and data to process, and the usual approach is to treat the data as though all the assumptions needed to justify the statistical and measurement procedures have been met. This is a dangerous procedure, because assumptions in statistics and measurement are not to be taken lightly—and certainly not ignored. Thus, this raises the serious problem of how to evaluate the assumptions underlying one's statistical and measurement procedures. Fortunately, there is a solution available to the experimental researcher: namely, repeat the experiment a second time and see if the same results can be obtained as in the first study. If so, this is sufficient evidence to conclude that all the assumptions underlying the procedures have been adequately met.

Rationale for Repeating an Experiment

It is worth spending a few moments to present the rationale for the conclusion given above. There are two bases for this conclusion, one theoretical, the other pragmatic. First, the theoretical rationale. The numbers we use when we measure something along an interval scale are supposed to represent real events in the world of biology or behavior. If the measurement scale has a one-to-one relationship to our endpoint, then our measurements are taken with almost no error, and a replication of the experiment would be expected to yield findings similar to the original experiment. Now consider the other extreme, where there is no relationship between the scale of measurement and our topic of experimental interest. In such a situation the numbers we obtain from our study would be distributed essentially at random. Therefore when we repeated the experiment, it is very unlikely that the same pattern of findings would be found, because we are working with chance variability. In actuality, our scale of measurement will fall somewhere between these two extremes, and if our independent replication gives the same results as we obtained the first time, we infer from this finding that the assumptions involved must have been sufficiently satisfied, or else we would not have been able to repeat the experiment successfully.

The pragmatic rationale can be stated more succinctly: The essence of experimental research is to be able to reproduce one's findings in an independent replication. If a researcher has a scale that he assumes to be an interval scale, if he carries out an experiment and obtains significant findings using a particular statistical procedure, and if, upon an independent replication, he repeats his experimental findings, this is sufficient evidence to establish the validity of the statistical and measurement assumptions.

THE PLAN OF THE REST OF THE BOOK

The great majority of research involving behavioral and biological experiments use a measurement scale based on interval assumptions. That is, we somehow devise a system of measurement such that we assign numbers to our observations, whether these be number of trials it takes for an animal to learn a maze, the amount of activity in a 24-hour period in a circular wheel, the time it takes an animal to retrieve a pup, the amount of corticosterone found in the blood after exposing an animal to a novel stimulus, the amount of acetylcholine esterase found in the brain after early environmental enrichment, and so on. Because this is the most common measurement scale used in biological and behavioral research, we shall discuss first descriptive and inferential statistics in terms of an interval scale of measurement. Chapter 3 will cover frequency distributions, measures of central tendency such as the mean and the median, and the measure of variability called the standard deviation. This will be followed by discussions on the normal curve, the t test, the analysis of variance, and correlational procedures in Chapters 4 through 10.

Chapter 11 will be concerned with the methods of analysis of nominal data—data based on classification schemes alone. Finally, Chapter 12 will examine some of the nonparametric procedures used with ordinal or rank data.

SUMMARY OF DEFINITIONS

Interval scale   A measurement scale in which numerical values that can be added and subtracted are assigned to the subject matter.

Measurement   A set of procedures or operations whereby a "score" (which can be quantitative or qualitative) is assigned to each element under study.

Nominal scale   A measurement scale that classifies the subject matter into two or more mutually exclusive and totally inclusive categories.

Nonparametric statistics   Any statistical procedure based on nominal or ordinal measurement data and that does not make any assumptions about the nature of the population distribution.

Ordinal scale   A measurement scale in which the subject matter is ranked in order on some dimension or attribute of interest to the researcher.

Parametric statistics   Any statistical procedure that assumes the numerical data have the property of additivity (i.e., are on an interval scale) and that also makes one or more assumptions about the nature of the population to which inferences are to be drawn (e.g., that the population is normally distributed.)

Ratio scale   A measurement scale that contains all the properties of an interval scale and also has an absolute zero.

CHAPTER 3

Descriptive Statistics: Frequency Curves, Averages, and Variability

THE NATURE OF OUR DATA

Usually there are several possible endpoints that may be used in an experiment. If we are in the midst of trying to develop these endpoints, there are, as indicated earlier, a number of criteria that we must apply in deciding whether or not a particular dependent variable will be useful. On the other hand, we may have decided already on a specific endpoint because it has been widely used by other researchers who have published in the literature. In either case, it is still necessary to determine the nature of the statistical distribution of the data.

For example, suppose that we are interested in studying avoidance learning of rats in a shuttlebox situation using electric shock reinforcement. We have purchased a commercial shuttlebox with the appropriate programming and control system, so our apparatus is the same as that of some of the other researchers in the field. However, our animals certainly differ to some extent from other animals around the world which have been tested in shuttleboxes, and our housing conditions are also different. Therefore, we must run a sample of our animals in the apparatus to get some idea of the nature of the distribution of learning scores.

We shall be concerned in this chapter with the methods for *describing* the data we may get from such an investigation. From this kind of study a researcher develops a feel for the behavior of the animals and their capabilities in the particular test apparatus being used. The researcher is then better able to appreciate the effects of an experimental intervention on this behavioral process. We shall first be concerned with a simple tabular method of portraying the data. Then we shall examine graphic methods of doing the same. After that, we shall discuss methods of determining an average score for all the animals, and we shall finish the chapter by discussing how to determine the amount of variability among the animals.

TABULAR DESCRIPTION: THE FREQUENCY DISTRIBUTION

Consider the shuttlebox example. Assume that we have randomly (this is a very important concept to which we shall return later in the text) selected 50 male rats from the colony and have tested them in an automated shuttlebox situation. Each animal is run in one session consisting of 90 trials. Very briefly, the situation is as follows: The apparatus is an enclosed box approximately 24 inches long by 10 inches wide, the floor of which is a grid that can be electrified. A signal light goes on 5 seconds before an electric shock occurs on the grid floor. If the animal does not move to the opposite side of the box during that 5-second interval, the electric shock is automatically turned on and remains on until the animal moves across to the other side. As soon as the animal escapes to the other side, both the light and the shock are terminated. After an intertrial interval of 60 seconds, the light again goes on and the animal is again required to shuttle over to the opposite side of the box to avoid shock. As before, it has 5 seconds in which to do this. After a number of such pairings of light followed by shock, many rats will learn to associate the light with the shock, and they will move over to the other side during that 5-second interval between the onset of the light and the occurrence of the shock. If this happens, the shock does not occur. Thus, the animal has learned to avoid the shock by making the appropriate response in time.

A common measure obtained in this apparatus is the total number of avoidances that the animal makes over a standard number of trials. In this example we test each animal for 90 trials in one continuous session, and the score obtained is the total number of avoidance responses out of the possible 90 responses. Hypothetical scores for the 50 animals are given in Table 3.1.

Our first task is to develop a set of procedures that will allow us succinctly to summarize and describe these data. It is quite obvious that listing the 50 scores as they are obtained is an inefficient way of doing this. The first procedure involves a tabular presentation of the data and is a simple way of providing an overview of all of the information of the experiment. Table 3.2 is such a presentation. In the left-hand column we list the scores obtained, starting with the lowest score at the top of the page and going down in order until we get to the highest score. Next we go through our data, putting down an X in the Tally column beside a score each time it occurs, and then summarize the tallies in the Frequency column. When this is completed we have what is formally called a *frequency distribution.* The orderly arrangement in Table 3.2 certainly gives us a better feel for our data than the haphazard tabulation in Table 3.1. We can see that the scores cluster around the middle of the distribution with a tapering off at both extremes. However, the presentation in Table 3.2 is still unsatisfactory because the occurrence of zero frequencies at some score points (e.g., 23, 24, 27, 33) and the relatively low number of cases even for the most frequent score (which is 44) do not allow us to form a good idea of the distribution. This is the usual case with research data, and this difficulty is resolved by increasing the size of the score interval, thereby reducing the number of different "scores" and increasing the frequency within each interval. Table 3.3 shows such a rearrangement of the data from Table 3.2.

TABLE 3.1 Number of avoidance
responses in shuttlebox apparatus,
with scores listed in order obtained

44	37
35	34
32	31
48	42
41	35
48	46
45	41
38	49
50	41
56	22
38	54
46	42
54	37
25	28
32	25
44	45
42	34
31	40
44	47
26	50
35	29
53	29
32	40
44	30
51	59

A few comments are in order about how we obtained Table 3.3. First, a good rule of thumb is that there be approximately 12 to 15 intervals for the data. The question then becomes: How do we choose the size of the interval? This is found by taking the range of raw scores (the highest score minus the lowest score, plus one) and dividing that range by either the number 12 or the number 15, the number of desired intervals. We round the answer to the nearest whole number and use that as the size of the class interval. For this particular example the lowest score is 22 and the highest is 59. This encompasses a range of 38 raw score points. When 38 is divided by 12 we get $3\frac{1}{6}$ as an answer; when divided by 15, the result is $2\frac{8}{15}$. In both cases the nearest whole number is 3, and this becomes the size of the class interval. We then prepare a Score Interval column like the left-hand column in Table 3.3 and tally the data again as shown in Table 3.3.

Now compare Table 3.2 and Table 3.3. These are the same data, but the arrangement in Table 3.3 makes a much more coherent picture that the arrangement in Table 3.2. We can get a lot of good descriptive information about the data just by examining Table 3.3: The distribution has one major peak

TABLE 3.2 Frequency distribution
of number of avoidance responses
in the shuttlebox for the 50 rats

Raw score	Tally	Frequency
22	X	1
23		0
24		0
25	XX	2
26	X	1
27		0
28	X	1
29	XX	2
30	X	1
31	XX	2
32	XXX	3
33		0
34	XX	2
35	XXX	3
36		0
37	XX	2
38	XX	2
39		0
40	XX	2
41	XXX	3
42	XXX	3
43		0
44	XXXX	4
45	XX	2
46	XX	2
47	X	1
48	XX	2
49	X	1
50	XX	2
51	X	1
52		0
53	X	1
54	XX	2
55		0
56	X	1
57		0
58		0
59	X	1
Total	50	50

(technically it is "unimodal") at the score interval 40–42, and the scores drop off in a fairly symmetrical fashion on either side of the peak.

If we wished, we could leave the data in the form of Table 3.3. However, at times it is convenient to portray the data in graphic form. There are two standard ways of doing this: the histogram and the frequency polygon.

TABLE 3.3 Frequency distribution of the
data in Table 3.2 grouped into class intervals
of three units

Score interval	Tally	Frequency
22–24	X	1
25–27	XXX	3
28–30	XXXX	4
31–33	XXXXX	5
34–36	XXXXX	5
37–39	XXXX	4
40–42	XXXXXXXX	8
43–45	XXXXXX	6
46–48	XXXXX	5
49–51	XXXX	4
52–54	XXX	3
55–57	X	1
58–60	X	1
Total	50	50

GRAPHIC DESCRIPTION: THE HISTOGRAM

The histogram directly portrays the information of the frequency distribution. It is often called a bar diagram. To prepare a histogram, draw a set of coordinates representing the X and Y axes as shown in Figure 3.1. Mark off the X axis into intervals corresponding to those used in the frequency distribution. Note that the intervals in Figure 3.1 are the same as those in Table 3.3. The values along the X axis should start at one class interval below the lowest interval that contains a score and extend to one class interval beyond the highest score (the reason for this will become apparent in the following section). The Y axis represents the number of cases present at each score interval and must always start at 0. Next, draw a bar equal in height to the number of cases at each interval. This is shown in Figure 3.2, which is the complete histogram.

The information conveyed in Figure 3.2 is identical to that in Table 3.3, but Figure 3.2 is more acceptable because this has been one of the conventional procedures used for portraying frequency distribution information.

GRAPHIC DESCRIPTION: THE FREQUENCY POLYGON

The disadvantage of the histogram is that the vertical bars tend to interfere with one's visual scan of the figure. The frequency polygon avoids this problem. In order to construct a polygon (meaning many-sided figure), the center of the class interval is taken as the most representative point for all the scores within an interval, and lines are drawn from the center of one interval to the center of the next interval. Also, the curve is brought to the baseline (i.e., zero frequency) at

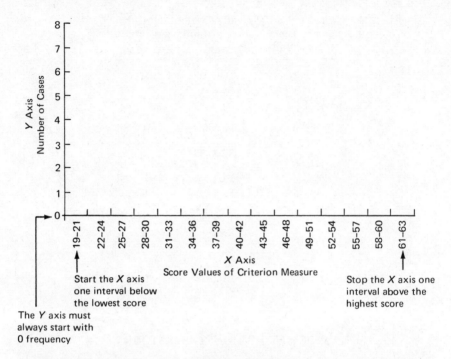

FIGURE 3.1 General layout for the histogram. The X axis always represents the endpoint or criterion measure and has the same class interval size as the corresponding frequency distribution (see Table 3.3). Start the X axis one class interval below the lowest score and end it one class interval above the highest. The Y axis always represents the number of cases (or frequency of occurrence) and must begin with 0 frequency.

both ends of the distribution (this was the reason for having the two terminal intervals when constructing the histogram). Figure 3.3 presents the polygon for the data of Table 3.3. The histogram in Figure 3.2 and the polygon in Figure 3.3 are geometrically identical, but the polygon is the easier figure to scan visually to get the best feel for the nature of the distribution of your data.

Figure 3.3 was constructed with the Y axis representing Number of Cases. This restricts the ease with which one can generalize to other situations involving different numbers of subjects. For this reason it is conventional to divide all the frequencies by the total number of subjects involved to yield a *relative frequency* or *proportion* value. This is illustrated in Table 3.4 for the data of Table 3.3 Figure 3.4 presents these data in graphic form. Figures 3.3 and 3.4 are identical in form, but the area under the curve in Figure 3.3 sums to 50 whereas the area under the curve in Figure 3.4 sums to 1.00. This is an obvious yet important observation which we shall see developed more fully when we get to the chapter on the normal curve.

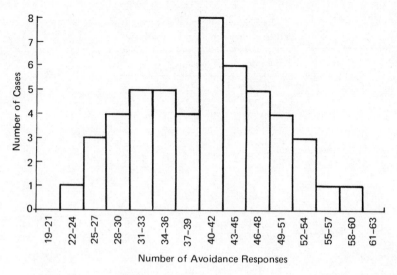

FIGURE 3.2 Completed histogram for the data in Table 3.3.

TYPES OF FREQUENCY POLYGONS

The frequency polygon shown in Figure 3.4 is a unimodal distribution which tends to be symmetrical. Roughly speaking, this approximates the normal curve. An ideal normal curve is shown in Figure 3.5. The vast majority of scores are bunched in the center of the distribution and there is a rapid drop-off in the proportion of occurrence of events as one moves toward either tail. In addition,

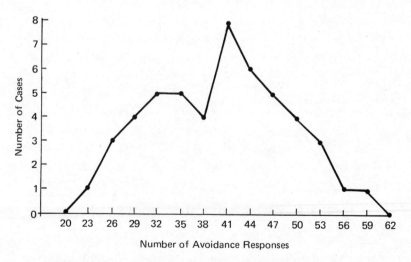

FIGURE 3.3 Frequency polygon for the data in Table 3.3.

TABLE 3.4 Relative frequency distribution of the data in Table 3.3

Score interval	Frequency	Proportion
22–24	1	.02
25–27	3	.06
28–30	4	.08
31–33	5	.10
34–36	5	.10
37–39	4	.08
40–42	8	.16
43–45	6	.12
46–48	5	.10
49–51	4	.08
52–54	3	.06
55–57	1	.02
58–60	1	.02
Total	50	1.00

the curve is symmetrical around its center. There are many examples in biological, behavioral, and psychological literature of data that distribute themselves in a form approximating the normal curve. This is tremendously convenient because the mathematical characteristics of the theoretical normal curve are well known, and the model of the normal curve may then be applied to these empirical distributions as we shall discuss in Chapter 5.

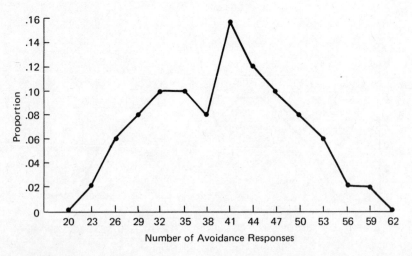

FIGURE 3.4   Relative frequency (proportion) polygon for the data in Table 3.4.

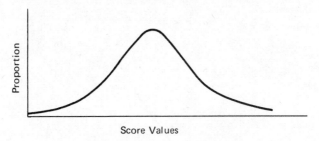

FIGURE 3.5   A normal curve.

However, there is nothing sacrosanct about the normal curve, and there are many instances in which data distribute themselves in a fashion other than a normal distribution. Figures 3.6 and 3.7 show two typical examples of *skewed* distributions. Figure 3.6 shows a distribution skewed to the right, called positive skew because the tail is pointed to the right, whereas Figure 3.7 is skewed to the left (negative skew).

Positively skewed distributions are commonly found when one is taking time measurements of an event. Suppose that we are recording the latency in seconds before a female rat starts retrieving her young after the pups have been removed to the front of the cage. There is a certain minimal time involved before she can make contact with the young, and this establishes a minimal value for any latency measure. If she is a motivated female, which will be the case if she is newly parturient and is nursing her young well, in general she will retrieve them near this lower limit. However, even for the most motivated, well-intentioned female, there often are other stimuli in the environment that may catch the animal's attention or distract her from her interest in the young. If that occurs,

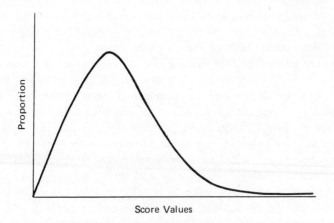

FIGURE 3.6   A positively skewed curve.

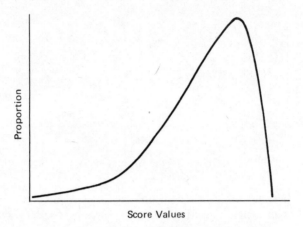

FIGURE 3.7   A negatively skewed curve.

latency will be prolonged. Because there is no upper limit with respect to time although there is a lower limit, the curve will be of the form shown in Figure 3.6, where most scores will be toward the lower time value but where a distraction or some other form of inference or a reduced level of motivation will allow a few of the scores to be abnormally high.

An example of a negatively skewed curve similar to that shown in Figure 3.7 is commonly found when one has a test in which all animals perform near capacity. For example, is most animal laboratories in which rats are born and reared there is selection for good mothers. Those females that have poor litters or that do not maintain their litters are generally not bred again. A common measure is the number of young from a litter that survive until weaning. Suppose that all litters are standardized at 8 animals at the time of birth, and the measure taken is the number of animals alive at the time of weaning. The vast majority of females will end up with 6, 7, or 8 animals alive at weaning, but there will be a few who destroyed all, or almost all, of their young. This would result in a frequency distribution similar to that shown in Figure 3.7.

Figure 3.8 presents what is called a *bimodal* distribution, because there are two modes or peaks. If you have data like this, be wary as to whether this is indeed a single distribution with two peaks, or whether there are two different distributions superimposed upon each other. Figure 3.9 presents the situation where there are two different, overlapping distributions which may be falsely interpreted as a single bimodal distribution.

An example of the latter situation would be seen if you were to record activity of male and female rats in an open field. An open field is a large black box (generally 3 to 4 feet square) the floor of which is marked off by thin white lines into squares for convenience in counting the number of squares traversed by an animal during some standard time interval. It has been found that females are markedly more active than males in such an apparatus. If you were to run a

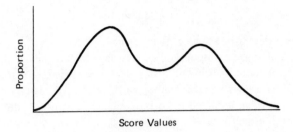

FIGURE 3.8   A bimodal distribution.

sample of animals of both sexes and ignore the sex variable when plotting the data, you would have a set of data that looks like Figure 3.8 which would, in reality be based on two separate distributions as described in Figure 3.9, with the females having greater activity than the males. It may be that there are some true bimodal distributions in biological or behavioral research, but this is a highly unlikely event. Normal curves and skewed curves are the more usual situation. Always treat bimodal data with serious suspicion until there has been some verification that the distribution is indeed bimodal.

DETERMINING AN AVERAGE VALUE

The frequency polygon provides a good visual portrait of a frequency distribution, but this is obviously of a qualitative rather than a quantitative nature. There are two major quantitative parameters needed to describe a frequency distribution: a measure of an average value (also called central tendency) and a measure of the amount of variability around this average score. This section will be concerned with two methods of determining averages.

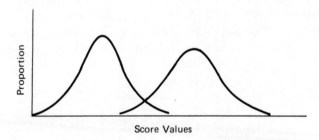

FIGURE 3.9   Showing how a bimodal distribution can result when two overlapping normal distributions are measured. By ignoring the characteristics of the separate distributions and pooling all the data, a curve similar to Figure 3.8 is obtained.

The Arithmetic Mean

An average value is one that is most representative of all of the data of a distribution. The most commonly used average, and the most powerful one statistically, is the one called the *arithmetic mean*. Here, for the first time, we shall introduce a formula. The symbol for the mean is $\bar{X}$. The symbol X refers to any raw score value obtained in an experiment. The bar over the value means "the average of." To obtain the mean one adds all the raw score values together and divides by the total number of observations. In formula form this is

$$\bar{X} = \frac{\Sigma X}{n} \tag{3.1}$$

where Σ = The Greek symbol, sigma, standing for the "summation of"
   X = raw score
   n = number of cases involved
For our shuttlebox example in Table 3.1, the sum of the raw scores is 2001, and there are 50 cases. Therefore, $\bar{X} = 2001/50 = 40.02$.

The mean has certain mathematical properties that make it uniquely suited to be the one value most representative of all values of a distribution. First of all, the mean is the "center of gravity" of the distribution. This may be illustrated by using the data in Table 3.1 in a physical analogy. Imagine that you have 50 wooden blocks, all of equal weight, and also a wooden plank marked off in 38 equal units. Assign the number 22 to the unit at one end of the plank and count up in units of one until you reach the last interval at the other end of the plank, which will have the value 59. (The score of 22 refers to the lowest avoidance learning score in Table 3.1 whereas the score of 59 is the highest). Now place a wooden block on each interval on the plank corresponding to the values in Table 3.1. Such an arrangement is shown in Figure 3.10. You will remember from elementary physics that a point may be found on the wooden plank where the force on one side exactly balances the force on the other side. If you were to place a pivot at that point, the plank would be perfectly balanced. The arithmetic mean of the distribution is that pivot point. Thus, if a knife edge were

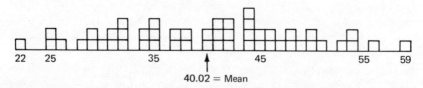

FIGURE 3.10 Showing the scores from Table 3.1 as though they were wooden blocks distributed along a wooden plank. The mean of the distribution is that point where the plank is in perfect balance.

placed under the plank in Figure 3.10 at the exact value of 40.02, the plank would be in complete balance.

We now wish to derive a corollary from the center of balance idea. To do so we introduce a second definition, which is concerned with a deviation from the mean. This is defined as follows:

$$x = X - \bar{X} = \text{a deviation from the mean} \qquad (3.2)$$

This states that a deviation from the mean is equal to a raw score minus the mean value itself. This deviation value is one we shall be using in a number of our subsequent statistical formulas. Because of the unique characteristics of the mean, which specifies that it is the balance point of all values, it follows that the scores above the mean must exactly balance the scores below the mean. Therefore, the summation of all these positive and negative forces operating around the mean must equal zero. This may be put into a simple mathematical formula by stating that the sum of the deviations around the mean equals zero, as expressed below.

$$\Sigma x = \Sigma (X - \bar{X}) = 0 \qquad (3.3)$$

Another mathematical property of the mean as a measure of average is its relative consistency over successive samples. For example, if we were to draw a second random sample of 50 animals from our colony and run them on the avoidance learning task, we could compute a second mean value. We could do this with a third group of 50 rats, a fourth group, and so on until we had used all the animals in our colony. If we were to compare the mean values obtained from our successive samples, we would find that there is less variability among these means than there would be among any other index of average that we could compute. In other words, the mean is the *most stable* of all of the possible average values we could obtain.

The Median

A second statistic commonly used as a measure of average value is called the median. It is defined as the score below which fall 50% of the observations. The median has also been defined as the middlemost score of the distribution. In the example of the avoidance conditioning situation, there were 50 animals involved and, thus, 50 scores. Therefore, the average of the scores of the 25th and 26th animals would be the middlemost score below which fall 50% of the cases and would define the median in this distribution. In this particular example both the 25th and 26th animals had scores of 41, and so the median is 41. There are formulas that enable us to pinpoint the median more exactly within the class interval of 41 (which really ranges from 40.5 to 41.5), but this is a form of refinement that is virtually never used in actual research. Therefore it is eliminated here. As far as we are concerned, the middlemost score may be taken as the median. If it happpens that the 25th score has a value of 41 and the 26th score has a value of 42, then the median would be 41.5.

At times the median is a value for which there is no observed score. For example, consider the following set of data:

$$6, 9, 11, 17, 18, 19$$

The middlemost score falls between the third and fourth values of this distribution and in this instance equals the value 14.

Comparisons Between the Mean and the Median

Given a completely symmetrical distribution, the mean and the median will be mathematically identical. As the degree of skewedness of a distribution increases, the mean and the median diverge more from each other. These relationships are shown in Figures 3.11 and 3.12 for positively and negatively skewed distributions. The mean, which is the balance point of the distribution, gives greater weight to the more extreme scores, and therefore is pulled more toward the tail of a skewed distribution. On the other hand, the median disregards the absolute values of the scores and treats each one as though it were above or below a certain hypothetical point (actually, the median) that separates the area under the curve into two equal areas.

If a distribution is roughly symmetrical, always choose the mean as the appropriate statistic to describe the average value of the data. In fact, use the mean unless there are very strong reasons to turn to the median as the appropriate measure. If the raw score distribution is extremely skewed or is bimodal, then you might consider using the median. If so, then you would have

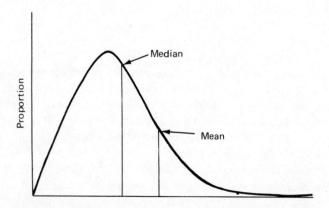

FIGURE 3.11 The relationship between the mean and the median for a positively skewed distribution.

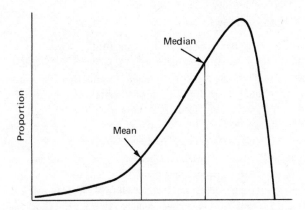

FIGURE 3.12 The relationship between the mean
and the median for a negatively skewed distribution.

to use some nonparametric statistic to evaluate your data (as, for example, in
comparing the medians from an experimental and a control group to see if they
differed significantly).

In practice we rarely have sufficient data about a distribution to make a
decision concerning the appropriate average statistic. In an experiment there are
usually anywhere from 5 to 15 or 20 cases for a particular treatment. Those
numbers are too small to give useful information about the shape of a
distribution, and the mean is the statistic to use for an appropriate description of
the data.

Another reason for choosing the mean is a very simple and pragmatic
one: we are then able to use powerful parametric techniques such as the t test,
analysis of variance, regression analyses, and correlational procedures.

There is one instance where a median may be used in the context that has
been specified for the mean. This is where there are a number of observations on
a particular experimental subject and we wish to obtain an average value from
the series of observations. Consider an experiment in which we are interested in
the amount of food ingested daily. For a period of one week we keep records to
the nearest 10th of a gram of the amount of powdered food that each of a group
of animals eat. In looking over the seven scores for each animal, we notice that a
number of them have one or two scores that seem to be quite divergent from the
others in that they are much lower than the typical amount ingested. Our feeling
is that these scores represent some sort of effect in the environment that has
nothing to do with ingestive behavior. In other words, we believe that those
scores are not representative of the amount of food the animal typically eats. In
such an instance we may choose to use the median of the animal's seven scores
rather than the mean, because in this fashion we eliminate the extreme weight
that one or two deviant scores would have in modifying the mean.

Variability of Means and Medians

Recall that, in the previous section of the mean, the statement was made that, in successive samples, the mean is the most stable average statistic. If we were to compute the mean and median for each sample, do this for a large number of samples, and then determine the amount of variability among the means and among the medians, we would find that the variability among the medians was 56% greater than among the means. In other words, to have a median equal in stability to a mean, it is necessary to have 56% more cases. Considering the shuttlebox example again, a mean score based on 50 animals is as stable an estimate of the average value in the population as a median score based on 78 animals. Now, there may be times when all we are interested in is a stable estimate of the population average, where there are good reasons to be concerned with the lack of symmetry of the distribution, and where the cost of obtaining additional values is inexpensive. Under these conditions the median is probably the statistic of choice. However, these particular conditions rarely occur in experimental research, and so the recommendation is to use the mean if it is at all reasonable.

A MEASURE OF VARIABILITY

As indicated earlier, two parameters are necessary to describe a distribution: an average and an index of variability. We have developed the logic which indicates that the arithmetic mean is the best measure to use to describe the average score. We now have to consider an appropriate indicator of variability.

First, though, let us consider an example in which the primary interest is in the measure of variability rather than the average. A behavioral geneticist has two different strains of mice that have the same distribution of scores on an activity measure, including the same mean score. Because activity is known to have a genetic basis, she is curious as to whether these two different strains happen to have the same genes influencing activity. As a method of testing this hypothesis she cross-breeds the two strains to get the F_1 generation, and she then breeds randomly within the F_1 to get an F_2 generation. The data for the parent generations and for the F_2 generation are shown in Figure 3.13.

As can be seen, the means for the parental and F_2 generation are identical, but the variability of the F_2 has been very markedly increased relative to their parents. The range of the parents is from 85 to 115, a 30-point range, whereas the activity range of the F_2 offspring is from 70 to 130, a 60-point range. These data establish that the genes of the two parent generations were not the same, although they did happen to give similar numerical scores. It was only by segregating these genes by means of the F_2 generation that the experimenter was able to establish this phenomenon, and the critical statistical index here was an indicator of variability rather than one of central tendency.

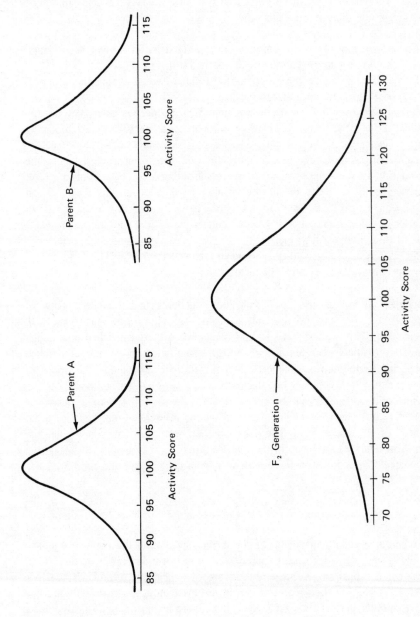

FIGURE 3.13 Example where the primary statistic is variability.

There are several statistics that could be used to measure variability. Before discussing any of these, however, we can specify certain characteristics that a measure of variability ought to possess in order to satisfy us as experimenters. Such a list would include the following:

1. The statistic should offer a unique solution. That is, for a given set of data, these should be a single, unique mathematical solution.
2. The statistic should be the most stable in that it would change its numerical value least from one random sample to another.
3. The statistic should have useful mathematical properties such that we can derive more information from using this particular statistic than from using any other statistic.
4. The statistic should take into account the quantitative characteristics of the data—that is, the actual numerical score should be used in the computations. After all, if we went to all the trouble to develop an interval scale of measurement so that we would be able to use numbers (rather than just classes or inequalities), then it is reasonable to expect our statistic also to make use of these numerical properties.
5. The measure of variability should be completely compatible with our measure of average, namely, the mean.

The list above should give you some general feel for the types of considerations that are involved in choosing among various statistics. It so happens that there is one statistic that meets all of these qualifications, and is especially valuable with reference to the third one listed above—namely, that it has special mathematical properties. This statistic is called the *standard deviation* and will be the only one discussed as a measure of variability.

When one considers variability, it is always with respect to a particular reference point. For a frequency distribution the reference point that we use is the mean of the distribution. The question then becomes: How can we measure the variability about the mean of the distribution?

We can get some intuitive feel for this by looking back to Formula 3.3:

$$\Sigma x = \Sigma(X - \bar{X}) = 0 \tag{3.3}$$

This formula states that the sum of the deviations around the mean is equal to zero. What happens if we square each deviation in Formula 3.3? If we do, all our values become positive, because the square of a negative number is positive. Therefore the sum of the squares of the deviations around the mean will be some positive number greater than zero. Now, if we were to substitute any value other than the mean into Formula 3.3, square the deviations around this other value, and add them up, we would find that all those sums would be numerically greater than the sum of the squares of the deviations around the mean. In other words, when we take deviations from the mean and square them, we get a minimum figure. This is expressed mathematically as follows:

$$\Sigma(X - \bar{X})^2 = \text{a minimum} \tag{3.4}$$

This is a demonstration of the *principle of least squares.* The principle of least squares in mathematics states that a unique solution is obtained if one finds a value such that the sum of the squares of the deviations around that value is at a minimum. The simplest example of this principle is seen in the standard deviation, although there are many other examples in statistics.

Formula 3.4 goes by a special shorthand title, namely, the *sum of squares,* or *SS.* This is all put together in Formula 3.5.

$$\Sigma(X - \bar{X})^2 = \text{sum of squares of the deviations about the mean}$$
$$= \Sigma x^2 = \text{sum of squares} = SS \tag{3.5}$$

One could use the numerical value obtained in the sum of squares formula as a measure of variability except for one obvious flaw: namely, the greater the number of cases involved the greater will be the numerical value. In other words, the sum of squares formula gives a total value (analogous to the ΣX part of the formula for the mean) rather than an average value. In order to get an average value, the intuitive thing is to divide by the number of observations, or n. This is almost correct. However, for reasons we shall not go into now, it is necessary to divide the sum of squares by the number of cases minus one rather than the number of cases itself. When this is done, the resulting statistic is called either a *mean square* (symbolized as *MS*) or else *variance* (symbolized as s^2).
This is shown in Formula 3.6.

$$\frac{\Sigma(X - \bar{X})^2}{n-1} = \frac{\Sigma x^2}{n-1} = \frac{SS}{n-1} = \text{mean square} = MS = \text{variance} = s^2 \tag{3.6}$$

Formula 3.6 is a satisfactory measure of variability, but it is cumbersome because it is expressed in square units and, therefore, is an area measure rather than a linear measure. However, we can easily convert to a linear measure by the simple procedure of taking the square root of the variance. This yields a statistic called the *standard deviation,* which is defined in Formula 3.7.

$$\sqrt{\frac{\Sigma(X - \bar{X})^2}{n-1}} = \sqrt{\frac{\Sigma x^2}{n-1}} = \sqrt{\frac{SS}{n-1}} = \text{standard deviation} = s \tag{3.7}$$

Determining the Standard Deviation When There Are Several Sets of Data

Suppose that you have data from several studies, all collected under comparable conditions, and you wish to combine these data to give you the one best estimate of the standard deviation. If the first study is called A, the second B, the third C, and so on, then each standard deviation separately is

$$s_a = \sqrt{\frac{SS_a}{n_a - 1}} \qquad s_b = \sqrt{\frac{SS_b}{n_b - 1}} \qquad s_c = \sqrt{\frac{SS_c}{n_c - 1}}$$

Our best estimate is obtained by taking the sum of squares within study A (SS_a), within study B (SS_b), and so on, and pooling these for our numerator. Similarly, we pool the separate divisors to give us our denominator. Because this standard deviation is obtained by combining the variability within each of our separate studies (or groups), we shall designate this statistic as s_{within} and define it by the following formula:

$$s_{within} = \sqrt{\frac{SS_a + SS_b + SS_c + \cdots}{(n_a - 1) + (n_b - 1) + (n_c - 1) + \cdots}} \qquad (3.8)$$

The square of this value is called s_{within}^2 or mean square$_{within}$ (MS_{within}), so that

$$s_{within}^2 = MS_{within} = \frac{SS_a + SS_b + SS_c + \cdots}{(n_a - 1) + (n_b - 1) + (n_c - 1) + \cdots}$$

Summary of Formulas About Variability

One of the confusing things about statistics, especially for the beginning student, is that statisticians have chosen to call the same thing by many different names. In part this is because they have approached a particular formula from different points of view, and in part this is determined by the particular symbols they had on their typewriters when they were preparing their texts for publication. So even though a variance is a variance is a variance, some people symbolize this by the letters MS whereas others symbolize this by the notation s^2. All of the usual symbols and formulas are summarized in Table 3.5 to eliminate confusion.

It is important that you study Table 3.5 carefully both to understand it and to memorize the various shorthand symbols given in the right-hand column. Note that under the first column the expression $(X - \bar{X})$ and the symbol x are both given. These are identical, as we showed in Formula 3.2. Both of these are presented here because textbook writers often will use one or the other of these expressions interchangeably in the same text and the student may be confused until he or she realizes that these are identical statements.

At the right-hand side of the table, the symbols are mostly self-explanatory. SS is a shorthand way of saying the sum of the squares of the deviation around

TABLE 3.5 Summary of formulas about variability

Formula	What it means	Shorthand symbol
$\Sigma(X - \bar{X})^2 = \Sigma x^2$	Take each deviation from the mean, square it, and sum them all together	Sum of squares $= SS$
$\dfrac{\Sigma(X - \bar{X})^2}{n-1} = \dfrac{\Sigma x^2}{n-1}$	Divide SS by number of cases minus one in order to get an average, or mean, value of variability	Mean square $= MS$ $= $ variance $= s^2$
$\sqrt{\dfrac{\Sigma(X - \bar{X})^2}{n-1}} = \sqrt{\dfrac{\Sigma x^2}{n-1}}$	Take square root of the above to convert from an area measure of variability to a linear measure	Standard deviation $= s$
$\dfrac{SS_a + SS_b + SS_c + \cdots}{(n_a - 1) + (n_b - 1) + (n_c - 1) + \cdots}$ $\dfrac{\Sigma x_a^2 + \Sigma x_b^2 + \Sigma x_c^2 + \cdots}{(n_a - 1) + (n_b - 1) + (n_c - 1) + \cdots}$	If you have data from several studies or groups, your best estimate of the mean square or variance is to add together the separate SS within each group to get the numerator and combine the separate divisors for the denominator	$s^2_{within} = MS_{within}$
$\sqrt{s^2_{within}}$	Take square root of above to convert from a variance to a standard deviation	s_{within}

the mean. MS indicates the mean square (this could have just as well been called the average sum of the squares if someone had decided to use that particular expression when writing a textbook in statistics) and implies that one is getting an average value by dividing by roughly the number of cases involved. Not only is the mean square symbolized by MS, but this expression is also called a *variance*, which is derived from the idea of variability. Probably the greatest puzzlement, however, is why variance should be symbolized by s^2. This can be understood if you look at the last set of symbols in Table 3.5, which refers to the standard deviation. This is symbolized by s, referring to the first letter of the term, standard deviation, so it is reasonable that when you square the standard deviation you end up with the expression entitled s^2. (Incidentally, engineers use the expression "root mean square"—which means to take the square root of the mean square—to designate the standard deviation.)

Although the various segments defy any simple semantic analysis, it is necessary that you familiarize yourself with these and learn the various definitions, as they will occur over and over again throughout the text.

TABLE 3.6 Computational formulas for variability

Definitional formula	Computational formula
$SS = \Sigma x^2 = \Sigma(X - \bar{X})^2$	$SS = \Sigma X^2 - \dfrac{(\Sigma X)^2}{n}$
$s^2 = \dfrac{\Sigma x^2}{n-1} = \dfrac{\Sigma(X-\bar{X})^2}{n-1}$	$s^2 = \dfrac{\Sigma X^2 - (\Sigma X)^2/n}{n-1} = \dfrac{n\,\Sigma X^2 - (\Sigma X)^2}{n(n-1)} = \dfrac{SS}{n-1}$
$s = \sqrt{\dfrac{\Sigma x^2}{n-1}} = \sqrt{\dfrac{\Sigma(X-\bar{X})^2}{n-1}}$	$s = \sqrt{\dfrac{n\,\Sigma X^2 - (\Sigma X)^2}{n(n-1)}} = \sqrt{\dfrac{SS}{n-1}} = \sqrt{s^2}$

COMPUTATIONAL FORMULAS FOR SUMS OF SQUARES, VARIANCE, AND STANDARD DEVIATION

The formulas given in Table 3.5 are definitional formulas used to define or specify a particular statistical concept. However, the formulas are not set up in a fashion that facilitates computation. The algebraically identical computational formulas are presented in Table 3.6.

Computational procedures involved in obtaining the mean, the sum of squares, the variance, and the standard deviation are shown in Table 3.7 using the data of the animals in the avoidance conditioning experiment.

TABLE 3.7 Illustrating computational procedures to obtain $\bar{X}$, SS, s^2, and s for a set of data, using the avoidance learning scores of Table 3.1

$\Sigma X\ = 2001$

$\Sigma X^2 = 84021$

$n\ = 50$

$\bar{X}\ = \Sigma X/n\ = 2001/50$

$\quad = 40.02$

$SS = \Sigma X^2 - (\Sigma X)^2/n = 84021 - (2001)^2/50 = 84021 - 80080.02$

$\quad = 3940.98$

$s^2 = SS/(n-1) = 3940.98/49$

$\quad = 80.4282$

$s = \sqrt{s^2} \ = \sqrt{80.4282}$

$\quad = 8.97$

SUMMARY OF DEFINITIONS

Bimodal distribution   A frequency distribution with two modes or major peaks.

Frequency distribution   A systematic way of ordering a set of data from lowest to highest score showing the number of occurrences (frequency) at each score point.

Frequency polygon   A graphic way of portraying a frequency distribution by plotting score values on the X axis, number of cases on the Y axis, and connecting the points to form a curve that is brought to the baseline (X axis) at both extremes.

Histogram   A graphic way of portraying a frequency distribution by plotting score values on the X axis and using bars to represent the number of cases on the Y axis.

Negatively skewed distribution   An asymmetrical and unimodal frequency distribution with the tail pointing toward the low score values. In such a distribution the median will be numerically greater than the mean.

Normal distribution   A special kind of frequency distribution that is unimodal, symmetrical, and is defined by a particular mathematical formula relating the mean and standard deviation of the normal curve to the proportion of cases under the curve. (This curve had also been called a bell-shaped curve.)

Positively skewed distribution   An asymmetrical and unimodal frequency distribution with the tail pointing toward the high score values. In such a distribution the median will be numerically less than the mean.

Principle of least squares   A mathematical principle which states that a unique solution results when the sum of the squares of the deviations from the mean is at a minimum value.

Proportion   A number ranging from 0 to 1.00.

PROBLEMS

1.(a) Compute the mean for the following set of data:

$$12, 4, 21, 17, 18$$

(b) Subtract the mean from each score to get $x = X - \bar{X}$. What is the value of Σx?

(c) Take each x value from above, square and sum these values to get Σx^2. In words, what is this summation called?

(d) When Σx^2 above is divided by $n - 1$, what is the resulting value? What is the statistic called?

(e) Use the computational formula for the variance from Table 3.6 to compute s^2 for the above raw scores. How does this value compare with the one immediately preceding?

2. Compute the mean, median, sum of squares, variance, and standard deviation for the following set of data. Based on the values for the mean and median, what kind of skewness is there?

$$62, 18, 66, 1, 89, 27, 78, 80, 18$$

3. An experimental psychologist tests 6 subjects to determine their reaction times to a light flash. Each subject is given 5 trials, and reaction time in milliseconds is recorded. The data are as follows:

Subject A: 250, 175, 320, 225, 780
Subject B: 197, 272, 478, 299, 290
Subject C: 343, 510, 396, 320, 370
Subject D: 407, 132, 134, 312, 203
Subject E: 223, 197, 174, 207, 387

Compute the mean and median reaction times for each subject. Based on the raw scores, means, and medians, which of the two measures of average is the better one to use?

4. An introductory biology instructor gave the same examination to three different sections. The statistics are as follows:

	Section 1	Section 2	Section 3
ΣX	2212	3151	3247
ΣX^2	178636	219427	280083
n	28	47	39

Find the mean, sum of squares, variance, and standard deviation for each section. Use formula 3.8 to obtain s_{within} on the assumption that the results of the three sections can be pooled to yield one standard deviation value.

5. The weaning weights of a group of rats, measured to the nearest gram, are as follows:

39, 42, 43, 46, 46, 48, 49, 50, 50, 51, 51, 52, 52, 52, 54, 54, 54,
55, 55, 55, 55, 56, 57, 57, 58, 58, 58, 58, 59, 60, 61, 61, 61, 62,
62, 64, 65, 65, 65, 66, 66, 68, 68, 69, 70, 71, 73, 73, 76,

Compute $\bar{X}$, SS, s^2, and s for these data. Then prepare a histogram and a relative frequency polygon of the data using a score interval of 3.

6. A class of college students viewed a movie, and each student rated the movie on a 9-point affective scale where 1 = Not Exciting and 9 = Exciting. The data are as follows:

Affective scale value	Frequency
1	2
2	4
3	8
4	13
5	21
6	20
7	22
8	17
9	14

Compute $\bar{X}$, SS, s^2 and s for these data. Then plot a histogram and a relative frequency polygon of the data using a score interval of 1.

CHAPTER 4

Making Inferences About Population Parameters: General Principles

Now that we know how to take a set of quantitative experimental data and obtain descriptive information concerning the form of its distribution, its mean, and its variability, we should pause for a moment to be certain that we realize why we are gathering this information. The data we have obtained from our experimental subjects are useful only if they give us meaningful information concerning the larger group or population of which our experimental subjects are but a sample. As an example, what general conclusions can an experimenter draw about the ability of a population of rats to perform in a shuttlebox, based on the information obtained on the 50 animals described in the prior chapter? Our question is: How do we make inferences about a population from a set of data obtained from a sample? This chapter is concerned with some general principles underlying this process of drawing inferences in experimental research.

POPULATIONS AND SAMPLES

First of all, if we wish to make generalized statements about a population, we have to define our population. A researcher defines a population in terms of the characteristics of the subjects and the measurement procedure. Once this has been done, a sample is drawn from the subject population by a process of random selection. *In order to qualify as a random sample, each element (i.e., subject) in the population should have an equal chance of being selected to be in the sample.* The researcher then performs a set of experimental operations on the sample, and from the sample data he or she may draw inferences concerning the population if (and only if) the sample has been randomly selected from that population. The continual emphasis in this paragraph on the concept of randomness is quite deliberate. This is *the* central concept in making inferences about a population. This appears to be such an elementary principle that many experimenters do not think deeply about it. And yet they should, because this is one of the key reasons why there is such great failure to substantiate research findings from one study to another, or from one laboratory to another.

Drawing a Random Sample

How do we draw a random sample? The procedure is to assign to each element in the population a numerical value starting with 1. We then enter a table of random numbers (see Table XII in the Appendix) at some haphazard point, and the first number chosen becomes the first element of the population that goes into the sample. We then proceed from this number in any direction we choose, and continue selecting numbers from the table until we have obtained the necessary sample size.

An important point to emphasize here is that randomization is a *procedure*, not a *product*. When we go through the operations of selecting a random sample, we are giving each element in the population an equal opportunity to be selected. This does *not* mean that we will end up with a sample that is a direct miniature representation of the population. Remember: By chance alone one can be dealt 13 spades from a deck of cards.

The reason for randomizing, from a statistical point of view, is to allow the laws of chance to operate. We know the mathematical and statistical characteristics of the laws of chance, and these are the bases for the formulas presented in this book which permit us to draw inferences about a population. These are in essence the same laws that tell us one is much less likely to win by drawing one card to an inside straight in poker than by keeping a pair and drawing three cards. Note in the poker example that on a few occasions one will draw an inside straight, but not very often. The laws of chance work *on the average over a long series*, whether in playing cards or doing experiments. By using a random procedure to draw our sample, we allow the laws of chance to operate. We may then use our knowledge of the mathematics of the laws of chance to draw conclusions about the population which will be correct over a long series of experiments. Admittedly, the conclusions we draw from any one experiment may be in error to a greater or a lesser degree.

Because we typically run only one or a few experiments, how do we know that our conclusions are correct in the long run? We are never certain of that, but what we can do is assign probabilities that we are right; we can also set limits or boundaries around our statistics indicating the amount by which we might be off in our estimations. We shall discuss most of these concepts and techniques in the next two chapters.

Experimental Research with Nonrandom Samples

As you are no doubt aware, there is a significant amount of empirical research in which samples are randomly drawn from a population following the theoretical definition given above. Those researchers doing polling, for example (Gallup, Roper, etc.) are very careful in their definitions of populations and their sampling procedures.

However, it is generally not possible to follow the definition for random sampling when doing experimental research in behavior or biology. For example, a very common situation is for researchers to purchase their experimental animals from a commercial supplier. Obviously, the experimenters are not in a position to assign numbers to all subjects in the colony of that supplier and then randomly select from the colony. They can make rather general specifications such as age, sex, weight, and whether or not the animals should be littermates, but that is about as far as they can reasonably go. Even if the experimenters have their own breeding colony, the theoretically defined randomization procedure may not be possible. Suppose, for example, that we are experimenters who want to test out an idea that occurred to us after reading a recent paper in the literature. We estimate that we need between 25 and 30 animals to run this experiment. We check through our colony and find that most of our animals are being utilized or have been set aside for other experiments, but we do find 28 animals that are available. These 28 animals have not been randomly selected from the total colony population, but they are there at the time that we need them. We, like all reasonable researchers, will take these animals and carry out our experiment.

The consequences of this discussion may be succinctly stated: When doing experimental research in behavior and biology, we typically do not have random samples from our populations. The procedures demanded by the theoretical statistician in order to assure that a sample has indeed been chosen randomly (which is a necessary condition in order to generalize to the population from which one has sampled) and the practical concerns of the day-to-day experimentalist are simply incompatible in most instances.

How to Draw General Conclusions from Nonrandom Samples

How, then, are we able to make generalizations in experimental biology and behavior? The answer, very simply, is that if we repeat the experiment a second time, drawing animals from the same population as we did the first time, and if we obtain essentially the same results as we did previously, then we have considerable confidence that we have a reproducible phenomenon. *The principle here is that we substitute a second independent replication of our experiment for our inability to draw a random sample.* It is extremely unlikely that the second sample we draw will have the same types of biases and oddities built into it as we obtained with the first sample, but it will have its own unique peculiarities. So if we have a phenomenon powerful enough to override the particular variabilities obtained from one sample to another, then we can feel quite confident that we have a real experimental finding.

It is always necessary to repeat the experiment a second time? The answer is "No." There is certainly no need to repeat the experiment if the first set of results is consistent with what has already been published in the literature, and the experiment either confirms or extends a set of established findings. For example, there is considerable evidence that electrical stimulation in certain regions of the brain will act as a positive reinforcer controlling an animal's learning behavior. If we took a species of animal that had not previously been studied for the effects of electrical brain stimulation, went through the appropriate surgical and experimental procedures, and found that this species also learned effectively when reinforced by stimulation, there would certainly be no need to repeat this experiment with another sample from this species, because the findings would confirm and extend those already existing in the literature.

But we *do* have to repeat the experiment (1) when a phenomenon is uncovered that has not been reported before, or (2) when the findings are contradictory to those in the literature. To publish the results of one experiment when either of the two conditions above obtain, even though the results are "highly statistically significant," is to publish at your peril. If the findings are potentially exciting or inconsistent with the currently accepted body of knowledge, you can fully expect that researchers from other laboratories will attempt to reproduce your data. If other laboratories cannot reproduce your findings, then you are in trouble. The only way to avoid this difficulty is to know that you can duplicate your own findings in your own laboratory.

When Randomization Must Be Employed

I am not cavalierly dismissing the concept of randomization. I am stating that in practice we cannot follow the dictates set forth by the theoretical statistician, but that does not mean that we should ignore the concept. Within the limitations set by our research subjects and our experimental methodology we should do the best we can to randomize. Moreover, *random assignment of subjects to groups is absolutely necessary when we are conducting an experiment involving two or more groups.* This point will be developed more fully in Chapter 7. For the moment the principle can be illustrated by the following example. It may be convenient to have animals that are housed in one rack of cages undergo an experimental treatment while animals in a second rack are made the control subjects. This is very bad experimental design, because the environmental events impinging upon one rack may differ markedly from those impinging upon a second. Instead, what the researcher should do is randomly assign animals within each rack to both experimental and control conditions.

Intralaboratory and Interlaboratory Repeatability

The emphasis so far has been on researchers being able to repeat their experimental findings in their own laboratories. It should be apparent that it is necessary to have intralaboratory replication before one can hope to obtain interlaboratory repeatability. However, at times these two types of verification have been confused. Some of the problems and principles involved here can be brought into focus by considering the shuttlebox investigation from Chapter 3. From what population are the rats a sample? We might define the population as "rats." This clearly is incorrect, because all elements of that population (i.e., all rats in the world) did not have an equal chance of being selected for the sample. If we purchased the animals from a commercial house, then we should specify the strain and the commercial house, and our definition would read something like the following: "All present and future male rats of the ABC strain purchased from the XYZ Production Laboratory."

There are a number of features of this definition on which we must comment. First of all, note the expression "all present and future rats. . ." This part of the definition is necessary if our conclusions are going to have any temporal generality. In addition, this part of the definition gives us essentially an infinite population, because we are talking about a boundless time dimension. As experimenters we are interested in making a generalization across time because our research findings would be particularly uninteresting if they applied only to all elements of the population at the specific time that we measured our sample. As statisticians we are interested in an infinite population because that simplifies some of our statistical assumptions and formulas, as will be seen in subsequent chapters. In order to make a generalization across time, it is necessary to assume that the genetic stock of the animals and the environmental conditions under which they are reared do not change.

We are now in the interesting position of having defined a population that is infinite in size, and that has no future temporal boundary, and yet we have also specified the strain of animals and the production laboratory with high exactitude. It should be apparent that it is only because we have precisely specified the strain and the production laboratory that we can talk about infinite populations over boundless time. In essense we are saying the following: "If you wish to reproduce my experiment any time in the future, order adult male rats of ABC strain from the XYZ Production Laboratory, follow the procedures I described in my publication, and I am quite confident that you will be able to obtain essentially the same result as I did, at least within sampling error."

Now it may well be that another researcher in another laboratory might use a different strain of rat obtained from some other commercial producer, and obtain the same results as we did in our experiment. If so, this is excellent, and it strengthens the generality of the data because we now know that the phenomenon is not restricted to one strain of animals from one production center. But suppose that another experimenter using a different strain of animal from another production center cannot repeat the findings of our original experiment? Are we to be held in disrepute because another experimenter cannot replicate our findings? Logically, the answer is "no," because the second investigator did not repeat our original experiment, because he did not draw his animals from the same population.

You may well be saying to yourself now, "These distinctions are quite picayune. If the experimental phenomenon does not generalize from one strain of rats to another or from one production facility to another, then it is probably not worth studying in the first place." Such a consideration would be valid if there were only trivial reasons why a change to a different strain or to another production facility brought about a failure to replicate. However, there may be extremely important behavioral and biological reasons why the second investigator cannot repeat our original findings with a different strain: (1) There may be genetic differences between stocks of animals produced by different commercial laboratories. We know, for instance, that even animals referred to by the same stock name (such as Sprague Dawley or Long Evans rats) differ genetically among commercial producers. (2) The diets fed to animals in different commercial production centers may differ markedly from one another; this can affect the animals directly and also indirectly by influencing the physiology and the behavior of the mother during gestation as well as during the nursing period. (3) The animal husbandry conditions will almost certainly differ from one production center to another, and we know that these conditions can have strong effects on the animals' later behavior and biology. (4) The number of animals housed per cage, the light cycle, variations in temperature and humidity, and other characteristics of the environment can also affect the physiology and behavior of the animals.

None of the variables enumerated above is trivial. Indeed, major research programs have been carried out to investigate each of the classes of variables described above. Therefore, we as researchers should be sensitive to the matter of defining our research population so that those who wish to replicate our experiment have sufficient information to obtain a sample from the same population that we investigated.

It is this failure on the part of experimenters to give a sufficient definition of their population, coupled with the failure of subsequent investigators to sample from the same population as the original researcher, that has caused so many reports in the literature of inability to repeat the findings of other researchers.

Another problem may have occurred to you by this point. Suppose that the population you are investigating is unique? For example, many biologists and behaviorists have their own colonies of animals which they have maintained by closed random breeding or by inbreeding for many years. If we have such a colony and draw our subjects from within that colony, then it is obviously impossible for another researcher to duplicate our experiment exactly unless we send him some of our own animals to investigate. On rare occasions a second investigator will request animals from the laboratory of another researcher in order to study a particular problem, but in general this does not occur.

Then how does one handle the problem of replicability of findings? Logically, nobody can duplicate your experiment except yourself, and therefore it is your responsibility as a researcher to be certain that you can repeat your own experiment in your own laboratory before you publish your results. This is as much an experimental argument as it is a statistical argument. When you publish a set of results in a research journal, you are in essense stating publicly, " I have obtained a significant finding from my experiment and I am very confident that I can reproduce that finding in my laboratory any time in the future that I wish to." In more formal language, you are saying that if you draw another random sample from the same population that you used in your first study and apply the same experimental procedures that you used previously, you expect to obtain essentially the same results within the appropriate sampling error. You should not consider publishing your paper unless you have confidence that you can do this. If you are not confident, then set aside the data until you can carry out a replication of the experiment.

Consider what might happen if you publish a paper and another researcher writes to you several months later and says he tried to repeat your experiment and was unable to do so. Do you have sufficient confidence in your research to write back and invite him to come to your laboratory where you will set up the experiment and show him how you obtained your significant results? If you do not have this degree of confidence in your data, then you should not have published the research in the first place.

POPULATION PARAMETERS AND SAMPLE STATISTICS

It is now time to turn to some formal considerations relating our sample to our population. In order to make these formal considerations, it is necessary to assume that our sample has been randomly drawn from our population. We know from Chapter 3 that we can characterize the data we obtain from a sample by three descriptive procedures: a frequency polygon portraying the shape of our distribution, the mean representing the average value of our data, and the standard deviation indicating the degree of variability about that mean. Similarly, our population can also be characterized by a frequency distribution, a mean, and a standard deviation. Indeed, the objective of our experiment is to allow us to make inferences concerning the nature of the frequency distribution in the population and the numerical values of the population mean and standard deviation. Figure 4.1 shows the logical procedure involved in these inferences.

We see in Figure 4.1 that we start with a population that is defined in an arbitrary but consistent fashion. A sample is drawn from that population through a random process and a set of experimental operations is imposed on the elements (e.g., measuring rats in a shuttlebox, determining amount ot DNA in certain brain tissue, recording speed of conduction of an electrical impulse down a nerve). The measurements we obtain following our experimental operations allow us to plot a frequency polygon, and to compute a mean and standard deviation. We can then use this information to draw certain reasonable inferences concerning the characteristics of our population from which we drew our random sample.

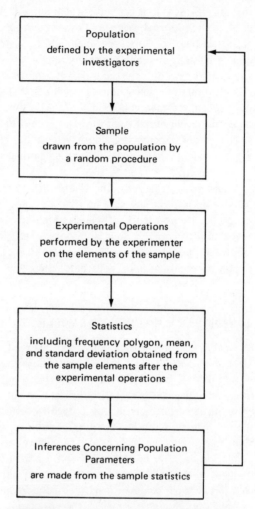

FIGURE 4.1 Showing the logical sequence
of going from a defined population to a
sample in order to draw inferences about
the population.

What would we like to know about our population? We would like to know
what is (1) the population distribution, (2) the population mean, and (3) the
population standard deviation. Now we must introduce certain conventions that
statisticians have agreed on. A numerical value obtained from a sample is called a
statistic, whereas a value that characterizes a population is called a *parameter.*
The symbols used to designate sample statistics are always Roman letters,
whereas the symbols used to designate population parameter are always Greek
letters. Table 4.1 shows this for three sample statistics (mean, variance, and
standard deviation) and their corresponding population parameters.

TABLE 4.1 Symbols and formulas for the mean, variance, and standard deviation for the population parameter and the sample statistic

Type of measure	Population parameter	Sample statistic
Mean	$\mu = \dfrac{\Sigma X}{n_{\text{population}}}$	$X = \dfrac{\Sigma X}{n_{\text{sample}}}$
Variance	$\sigma^2 = \dfrac{\Sigma(X - \mu)^2}{n_{\text{pop}}}$	$s^2 = \dfrac{\Sigma(X - \bar{X})^2}{n_{\text{sam}} - 1}$
Standard deviation	$\sigma = \sigma^2$	$s = s^2$

The symbol designating the mean of the sample is $\bar{X}$, as we have indicated previously. The corresponding symbol for the population mean is the Greek letter μ (pronounced *mu*), which is defined as follows:

$$\mu = \frac{\Sigma X}{n_{\text{population}}} \tag{4.1}$$

The symbol X signifies a measured score on an element in our sample or our population. Thus, both formulas state that we are to sum the elements that we have measured (ΣX) and divide that sum by the total number of observations. Thus, to obtain the sample mean we would take all the elements in our sample, sum the scores obtained, and divide by the sample size, designated as n_{sample}.

For our population means, μ, the definition is a theoretical one because we are never able to measure all elements in the population because our population is usually infinite, or at least extremely large. Note that this formula is a direct parallel of the formula for the sample mean: It instructs us to sum the numerical values obtained on all elements of our population and divide by the total population size. (The symbols n_{sample} and $n_{\text{population}}$ are used here only to avoid any ambiguity or confusion. In general practice n refers to the sample size, and that will typically be the way it is used throughout this text. Only where it is necessary to draw a distinction between the sample size and the population size will subscripts be introduced.)

Consider now the formulas for the variance given in Table 4.1. The symbol for the sample variance is s^2, whereas the lowercase Greek σ^2 (sigma squared) is used to designate the population variance. The population variance is defined as

$$\sigma^2 = \frac{\Sigma(X - \mu)^2}{n_{\text{pop}}} \tag{4.2}$$

These formulas tell us to take the deviation of each score from the mean of the group, square that deviation, and obtain the sum of these squared deviations. That accounts for the numerator of both formulas. The denominators differ in that we are to divide by one case less than the sample size ($n_{sam} - 1$) when obtaining the variance for the sample, whereas we are instructed to divide by the total population size (n_{pop}) when working with the population. The obvious question here is: Why do we subtract 1 from the denominator of the sample size when computing the variance but we did not do the same for the population? This involves the concept of unbiased estimates of population parameters.

Unbiased Estimates of the Mean and Variance

A point that has been implicit in the above discussion, and that should now be stated explicitly, is that the mean, variance, and standard deviation *of the population* are fixed values. That is, there is one and only one numerical value for the population mean, μ; the population variance, σ^2; and the population standard deviation, σ. This uniqueness rests on the assumption that our measurements are taken without error upon all elements of the population. Thus, if we were to repeat our measurements a second, third, fourth, etc., time on all elements of the population, we would obtain identical numerical values and thus the mean, variance, and standard deviation would be the same.

We also know that the mean and variance obtained from our sample will be *variable*. That is, if we draw a second random sample from our population and compute the mean and variance, we will almost certainly get values that differ numerically from the first set of values we obtained. The third sample would also be expected to yield somewhat different numerical values, and so on. This difference in the numerical values of our statistics from one sample to another is called *sampling error*. If we were to plot the distribution of the means and variances obtained from successive samples of the same size, we would have what is called the *sampling distribution* of that statistic.

Suppose that we were to repeat our experiment a very large number of times, and each time we computed the mean and variance by the formulas given in Table 4.1. If we obtained the sampling distribution of the mean and determined the mean of that sampling distribution (i.e., the mean of all those means), we would have a value that would be a very close approximation to the population mean, μ, and the mean of our sample variances would be very close to the population variance, σ^2. Indeed, if you can imagine that we have taken an infinite number of random samples of the same size and then computed the overall average mean and variance for that infinite number, then those values would be identical with μ and σ^2. If the conditions specified above are fulfilled—that is, if the average mean or variance obtained from an infinite number of random samples is numerically equal to the population mean and variance—then we say that our sample statistics are *unbiased* estimators of the population parameters. By unbiased we mean that the sample statistic (whether mean or variance) will have an average value that will, in the long run, coincide with the value of the population parameter that is being estimated.

The fact that the sample mean is unbiased is readily and intuitively seen by examining the formulas in Table 4.1. Our random sample should have a mean value that is somewhere near the population value, μ (we can quantify the statement "somewhere near," and will do so in Chapters 5 and 6), but it is most unreasonable to expect the sample mean to be numerically the same as the population mean. If we draw a second random sample, the mean we obtain might again be above or below the population mean, but again it should be somewhat close to our population value. If all our samples are randomly drawn and we have drawn an infinite number of random samples, then the mean of all those samples is bound to be the population mean, μ. Thus, we may state that the mean, $\bar{X}$, obtained from a randomly drawn sample is an unbiased estimator of the population mean, μ.

Consider now the variance. We start by examining the definitional formula for the population variance, σ^2. The formula instructs us to take all of our deviation around the population mean, μ, and then square those deviations. Now, we almost never have knowledge about the population mean. However, in order to develop the following argument, let us suppose that we do have a numerical value for the population mean in hand. If so, the definitional formula for the sample variance would be as follows:

$$s^2 = \frac{\Sigma(X - \mu)^2}{n_{\text{sam}}} \tag{4.3}$$

The formula given above now parallels the formula for the population variance except that we have in the denominator the sample n when working with the statistic, whereas we have the population n when working with the parameter. If this formula were applied to a very large (infinite) number of samples, the overall average would be the same as the population variance, σ^2.

If we take the above formula and substitute $\bar{X}$ for μ, we obtain

$$s^2 = \frac{\Sigma(X - \bar{X})^2}{n_{\text{sam}}} \tag{4.4}$$

This is the formula one often finds in statistic textbooks to define the variance of the sample. And this is the correct formula to describe the sample variance. However, this formula is a *biased* estimator of the population variance, and thus you will not encounter it at any other point in this book. It is introduced here merely to show you how it is biased and why we have to use Formula 3.6 given in Chapter 3 and in Table 4.1. Formula 4.4 is biased in that it will yield a numerical value that is, in the long run, less than the population variance, σ^2. This is because we use $\bar{X}$ rather than μ in the formula. The reasoning behind this is as follows. Recall Formula 3.4 in Chapter 3, which is repeated below:

$$\Sigma(X - \bar{X})^2 = \text{a minimum} \qquad (3.4)$$

This formula states that the sum of the squares of the deviations taken around their own mean is a minimal value. If any other numerical value except the sample mean is substituted in that formula, the result will be a higher figure. Now, in order to obtain an unbiased estimate of the population variance, it is necessary to use the population mean, μ, in the formula for the variance as given in Formula 4.2. Because μ will almost never be numerically equal to $\bar{X}$, the numerical value of the sum of squares that we would obtain if we used μ would have to be numerically larger than the value we obtain when using $\bar{X}$. Thus, because we use $\bar{X}$ (which in reality is all that we ever have in hand, because we do not know what μ is), the numerator of our formula to estimate the variance is numerically smaller than it should be.

How do we adjust for this? We can show by appropriate mathematics that we may compensate for this bias by reducing the denominator sample size by 1. Thus, Formula 3.6 for the sample variance given in Table 4.1 is an unbiased estimator of the population variance. That is, in the long run the average of all our sample variances will be numerically equal to the population variance, σ^2.

With this background, we are now ready to take up a discussion of two types of statistical distributions—the normal curve and the t distribution. These distributions are the topics of our next two chapters, and they are important because an understanding of them is necessary in order for us to follow the logical steps involved in drawing inferences about a population from a sample. Following this, we shall get to the heart of statistical application to experimental research by taking up, in Chapter 7, the question of how to compare the means of an experimental and a control group to see if they are "statistically significantly different."

SUMMARY OF DEFINITIONS

Parameter A numerical value that characterizes a population figure (e.g., the population mean μ).

Random sample A sample obtained by following a procedure in which all elements of a population have an equal chance of being chosen to be in the sample.

Sampling distribution A distribution of a statistic such as the mean or the variance.

Sampling error The difference in numerical value of a statistic from one sample to another.

Statistic A numerical value obtained from a sample (e.g., the sample mean $\bar{X}$).

Unbiased estimator A statistic obtained from a sample whose value, in the long run, will be the same as the population parameter being estimated.

CHAPTER 5

The Normal Curve

In the previous chapter we discussed some of the general principles involved in making inferences about population parameters from sample statistics. The general logic involved is summarized in Figure 4.1. In this chapter we wish to consider the special case of drawing inferences from a sample of data to a larger population when we have a normal distribution curve.

When we talk about the normal curve, we move out of the world of empirical data into the world of abstractions. The normal curve does not exist as a real event. It is a mathematical expression for a frequency distribution with two parameters, μ and σ. The curve can be depicted geometrically as shown in Figure 5.1.

The X axis represents all possible values of a variable and the Y axis represents the number of cases, or density of the curve. The X axis is continuous and ranges from minus infinity to plus infinity. It is for this reason that the tails of the normal curve in Figure 5.1 never touch the X axis. The curve is completely symmetrical and unimodal; and the mean, median, and mode all have the same numerical value. By convention, the total area under the normal curve is 1.00 or 100%.

What relationship is there between the mathematical nature of the normal curve and data obtained in the real world? None whatsoever. However, by a happy *coincidence* many distributions obtained in research activities have properties similar enough to the formal mathematical properties of the normal curve so that the normal curve can be used as a model for generating statements about the data. In actuality it is incorrect to say that a set of data is "normally distributed," because we are referring to a mathematical formula when we use that expression. However, it is much more convenient to use that expression than to use the more formally proper one, to wit: "the distributional properties of the data are a close enough approximation to the mathematical properties underlying the normal curve that the model of the normal curve may be used as a generalized way of describing those data."

What are these mathematical properties of the normal curve which are so important to us as empirical researchers? We have continually emphasized that a frequency distribution, a mean, and a standard deviation are needed to describe

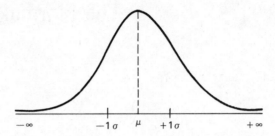

FIGURE 5.1 The normal curve.

the data. Although it is apparent that these three statistics are inherently related, we have not explicitly indicated what that relationship is until now. Remember that the area contained within the frequency distribution is the geometrical representation of n, or the number of cases in our sample. *Given a normal curve, we can specify the proportion or percentage of cases occurring between any points within our frequency distribution by using the mean as our reference point and the standard deviation as our unit of measurement.*

The principles involved here will be illustrated by the example below. We shall assume that we have a theoretical normal distribution with a population mean (μ) of 50 and a standard deviation (σ) of 8. This distribution is shown in Figure 5.2.

Figure 5.2 has a mean of 50, and the raw score values range from 26 to 74 along the X axis. Because the standard deviation is 8, the score value, 42, is one standard deviation below the mean whereas the score value of 58 is one standard deviation above the mean; a score of 34 is two standard deviations below the mean whereas the value of 66 is two standard deviations above the mean; and the scores of 30 and 70 are, respectively, two-and-a-half standard deviations units below and above the mean. These are also shown in Figure 5.2.

When talking about the normal curve, it is conventional to talk in standard deviation units rather than in raw score units in order to maintain generality. Thus, for all normal curves, 34.13% of all cases will fall between the mean and plus one standard deviation from the mean. (We shall see how we got this figure in a moment.) Because the normal curve is symmetrical, 34.13% of the cases will also fall between the mean and minus one standard deviation. Therefore, we see that the mean plus and minus one standard deviation ($\mu \pm 1\sigma$) will encompass 68.26% of all cases, or approximately two-thirds of the population. Between 1 and 2 standard deviations on either side of the mean will be found 13.59% of the cases. Between 2 and 2.5 standard deviation units from the mean will be found 1.66% of all cases, thus leaving 0.62% of the cases remaining beyond 2.5 standard deviations. Another way of expressing this is to state that the mean plus and minus two standard deviations ($\mu \pm 2\sigma$) will include approximately 95% of all cases in the population, whereas the mean plus and minus 2.5 standard deviations ($\mu \pm 2.50\sigma$) will encompass approximately 99%.

In Figure 5.2 the standard deviation units are rather gross. This is simply for convenience to portray the concept that there is an exact relationship between distance along the X axis in standard deviation units from the mean and area under the normal curve (or percent of cases in the population). The normal curve is continuous, and these units may be subdivided as finely as we wish. In Figure 5.3 the same information as shown in Figure 5.2 is presented but broken down into fractions of standard deviations. To the right of the mean the curve is broken into quarters of standard deviations, whereas to the left of the mean it is broken into halves. The percent of cases within each quarter- or half-unit is shown.

As indicated earlier it is usual to talk in standard deviation units rather than raw scores units when discussing the normal curve because the former is

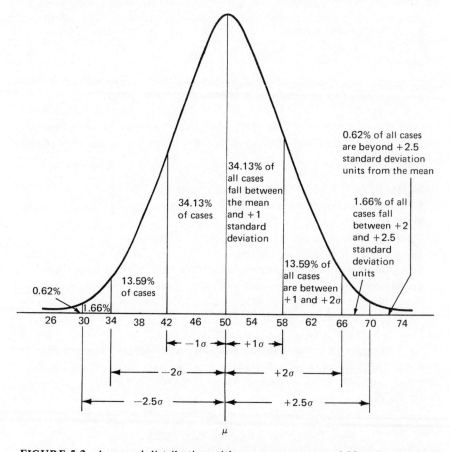

FIGURE 5.2 A normal distribution with a raw score mean of 50 and a standard deviation of 8. Beneath the raw score values are listed selected standard deviation units. The percentage of cases occurring within these standard deviation units are shown within the normal curve.

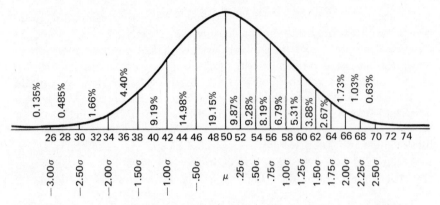

FIGURE 5.3 The same normal distribution as in Figure 5.2 with the X axis sub-divided into quarters of standard deviation units. The percent of cases are shown that occur within each segment specified by the standard deviation units.

universal whereas raw scores are unique to a particular problem. We can easily illustrate this point by considering the following example. Suppose that we had a normal distribution with a mean of 65 and standard deviation of 20 ($\mu = 65$, $\sigma = 20$). Answer the following questions. (As a suggestion, first convert the raw scores to standard deviation units and then use Figure 5.3 to help you find the answer.)

1. What percent of cases would have scores greater than 90?
2. What are the score values within which we could expect approximately two-thirds of the cases to fall?
3. What percent of the cases should get scores of 50 or more?
4. For a particular experiment we only want to work with the upper and lower 10% of the population. What raw score values would this represent?
5. For another experiment we want animals as homogeneous as possible and therefore decide to take the middlemost 55% of the population. What raw score values does this encompass?

Once you have answered these questions, and understand the principles involved in getting the answers, you have essentially mastered the basic concepts underlying the normal curve. The answers to these questions are given at the end of the chapter.

TABLE OF THE NORMAL CURVE

The percentages in Figures 5.2 and 5.3 were obtained from the table of the normal curve. This table is to be found in Table I in the Appendix. We shall now consider how to make use of this table with the help of Figure I in the Appendix.

Recall our basic principle that there are certain exact relationships between linear distances along the X axis in standard deviation units and areas under the normal curve. Therefore, we need to make our measurements along the X axis in units based on the standard deviation. This unit is called a *standard score* or *z score* and is defined as follows:

$$\text{Standard score} = z = \frac{X - \mu}{\sigma} = \frac{x}{\sigma} \tag{5.1}$$

What this formula does is place the raw scores of the X axis into standard score units. For example, look at Figure 5.2. The mean of that distribution is 50 and the standard deviation is 8. If we let $X = 42$, we may substitute in the above formula and get

$$z = \frac{42 - 50}{8} = -1$$

The z score of -1 indicates that the raw score of 42 is one standard deviation unit below the mean.

As another example, consider the raw score of 70 in Figure 5.2. Its standard score equivalent is 2.5, as the following formula shows:

$$z = \frac{70 - 50}{8} = 2.5$$

Therefore, the first thing to do when working with the normal curve is to translate the raw scores into standard scores, or z scores. Then enter the normal curve table with the z score unit in the first column of Table I in the Appendix. The next column, labeled A and titled "Area from mean to x/σ," gives the proportion of cases that will occur between the mean and the standard score value (see also Figures I(a) and I(b) in the Appendix).

All the data in Figures 5.2 and 5.3 were obtained from the first two columns of Table I in the Appendix. For example, in Figure 5.3 consider the value of 52 along the X axis: This is two raw score units above the mean, and with a standard deviation of 8 this represents a z score value of .25. If you look in the standard score column in Table I for the value 0.25, you will find that .0987 proportion of the cases (or 9.87%) will lie between the mean and that value, which is what we have entered in Figure 5.3. Look now at the value of 46 in Figure 5.3. This falls below the mean and therefore has a standard score value of $-.5$. The normal curve is perfectly symmetrical; therefore, negative standard score units will be numerically identical to positive standard score units except that they will be below the mean. Table I in the Appendix gives only half of the normal curve—that portion above the mean. If we enter Table I with the standard score value of 0.50, we find that .1915 proportion of all cases will fall within this area, and that is the value which we place in Figure 5.3.

The third column in Table I, labeled *B* and titled "Area in larger portion," contains the same information as in the second column except that 50% has been added to it—also see Figures I(c) and I(d). This includes the information from the other half of the normal curve and it gives the proportion of cases included from the extreme tail of the normal curve (infinity) up to the particular *z*-score value in which you are interested.

The fourth column, labeled *C* and titled "Area in smaller portion," indicates what proportion of cases are still to be found beyond your particular *z* scores (see Figures I(c) and I(d) as well). For example, look at a standard score value of 1.00, and you will find that beyond one standard deviation from the mean will be found .1587 proportion of all cases. Suppose that we wanted to find that standard score value that included all but 5% of the population. This means that the extreme 2.5% in each tail will be beyond this particular standard score value. If you look in column *C* of Table I, you will find that a *z* score of 1.95 will leave 2.56% of cases remaining in the smaller portion. The actual value of *z* required to encompass all but 2.50% in either tail is 1.96. If we wished to include all but the extreme 1% of our population, we would have to go out 2.58 *z* score units. These two values, 1.96 and 2.58, will be used later in the chapter when we talk about confidence limits.

The last column in Table I, labeled *y*, is rarely used by experimental researchers but is included in order to complete the normal curve table. This is the height of the normal curve (i.e., the *Y* axis) at any specified standard score value.

PROBABILITY AND THE NORMAL CURVE

We have now arrived at the central concept concerning the importance of the normal curve, which is that *it allows us to make probability statements about our data.* A probability figure is a numerical value between 0 and 1 which specifies the likelihood of an event occurring. In essence, it tells us the percentage of times we may expect something to occur. For example, an event that has a probability value of .75 is one that has a 75% probability of occurrence.

Because our normal curve table is set up in proportions, we may use it to make probability statements. Let us go back to our example of the normal distribution with a mean of 65 and standard deviation of 20 ($\mu = 65$, $\sigma = 20$), which we discussed on page 56, and rewrite those questions in the form of probability statements.

1. What is the probability of obtaining a score greater than 90?
2. What is the probability of obtaining scores either greater than 85 or less than 45 (i.e., what is the combined probability for both of these events)?
3. What is the probability of obtaining a score of 50 or more? Of less than 50?

4. What is the probability of obtaining a score of 40 or lower from this population?
5. If we sampled randomly from this population, what is the probability that we would draw cases within the score limits of 50 to 80?

    If you will compare your answers to these five questions with your answers to the questions on page 56, you will see that they are in perfect correspondence. This simple exercise should suffice to show you that a proportion value obtained from the normal curve table is identical to a probability statement. Thus, if we can justify the assumption of normality, we have at our disposal a very powerful tool which permits us to make exact quantitative statements about our data.

APPLICATION OF THE NORMAL CURVE: MAKING INFERENCES ABOUT THE POPULATION FREQUENCY DISTRIBUTION

    Let us now see how we can apply normal curve concepts in our research endeavors using our shuttlebox data to illustrate the principles. One general question concerns the distribution characteristics of future rats tested in the shuttlebox. You will recall that our 50 rats yielded a mean $(\bar{X})$ avoidance score of 40.02 with a standard deviation (s) of 8.97. These are sample statistics, and we do not have any information on the respective population parameters, μ and σ. However, these are all the data we have available to us, and so we proceed to use them to draw inferences about the distribution in the population. We do so by the use of the following logic, which employs the "If. . ., then. . ." type of sentence.

    If we are sampling from a normal population with $\mu = 40.02$ and $\sigma = 8.97$, *then* what will be the nature of the distribution that we expect to get in future samples? Note the arrangement of that sentence. We are making the assumption that μ and σ in the population are identical with $\bar{X}$ and s in our sample. If $\bar{X}$ and s are unbiased, then it it not unreasonable to expect that their numerical values will be somewhere near the population values, although it is most unreasonable to expect that these sample statistics will correspond identically with the population parameters. However, we make this assumption because it enables us to make some fairly general statements about the behavior of future rats in our shuttlebox. The validity of these statements can be tested as more research data are collected.

    Figure 5.4 is another normal curve in which we have marked off certain areas (i.e., percent of cases) under the curve. Along the X axis we have listed raw scores values based on the assumption that $\mu = 40.02$ and $\sigma = 8.97$ and, below them, some z score equivalents. We can see that the middlemost 50% of our population will have learning scores between the values of 34 and 46 (rounded off to the nearest whole number). Thus, if we wished a homogeneous group from our population for some particular experiment, we would be able to find about half of our animals with scores within 13 units of each other.

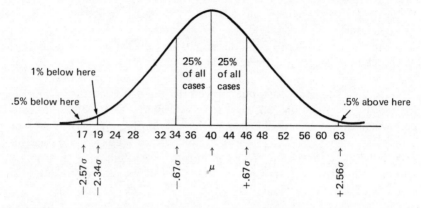

FIGURE 5.4 Theoretical normal curve for the shuttlebox data based on $\mu = 40.02$ and $\sigma = 8.97$. Both raw score values and some z score equivalents are shown on the X axis as well as the percent of cases at certain score limits.

Only rarely would we expect to find an animal with a score less than 19. From Figure 5.4 we see that this would occur only 1% of the time. If several animals gave us scores this low or lower we might wonder whether they are ill or whether our animal caretakers have been feeding and watering them regularly. We should recognize that by random sampling we shall get 1% of our animals with scores of 19 or lower, but if a larger number than 1% occurred this would justify suspicion.

We would almost never expect to find animals with scores as low as 17 or as high as 63, because these two groups together include 1% of our total population (one-half of 1% occurs in each tail). Thus if we had need for a very "bright" group of rats who demonstrated excellent learning ability—say we wished a group of animals all of whom scored 63 or higher—it would not be reasonable to try to select them from this population. Instead, we should change the characteristics of our population. How can we do that? Remember that our definition of a population must also include the operations in which we engage when measuring the animals as well as the method employed to select the sample of subjects. Thus part of our population definition includes the shock level used in the shuttlebox, the length of time the conditioned stimulus is on, the amount of time between trials, and so on. Changing any one of these conditions can lead to significant changes in our learning scores. Therefore, if we wanted a group of high-scoring rats, we would be more successful by manipulating the conditions in our learning experiment than by trying to select from this population a group of rats that met our demands.

APPLICATION OF THE NORMAL CURVE: MAKING INFERENCES ABOUT THE POPULATION MEAN

We now turn to the matter of making inferences about the population mean from our sample data. If we have randomly sampled from our population, our best estimate of the population mean, μ, is our sample mean, $\bar{X}$. Given random sampling, our mean would be unbiased and would be as likely to be above the population mean as below it, so that in the long run the average of all our sample means would be very close to the total population value. Thus one way, at least in theory, of trying to determine our population mean would be to repeat our study a second time, a third time, and so on, each time computing the mean until we had drawn a sufficient number of samples to allow us to obtain a sampling distribution of the mean. We could take the mean of this sampling distribution (i.e., the mean of the means) as our best estimate of the population mean, μ, and we could take the standard deviation of that distribution of means as an indicator of the amount of variability we would expect to obtain from one sample mean to another. In actual practice we would not do such a set of experiments—first because no one has the time, animals, and money to work out an empirical distribution of means in this fashion, and second because it is not necessary to do this as it is possible to take the data of the first experiment and draw inferences concerning the population mean and the variability around it.

Thus, there are two ways to get information about the population mean, μ: (1) construct an empirical distribution by repeating the experiment a sufficient number of times, or (2) use the data we have from the first experiment to draw certain inferences via statistical models. Even though we shall almost always use the second method, we shall gain a better understanding of the second method if we look at a hypothetical experiment utilizing the first method.

An Empirical Sampling Distribution

Suppose that we had drawn a second sample of 50 animals from our colony, repeated the shuttlebox experiment that we discussed in Chapter 3, and computed a mean for that set of 50 animals. And now suppose we did this a third time, a fourth time, and so on until we had repeated the experiment 40 times in total, each time computing the mean based on 50 cases. Table 5.1 lists the 40 means, and Figure 5.5 is a frequency distribution of those means. (You will recall from Chapter 4 that this frequency distribution is called a sampling distribution.)

The mean of the distribution of scores in Table 5.1 is 38.99 and its standard deviation is 1.19. Now this standard deviation is quite different from the standard deviation that we have encountered previously. Up until now all the standard deviations were based on variability among raw scores, whereas this standard deviation is based on variability among means. In order to avoid confusion between these two kinds of standard deviations, the standard deviation based on means is called the *standard error of a mean,* and is symbolized as $\sigma_{\bar{x}}$ when we are talking about a population of means, and as $s_{\bar{x}}$ when we are working with sample statistics.

There are several interesting features about Table 5.1 and Figure 5.5 which should be noted. First, the standard error of the mean is numerically much

TABLE 5.1   Distribution of 40 shuttlebox
means, each based on 50 cases

36.31	38.25	39.11	39.84
36.72	38.35	39.20	39.92
37.11	38.41	39.23	40.02
37.50	38.41	39.24	40.17
37.52	38.54	39.30	40.20
37.65	38.60	39.44	40.32
37.90	38.70	39.56	40.56
37.99	38.80	39.59	40.99
38.03	38.99	39.68	41.20
38.10	39.00	39.73	41.55

smaller than the standard deviation obtained in our first sample of animals in the shuttlebox. Next, the distribution of means is more normally distributed than was the distribution of raw scores in our shuttlebox example. These points are both illustrated in Figure 5.6, which shows the distribution of raw scores in our shuttlebox example (taken from Table 3.2); superimposed on this distribution is the distribution of means taken from Figure 5.5.

The difference in variability between the two distributions is quite outstanding. The variability among a group of means will *always* be less than the variability among a group of raw scores because the mean is a more stable figure and will vary less from one sample to another than will scores of individual animals. What will happen to the variability among sample means as the number of cases in the sample is increased? Intuitively we realize that the more cases that we have per sample the more stable is our mean, and therefore we would expect the variability among means to decrease as sample size increases. This is

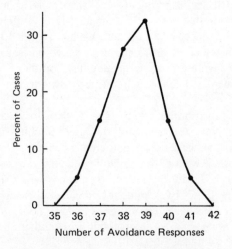

FIGURE 5.5   Distribution of 40 shuttlebox means based on $n = 50$ for each.

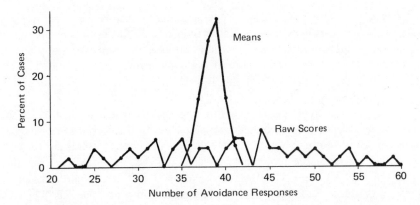

FIGURE 5.6 Comparison of the distribution of shuttlebox raw scores from Table 3.2 and the distribution of the shuttlebox means from Figure 5.5.

indeed the case and, in fact, there is a formula that will allow us to predict the variability among sample means if we know the standard deviation of the raw score distribution and the sample size. This leads us to a discussion of how to use the data of the first experiment to draw inferences about the population mean.

The Standard Error of the Mean

The formula for the standard error of the mean, if we know the population standard deviation, is

$$\sigma_{\bar{x}} = \frac{\sigma}{\sqrt{n}} \tag{5.2}$$

If we do not know the population standard deviation (which is the usual case), but we do know the sample standard deviation, the formula becomes

$$s_{\bar{x}} = \frac{s}{\sqrt{n}} \tag{5.3}$$

In formula 5.3 the value of s would be from our first experiment because that is all we usually have to go on. However, if we had information from more than one study, we know from Formula 3.8 that our standard deviation obtained by combining our data, s_{within}, is our best estimate. Our formula now becomes

$$s_{\bar{x}} = \frac{s_{within}}{\sqrt{n}} \tag{5.4}$$

Thus we see that the variability among means is directly dependent on the variability of the raw score distribution (which makes sense because the more variable are the raw scores the less stable is our estimate of the mean), and is inversely related to the square root of the number of cases (which also makes sense because, as discussed above, the larger the number of cases the more stable we expect our mean to be).

We saw in our hypothetical example in Table 5.1 that if we had drawn 40 samples each containing 50 animals and computed the mean, our standard error of the mean would have been 1.192. As indicated previously, this is not the way we would approach the problem in research. Instead, we would take the data from our first experiment and use that information to determine the standard error of the mean by substituting that information into Formula 5.3. When we do this, we obtain the following:

$$s_{\bar{x}} = \frac{8.97}{\sqrt{50}} = 1.269$$

We see, then, very close correspondence between the empirical value of the standard error based on our hypothetical 40 sets of samples ($s_{\bar{x}} = 1.192$) and our estimate of that value based on the information in our first sample ($s_{\bar{x}} = 1.269$). They are not expected to be exactly equal because of sampling error, but we would expect them to be numerically similar to each other, as they are.

Another feature of Figure 5.6 that has not been commented on previously is that the mean of our 40 samples (38.99) is not the same as the mean of our first sample (40.02). Again, we do not expect these values to be numerically identical because of sampling error. The value of 38.99 is a better approximation of the population mean, μ, than is the value 40.02, because it is based on more cases. However, in practice we would not have the value of 38.99 at our disposal because we would not have carried out so laborious a set of experiments. Instead, we would use the value 40.02 based on our first experiment. Even though 40.02 is our best estimate of the population μ, we are quite certain that μ differs from our sample mean by some numerical amount. What we now wish to do is use our sample mean, $\bar{X}$, to allow us to draw inferences as to possible reasonable values for the population mean, μ. To put this another way, we would like to set limits within which we would feel confident that the population mean, μ, fell. What do we mean by the expression "feel confident"? Statisticians have agreed on two reference points for this statement, which they call the *95% confidence limit* and the *99% confidence limit*. By confidence limits we mean that in repeated application of the procedure to independent samples, we shall be right (i.e., our limits will encompass μ) 95% or 90% of the time. We can now phrase our question in a more formal and exact form: Given a sample mean and a population standard deviation, how may we set a 95% (or 99%) confidence limit on our population mean, μ?

Setting Confidence Limits on the Population Mean

In order to set a 95% confidence limit on the population mean, we have to include all but 5% of our cases. From the table of the normal curve we find that when we go out 1.96 standard error units (i.e., 1.96 $\sigma_{\bar{x}}$) this will encompass all but 2.5% in each tail of the distribution. Therefore, the limits set by $\mu \pm 1.96\ \sigma_{\bar{x}}$ will include 95% of all means obtained from samples of size n. We can express this as follows:

The probability is .95 that

$$-1.96\ \sigma_{\bar{x}} \leqslant \mu - \bar{X} \leqslant +1.96\ \sigma_{\bar{x}} \tag{5.5}$$

If we add $\bar{X}$ to each term of the above inequality, we do not affect the relationship. This results in the following:

$$\bar{X} - 1.96\ \sigma_{\bar{x}} \leqslant \mu \leqslant \bar{X} + 1.96\ \sigma_{\bar{x}} \tag{5.6}$$

In words, over all possible samples of size n the probability is .95 that $\bar{X} \pm$ 1.96 $\sigma_{\bar{x}}$ will include the population mean μ. Therefore,

$$95\% \text{ confidence limit on } \mu = \bar{X} \pm 1.96\ \sigma_{\bar{x}} \tag{5.7}$$

The same logic is involved in obtaining the 99% confidence limit as with the 95% limit except that we must go out farther along our X axis to encompass all but 1% of our cases. From the table of the normal curve we find that 2.58 standard error units will include all but one-half of 1% of the cases in each tail of the distribution. Therefore, in order to obtain the 99% confidence limit we use the following formula:

$$99\% \text{ confidence limit on } \mu = \bar{X} \pm 2.58\ \sigma_{\bar{x}} \tag{5.8}$$

Interpretation of Confidence Limits

In order to apply Formula 5.7 to our shuttlebox data, we have to assume that the population standard deviation is known. We shall assume that $\sigma = 8.97$, so that $\sigma_{\bar{x}} = 1.27$ when $n = 50$. Substituting into Formula 5.7, we obtain

$$40.02 \pm (1.96)(1.27) = 37.53 - 42.51$$

Now that we have obtained these confidence boundaries, let us be certain that we know exactly what they mean. When we state that the 95% confidence limit on the population mean, μ, is 37.53 to 42.51, we are stating that we are subjectively 95% confident that the true population value, μ, falls within these limits. This is *not* a probability statement. The population mean, as you remember, is a fixed value and does not vary from one situation to another. Therefore either μ does fall within the confidence boundaries we have set, or it does not. The reason we feel 95% confident that the mean is within the limits we have set is based on the logic that we have used in obtaining these limits. It is quite important to understand the logic involved here rather than merely to apply a formula blindly.

There is one instance in which we can make a probability statement about our population mean. If we had repeated our experiment a number of times, and for each of our samples we had computed a 95% confidence limit, we could state that, for all of our samples, in 95% of the cases our population mean would fall within the confidence boundaries set. Note, however, that it is necessary for us to repeat the experiment a number of times in order to make a probability statement. On any one occasion all we can do is make a confidence statement based on the logic we have already discussed concerning our subjective feelings about how right we are.

TESTING HYPOTHESES ABOUT THE POPULATION MEAN

In setting a confidence interval our approach is to specify a band width within which, we believe, our population mean lies. All scores within that band are considered to be equally likely values for our population mean. But suppose that we wished to test a particular hypothesis about the value of our population mean. For example, returning to our shuttlebox problem, suppose that a research colleague from another laboratory had the same apparatus as we had and also purchased his animals from the same supplier as we did. We would be curious to know whether our animals had similar learning scores to his. Let us assume that he had tested many animals in his shuttlebox over an interval of several years, and had obtained a mean learning score of 38.88. If we consider this value to be our best estimate of the population mean, how likely is it that our sample mean of 40.02 could have been drawn from such a population? We shall assume that our population standard error is known to be 1.27. To answer the question posed above, we set up our z score question from Formula 5.1 in the following form:

$$z = \frac{\bar{X} - \mu}{\sigma_{\bar{x}}}$$

Making the appropriate substitutions, we obtain

$$z = \frac{40.02 - 38.88}{1.27} = 0.90$$

On the assumption that we are sampling from a normal distribution, we may enter our table of the normal curve with the z score of .90. We find that a sample mean of 40.02 *or higher* could have arisen by chance 18.41% of the time if we were randomly sampling from a population with a mean of 38.88. Thus we may conclude that it is quite likely that the population from which we drew our sample could have a mean of 38.88.

In the statement above, the expression a score "of 40.02 *or higher*" was used. Whenever we test a hypothesis, we must also consider the possibility of drawing even more extreme scores from our population, and so it is necessary to determine the probability of obtaining our particular value as well as all values more extreme than it. Therefore, we use the column entitled "Area in smaller portion" in our table of the normal curve to evaluate our z score.

ASSUMPTIONS UNDERLYING THE USE OF THE NORMAL CURVE TABLES

There are three major assumptions that must be met in order for us to be able to use the normal curve tables and to assign probability values to our findings.

Random Sampling

Our first assumption is that we have a random sample. We have indicated that in the real world of day-to-day research we are probably going to violate that assumption, but we should still strive to sample randomly whenever we can. Two ways that we have of checking on the random sampling assumption are by repeating our experiment or by comparing our results with those of other researchers (as in the example above). If we get consistent results from one experiment to another within our laboratory, or from one lab to another, then we can feel confident about our assumption of random sampling.

Independence of Observations

Our second assumption is that we have independent (i.e., uncorrelated) observations. This means that the occurrence of one score in no way influences the probability of occurrence of the same or other scores on subsequent trials. Statistically, if each observation is not independent of every other observation, then the laws of chance will not operate to give us a normal distribution. And experimentally, if our measurements on different animals are not independent, then we have a worthless experiment.

Another way to describe independence from the experimenter's perspective is to state that the score obtained on an animal must not be systematically influenced by any event except the experimental variable. Sophisticated researchers are aware of these problems and sensitive to them. Thus, for example, in many biological and behavioral studies the animals are evaluated "double-blind," so that no one scoring the animal knows from which treatment group it came. For the same reason, many researchers prefer automated equipment so that any biases or idiosyncracies of the research technician are minimized or eliminated. Quite often researchers work up detailed procedures to balance out such variables as the sequence in which animals are tested, the time of day of testing, the place where testing is done (e.g., if two similar pieces of test apparatus are located in two different rooms), and other possible contaminating variables.

Normal Distribution

The third assumption we must make is that the statistic in which we are interested is drawn from a normally distributed population. The stringency of this assumption varies as a function of whether, as experimenters, we are interested in the form of the raw score distribution or the mean of the raw score distribution.

Normality and the form of the distribution If our experimental question concerns the form of the distribution—as in the previous section in this chapter entitled "Applications of the Normal Curve: Making Inferences About the Population Frequency Distribution"—then it is necessary that our raw scores be normally distributed. This is a very stringent assumption, because we know that many variables in biology and behavior are not normally distributed in the population.

Normality and the mean of the distribution As experimenters our research questions usually revolve around the mean of the distribution rather that the form of the distribution. For example, the usual research question we ask is: Did my experimental treatment change the mean value in the study, relative to my control group? When our primary interest is in the mean, rather than the form, of our distribution of raw scores, then the normality assumption is much less stringent. We may make two statements about the distribution of means in the population: (1) if the raw score population is normally distributed, then means from that population will also be normally distributed, and (2) even if the raw score population is *not* normally distributed, means from such a population *will be* normally distributed if the random sample is sufficiently large. This latter statement is known as the *central limit theorem.*

THE CENTRAL LIMIT THEOREM

The importance of the central limit theorem for experimental researchers cannot be overemphasized. Regardless of the form of the raw score distribution, we can show that means will be normally distributed if we have a random sample of sufficient size. Statisticians seem to agree that approximately 20 to 30 cases per group is sufficiently large so that we can use normal curve statistics in evaluating our means.

No attempt will be made to prove this theorem, but the reader can get an appreciation for the phenomenon by examining Figures 5.7 and 5.8. In Figure 5.7 we see a skewed distribution of 10 cases, whereas Figure 5.8 shows us a bimodal distribution of 10 cases. In both instances all possible samples of 2 cases were obtained, and the distribution of means of these cases are also shown in the tables. We can see that in both instances the distribution of means tends toward the normal.

DISTINCTION BETWEEN A DISTRIBUTION OF RAW SCORES AND A DISTRIBUTION OF MEANS

The beginning student is often confused by the distinction made between a distribution of raw scores and a sampling distribution of means. An examination of Figures 5.7 and 5.8 shows the differences clearly. If you were to draw random samples of 3 cases from those two raw score distributions, you would find that the distributions of the means would be even closer to a normal distribution than the distribution based on 2 cases per mean in the figures.

For experimental researchers the principles and procedures are clear: In most cases we are going to be interested in the mean of our group or groups. In order to evaluate those means we shall have to assume that the means are normally distributed. In order to meet that assumption we need independent observations from a random sample that (1) is drawn from a normally distributed raw score population, or (2) has approximately 20 to 30 cases per group. If we do not have a random sample, then the study should be replicated with independent observations.

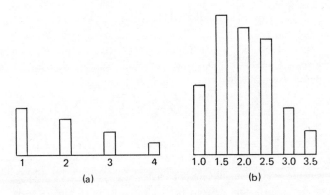

FIGURE 5.7 Demonstration that a distribution of means (Figure 5.7*b*) will tend to be normally distributed even though the raw score distribution is markedly skewed (Figure 5.7*a*).

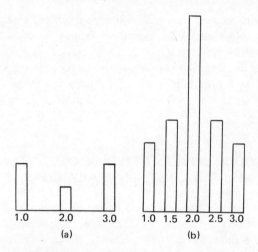

FIGURE 5.8 Demonstration that a distribution of means (Figure 5.8*b*) will tend to be normally distributed even though the raw score distribution is bimodal (Figure 5.8*a*).

Answers to Questions on Page 56

1. 10.56%
2. 45–85
3. 77.34%
4. Below 39 and above 91
5. Between 50 and 80

SUMMARY OF DEFINITIONS

Central limit theorem A theorem which states that, regardless of the distribution of raw scores in the population, the sampling distribution of means is approximately normal when the sample size is large.

Confidence limit An estimated range of values set around the sample mean with a given high probability (usually 95% or 99%) of including the population mean.

Independence Two or more events are said to be independent if the occurrence of one event in no way influences the probability of occurrence of the same or other events on subsequent trials.

Standard error of the mean The standard deviation of the sampling distribution of means.

Standard score A score based on standard deviation units. It is obtained by subtracting the mean of the distribution from a raw score and then dividing the resulting value by the standard deviation of the distribution. This is also called a z score.

   z score   Another name for a standard score.

PROBLEMS

1. Given a normal distribution with $\mu = 42$ and $\sigma = 8$, convert each of the following raw scores to standard scores and then determine the proportion of cases that are beyond each standard score value by using the table of the normal curve.

$$24, 30, 34, 40, 42, 46, 48, 56, 66$$

2. A number of national achievement tests are scaled so that the mean is 500, the standard deviation is 100, and the distribution is normal. Given these parameters, answer the following questions.
   (a) What is the probability of obtaining a score of 600 or better?
   (b) What percent of cases would fall below a score of 350?
   (c) What percent of cases would be expected to have scores greater than 750?
   (d) What proportion of cases will fall between 450 and 625?
   (e) An admission committee decides that it will only consider students who have scores in the upper one-third of the distribution. What is the minimal score needed for consideration?
   (f) A student is told that her score was such that only 5% of the population exceeded her. What was her score?

3. The mean weaning weight for rats in one laboratory was found to be 58 grams with a standard deviation of 8.5. On the assumption that these values are good estimates of μ and σ in the population, and that the distribution is normal, answer the following questions.
   (a) What is the weight below which will fall 5% of the population?
   (b) What is the weight above which will fall 10% of the population?
   (c) What percent of rats will fall between 55 and 64 grams?
   (d) The researcher is interested in obtaining a group of rats that weight no more than 35 grams at weaning. What proportion of animals from his colony will meet this criterion?
   (e) What will be the weight range of the middlemost 50% of his colony?

4. The researcher in the previous problem wishes to set confidence limits for the population mean weaning weight. He is willing to assume that the population $\sigma = 8.5$. Given the sample $\bar{X} = 58$ and $n = 25$, what is the value of $\sigma_{\bar{x}}$? Use this value to obtain the following confidence limits: 95%, 99%, and 90%.

5. In a small liberal arts college the average Verbal Graduate Record Examination (GRE) score of the 334 graduating seniors was found to be 575 and their mean Quantitative GRE was 547. On the assumption that the population σ for both tests is 100, what is the standard error of the mean $(\sigma_{\bar{x}})$ for this graduating class? Use this value to set the 95% and 99% confidence limits on the population means for the Verbal and Quantitative GRE scores.

CHAPTER 6

The *t* Distribution

In the previous chapter we introduced a statistic called a *z* score, or standard score, which we used to enter the table of the normal curve for purposes of determining a probability value. Let us examine certain characteristics of that statistic. The formula is reproduced below for the condition in which we are attempting to evaluate a sample mean relative to the population mean.

$$z = \frac{\bar{X} - \mu}{\sigma_{\bar{x}}}$$

If we examine the numerator of that formula, $(\bar{X} - \mu)$, we see that we have a variable, the sample mean, which will change from one experiment to another and a constant, the population mean, which is a fixed value. We have also learned that the distribution of sample means will be normal either (1) when the distribution in the population from which the mean was drawn is normal, or (2) when the sample size is sufficiently large. Therefore, when either of these conditions is true, the value we obtain in the numerator of our formula $(\bar{X} - \mu)$ will be normally distributed, because subtracting a constant from any set of data will in no way change the form of those data.

Consider now the denominator. You will recall that a statement was made in the previous chapter that we had to assume the population standard deviation was known in order to use the table of the normal curve. The reason for this assumption will now be explained. The population standard deviation is a fixed value, and *n* is also a fixed value, so the quantity $\sigma/\sqrt{n}$ is a constant. Therefore the numerator of our equation is now divided by a constant, and dividing by a constant will in no way change the form of the distribution.

In sum, therefore, we see that the standard score, *z*, is normally distributed because we have a normally distributed variable $(\bar{X})$ from which we have subtracted a constant (μ) and divided by a constant $(\sigma_{\bar{x}})$. It is for these reasons that we can evaluate *z* by use of the table of the normal curve.

But is a rare event in experimental research when the researcher knows the numerical value of the population standard deviation. In most circumstances all that we shall have is an estimate of that value by means of our sample standard deviation, s. When we have s available to us instead of σ, then we have a new statistic called the t statistic which is defined as follows:

$$t = \frac{\bar{X} - \mu}{s_{\bar{x}}} \qquad (6.1)$$

The t statistic is identical with the z statistic in the numerator, but the two denominators differ, and this makes a world of difference concerning the nature of the distribution we obtain. The difficulty, of course, is that $s_{\bar{x}}$ is variable rather than having a fixed value. For example, if we were to draw two independent samples with the same size n, and were to compute the means and standard deviations for these two samples, it is extremely unlikely that the numerical values of the two standard deviations would be identical.

What are the consequences of all this? First, the numerator of our t statistic, just like the numerator of our z statistic, will be normally distributed. However, the denominator of our t statistic is a variable and has a distribution of its own which is not normal (it is the square root of a chi-square distribution, although that is not an important piece of information at this point). When we divide the normally distributed numerator by a denominator that varies, the resulting values will not be normally distributed except when n approaches infinity.

Why should the t distribution approximate a normal distribution as n approaches infinity? The answer is that as our sample size gets larger, we are able to make a more accurate approximation of our population standard deviation, σ; and when our sample size is infinite, then we are able to compute σ exactly. When the sample size is very small, however, our estimate of σ from our statistic, s, is highly erratic; for this reason, the form of our t distribution changes as our sample size changes, being least normal in form when the sample size is small, and being exactly normal when the sample size is infinitely large. Figure 6.1 shows several t distributions of varying sample size.

DEGREES OF FREEDOM

In Figure 6.1 we see that the number of cases in our sample is a critical parameter in the definition of our t statistic. However, this parameter is not defined in terms of n but in terms of *degrees of freedom* which, in this instance, is equal to $n - 1$. Thus, in Figure 6.1 there are 2, 10, and 26 cases associated with the values of 1, 9, and 25 degrees of freedom, respectively. When degrees of freedom are infinite, we have a normal curve. (The letters *df* are an abbreviation for degrees of freedom.)

The concept of degrees of freedom is an important one, which will be with us for the rest of this book (and for the remainder of your research life if you use statistics in evaluating data), and we should give it some attention. The concept concerns the sample variance (or standard deviation, which is the square root of the variance), and specifies the number of ways our data are "free" to vary in

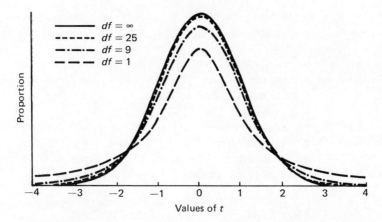

FIGURE 6.1 Distribution of the *t* statistic for various degrees of freedom. As the *df* becomes infinite, the distribution of *t* becomes normal. (From D. Lewis, *Quantitative Methods in Psychology*, New York: McGraw-Hill, 1960. Copyright 1960 by the McGraw-Hill Book Company, Inc. Reprinted by permission.)

determining the variance. This can be made clear by looking again at Formula 3.6 for the variance, which is shown below.

$$s^2 = \frac{\Sigma(X - \bar{X})^2}{n - 1} \qquad (3.6)$$

The important part of this formula concerns the numerator, which specifies that we are to take the deviations from the mean, square, and sum them. However, recall that Formula 3.3 specified that the sum of the deviations about the mean must equal zero, as indicated below:

$$\Sigma x = \Sigma(X - \bar{X}) = 0 \qquad (3.3)$$

The algebraic expression in Formula 3.3 is the same as the numerator of Formula 3.6 except that we have squared the value in Formula 3.6. However, it is apparent that whatever restriction we have on Formula 3.3 must also apply to Formula 3.6. There is one restriction on Formula 3.3, namely that the sum of the deviations around the mean must equal zero. We can satisfy that restriction by allowing *all but one* of our deviations to be whatever number we arbitrarily choose. However, that last number is fixed because its numerical value must be such that the sum of the deviations equals zero. Consider a simple example in which there are three cases. We can arbitrarily decide that the deviation from the mean of our first case will be +7. We can equally arbitrarily decide that the deviation from the mean of our second case will be −3. Once we have done this, then it is necessary for the deviation from the mean of our third case to equal −4 so that the sum of the deviations (+7, −3, −4) equals zero.

Another way of expressing the same concept is to state that one degree of freedom is used up in determining the mean of the sample. Once the mean is fixed for the sample, then the deviations around that mean are used to obtain the variance (or standard deviation), so that there are $n - 1$ degrees of freedom remaining for determining the variance. This is seen by examining the divisor in the formula for the variance or standard deviation—it will always be equal to the df.

As you go through this book you will find other statistics for which there will be formulas to determine the degrees of freedom. However, the principle is the same in all instances: df always specifies the number of ways that a particular statistic can vary and still meet the restrictions set down by our formulas.

CHARACTERISTICS OF THE t DISTRIBUTION

In inspecting Figure 6.1 we see that the t distribution is symmetrical, and that the major difference between the t and the normal is that the tails of the t distribution are raised more above the baseline than the tails of the normal curve. In consequence, we have to go out more standard errors (i.e., standard deviation units) along the X axis with the t distribution than we have to with the normal distribution in order to encompass an equivalent percentage of the cases. This is shown in Figures 6.2, 6.3, and 6.4 for the t distributions with 1, 9, and 25 degrees of freedom, respectively. Remember that with a normal curve we have to go out ±1.96 standard deviation units from the mean to include the middlemost 95% of the cases in our population, and ±2.58 standard deviation units to encompass all but the extreme 1% of our population.

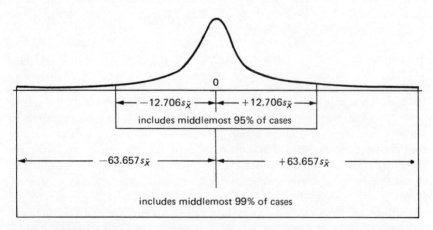

FIGURE 6.2 Distribution of the t statistic for 1 df, showing the number of standard deviation units we must go from the mean to include the middle 95% or 99% of the population. The standard deviation units on the X axis are not to scale.

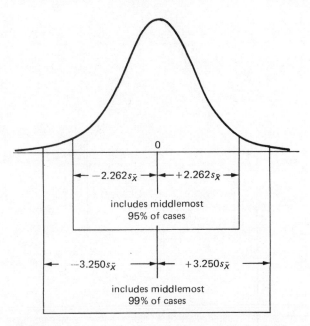

FIGURE 6.3   Distribution of the *t* statistic for 9 *df* showing the number of standard deviation units one has to go from the mean to include the middle 95% or 99% of the population.

In Figure 6.2 we see that if we were to conduct experiments using a sample size of 2, we would need to go out 12.706 standard errors of the sampling distribution on either side of the mean to encompass the middle 95% of cases in our population, and we would have to go out 63.657 standard errors in order to include all but one-half of 1% in each tail of the distribution. It is apparent that it is quite inefficient to try to conduct experiments with sample sizes of 2.

When our sample size increases to 10, we see from Figure 6.3 that the number of standard deviation units we must go out from the mean to include the middlemost 95% or 99% of the population is markedly less than for sample sizes of 2, in this instance being ±2.262 and ±3.250 standard errors, respectively. When we move to the situation where we have 25 degrees of freedom (Figure 6.4), we are beginning to approximate the normal distribution, and we find that we must move out ±2.060 standard deviations units to include all but 2½% in each tail, whereas ±2.787 standard errors will take in the middlemost 99% of our cases.

TABLE OF THE *t* DISTRIBUTION

The numerical values for the standard deviation units in Figures 6.2, 6.3, and 6.4 were obtained from the table of the *t* distribution in Table II in the

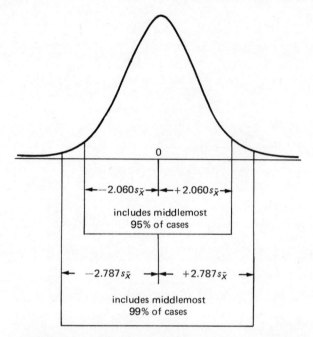

FIGURE 6.4 Distribution of the *t* statistic for 25 *df* showing the number of standard deviation units one has to go from the mean to include the middle 95% or 99% of the population.

Appendix. This table contains much of the same information that is contained in the table of the normal curve, but in somewhat different form. First of all, there is a *t* distribution for every possible degree of freedom, and so to tabulate these as we did the normal curve would take up a complete textbook by itself. What we do instead is take selected probability values along the *t* distribution and table them for differing *df*. Thus, in Table II the rows list degrees of freedom and the columns give you probability values. These are *two-tailed* probability values, and they tell us the total proportion of cases remaining in both tails of our distribution beyond the limits set by the standard error. This is in distinction to the normal curve table, which is *one-tailed*.[1] This can be illustrated by looking at the first line in the *t* table for 1 degree of freedom and moving over to the column where the probability value is .05. The numerical value in that cell is

[1] The distinction between one-tailed and two-tailed probability values has to do with the way that statistical tables are arranged, and is something which should be kept in mind when using tables containing probability values. One-tailed values give the proportion of cases remaining in the right-hand tail of the probability distribution after going out so many standard error units. This value must be doubled to give the equivalent two-tailed probability value.

12.706 and if you now look back at Figure 6.2, you will see that this is the number of standard deviation units you must go out on either side of the mean to encompass all but 5% of the cases. If you now look under the column where the probability is .01 with 1 degree of freedom, you will find the value of 63.657, which in Figure 6.2 is the number of standard errors that you must go out on either side of the mean to include the middlemost 99% of all cases.

The same situation applies for the cases where we have 9 degrees of freedom and 25 *df*, and you should compare the numerical values in Table II with the information given in Figures 6.3 and 6.4 so that their relationships are firmly understood.

The table of the *t* distribution is usually given in some detail for the first 30 *df* and after that only the .05 and .01 probability levels are listed. Look in Table II at the row listing 30 *df* and also the row just below it which gives comparable values for the normal distribution (i.e., infinite *df*). You will find that there is very good correspondence between the normal curve and the *t* distribution with 30 *df* in the middle of the curves (i.e., for high probabilities), but the discrepancy between these two distributions becomes greater as you move toward the extremes. If you examine the set of numbers at the bottom of Table II, you will find that it is not until you have 100 degrees of freedom that the numerical values for *t* at the .05 and .01 levels are close to the numerical values for the normal curve at the same probability levels. Therefore, unless you have very large degrees of freedom, you should always use the *t* table, rather than the normal curve table, in evaluating your data.

WHEN TO USE THE *t* DISTRIBUTION

At this point you may be confused because we stated in Chapter 5 that means will be normally distributed when *n* equals 20 to 30, and that we could use the table of the normal curve to evaluate the means. This is true for the situation in which the value of the population standard deviation, σ, is known. Because σ is usually unknown, normal curve tables generally cannot be used to assess data. However, the *t* table can always be used, because this is based on the use of the sample standard deviation, *s*. There are two general rules for the use of the *t* table, as follows:

1. When sampling from a population with a normal distribution of raw scores, the *t* table may be used, regardless of sample size.
2. When *n* equals 20 to 30 per mean, the *t* distribution can *always* be used, regardless of the form of the population distribution.

Thus, in the situation where we are interested in evaluating means, the same principles enumerated in Chapter 5, where σ was known and normal curve tables could be used, also apply here where σ is unknown and we have to use *s* and the *t* table. Because of their importance for experimental researchers, these principles are repeated below and modified slightly to take into account the t distribution:

Most researchers are primarily interested in the mean of their group or groups. In order to evaluate those means we must assume that they are normally distributed. In order to meet that assumption we need independent observations from a random sample which (1) is drawn from a normally distributed raw score population, or (2) has approximately 20-30 cases per group. If we do not have a random sample, then the study should be replicated with independent observations. In the case where σ is known, the normal curve table may be used to evaluate the mean or means. When σ is unknown—the usual case with experiments—s is used as an estimate of σ, and the t distribution is used in place of the normal distribution.

APPLICATION OF THE t STATISTIC: STANDARD ERRORS AND CONFIDENCE LIMITS

It is always important to keep in mind our experimental objectives when discussing statistical matters so that we never lose sight of the forest because of the trees. One prime objective is to get an estimate of our population mean from the sample data we have in hand. We know that our sample mean is our best estimate of the population mean, μ, for the particular conditions under which we drew our sample and made our measurements. However, we also know that if we repeat the experiment a second time, it is highly unlikely that we shall obtain the same numerical value for the mean as we did in our first study. An indication of how much we are likely to be off is given by the standard error of the mean, which is defined as $s/\sqrt{n}$ and abbreviated SE.

We see, therefore, that we have two *estimators* of our population mean: a *point estimator* based on our sample mean, and an *interval estimator* based on our standard error. In fact, it is conventional when we report means in the research literature to attach the standard error as well. Symbolically, we have $\bar{X}$ ± SE. As an example, in our shuttlebox situation we had a mean of 40.02 with a standard error of 1.27, and we would present these data in a research paper as 40.02 ± 1.27. This immediately indicates the best point estimate of the population mean, and the standard error gives us some idea of the variability we would expect in means of future samples. Thus, if the standard error is numerically very small, our estimate of the population mean is said to be *precise*. Under these conditions, we are much more likely to pick up a difference brought about by an experimental treatment than if our standard error was relatively large so that our estimation of the population mean is imprecise.

Although we can use $\bar{X}$ ± SE as a rough-and-ready way of evaluating the data of an experiment, we need a more exact quantitative method to specify limits within which we feel confident that the population mean falls. Therefore, we set confidence limits around the population mean following the same logical procedure as we developed in the previous chapter. In that chapter, where we had the normal curve as our reference point, we knew that if we went out ±1.96 standard errors from the mean we included the middlemost 95% of our cases, whereas 2.58 standard errors gathered in all but one-half of 1% in each tail of our distribution. We could only use this approach on the assumption that our population standard deviation was known. In the situation where we do not know the population σ, and have to use our sample statistic, s, as an estimate of it, we must use the *t* distribution rather than the normal. In this situation it is necessary to know the degrees of freedom of the *t* distribution as well as the percentage of cases we wish to include within our interval estimate. It is convenient to put these various considerations down into a concise expression as follows:

1. Let α (read *alpha*) designate the percentage of cases outside our domain of interest. In virtually all instances this will refer to the remaining 5% or 1% of the cases after we have set limits that include the middlemost 95% or 99% of the population. Therefore, α generally will equal .05 or .01.
2. Let *df* equal the number of degrees of freedom for the sample we have in hand.
3. Then the expression, t_{α} (*df*), contains in compact form all the information needed to determine the numerical value of *t*.

We can illustrate this by means of our shuttlebox example. The relevant statistics from that experiment are as follows:

$$\bar{X} = 40.02$$
$$s = 8.97$$
$$s_{\bar{x}} = 1.27$$
$$n = 50$$

If we wish to find the 95% confidence interval on our population mean, we must find the value of *t* that satisfies the expression, $t_{.05}(49)$. In Table II in the Appendix we see that *t* is tabled for 48 and for 50 *df*. If we interpolate, we find that the numerical value of *t* with 49 degrees of freedom at the .05 level is 2.009. This *t* value tells us that we must go out 2.009 standard error units on either side of the mean to include the middle 95% of our cases when we have 49 *df*. Our standard error is 1.27, and so we multiply that value by 2.009 and get 2.55. If we now add and subtract this to our mean, we have 40.02 ± 2.55 = 37.47 − 42.57. We can now state that we are 95% confident that the true population mean falls within these limits. Do not forget that a population mean either *does* or does *not* fall within these limits, because it is a fixed value, and that a confidence statement is not the same as a probability statement.

If we wish to be even more confident concerning the limits of the population mean, then we use the 99% confidence interval. In Table II of the Appendix we find that the value of t satisfying the expression $t_{.01}(49)$ is 2.680. Because our standard error is 1.27, we have the equation

$$40.02 \pm (1.27)(2.680) = 40.02 \pm 3.30 = 36.72 - 43.32$$

Let us look at this confidence interval through the eyes of an experimenter. It tells us that we can expect our sample means to fall between the limits of 36.72 and 43.32 in virtually all future experiments of sample size 50 drawn from our population in the same fashion as we drew this sample and using the same measurement procedures as we used in this experiment. As researchers we now have to decide whether this is precise enough for our investigations. For example, suppose that we are interested in studying a pharmacological preparation that we have reason to believe would improve learning scores by at least 10 points. This is clearly beyond the range of variability we could expect to get within our control groups (i.e., those animals not receiving the pharmacological preparation), and thus we would feel quite confident that we could demonstrate that our pharmacological preparation would have a "significant" impact on the learning behavior of our animals.

On the other hand, suppose that we are working with a drug that has rather sublte effects on the central nervous system, and that we do not expect to change the learning scores of our animals by more than 3 or 4 points. We see that the 99% confidence interval on our mean encompasses 6.60 points, and so it would be much more difficult to demonstrate a significant effect of only 3 or 4 points, because that amount of difference could occur by normal variation in random sampling from one experiment to another. In this situation, we should strive to make our estimation of the populations mean more precise before embarking on an experimental investigation of the drug. We already know one way to increase the precision of the experiment—increasing the sample size will decrease the standard error of the mean—but there are many other techniques also available, and one of the major purposes of Chapters 7, 8, and 9 is to show ways of designing experiments to increase their precision.

A GENERAL FORMULA FOR ESTIMATING CONFIDENCE LIMITS

Now that we have worked through the steps involved in determining confidence limits, we can put this all together into a general formula as follows:

$$\text{Confidence limits} = \bar{X} \pm [t_\alpha \, (df) \,] \, s_{\bar{x}} \qquad (6.2)$$

In order to obtain the 95% confidence limit, let $\alpha = .05$ in the above formula; for the 99% confidence limit, $\alpha = .01$.

Formula 6.2 is directly comparable with Formulas 5.7 and 5.8 in the previous chapter. For comparison, Formula 5.7 is reproduced below.

$$95\% \text{ confidence limits} = \bar{X} \pm 1.96 \; \sigma_{\bar{x}} \qquad (5.7)$$

In Formula 6.2 the term in brackets, $[t_\alpha \; (df)]$, is functionally equivalent to the value, 1.96, of Formula 5.7. Indeed, when the degrees of freedom are 1000 or more and $\alpha = .05$, the two values will be numerically equal. The last term in Formula 6.2, $s_{\bar{x}}$, is our best estimate of the population value, $\sigma_{\bar{x}}$, of Formula 5.7.

SUMMARY OF DEFINITIONS

Degrees of freedom Number of ways that a statistic can vary and still meet the restrictions placed on it. The degrees of freedom will always be some number less than *n*, the sample size.

One-tailed probability values The proportion of cases remaining in the right-hand tail of a distribution beyond the limit set by a standard error score. This probability value must be doubled to give the equivalent two-tailed probability value. The normal curve table in the Appendix lists one-tailed probability values (Table I).

Precision The numerical value of the standard error of the mean indicates the precision with which the mean is estimated. The smaller the standard error, the more precise is our estimate of the population mean.

Two-tailed probability values The proportion of cases remaining in both tails of the distribution beyond the limits set by a standard error score. The *t* table in the Appendix lists two-tailed probability values (Table II).

PROBLEMS

1. A clinical psychologist finds that the standard deviation on a personality questionnaire administered to introductory psychology students is 28 points. Assuming that he will find that $s = 28$ in future samples, what must his *n* be if he wishes his standard error of the mean ($s_{\bar{x}}$) to be equal to (a) 10, (b) 5, (c) 2.5, (d) 1.25, (e) 1?

2. (a) An endocrinologist shocks a group of 10 rats and then determines the amount of corticosterone in the blood of each animal. She finds that $\bar{X} = 38.6 \; \mu g/100$ ml with $s = 7.1$. What is $s_{\bar{x}}$?
   (b) Set the 95% and 99% confidence limits for the population mean.
   (c) The endocrinologist would like a standard error no greater than 1.75. What must her *n* be to achieve this value?

3. A researcher finds, in an experiment with 10 subjects, that $\bar{X} = 95.7$ and $s = 13.2$. He is willing to assume that the mean and standard deviation will be approximately the same in future samples, and he is interested in determining what effect sample size will have on his standard error and confidence limits. To answer these questions he sets up the following table. Fill in all the blanks in the table.

n	s	$s_{\bar{x}}$	$t_{.05}(df)$	Lower 95% limit	Upper 95% limit	$t_{.01}(df)$	Lower 99% limit	Upper 99% limit
10	13.2							
15	13.2							
20	13.2							
25	13.2							
50	13.2							
100	13.2							
200	13.2							

Note that increasing n does two things (1) it decreases $s_{\bar{x}}$, and (2) it decreases the numerical value of $t_\alpha(df)$.

4. A research paper reports the basic data as 107.68 ± 9.85 based on 25 cases. On the assumption that the researcher is presenting a standard error, what is the value of s?

5. An experimenter studying learning finds that the mean learning score for one group of 15 subjects is 38.2 with a standard deviation of 5.3. She repeats the study with a second group of 10 subjects drawn from the same population and obtains a mean of 37.0 with a standard deviation of 4.8. Because the two sets of data are very similar, she decides to pool the results.
   (a) What is the mean of the pooled data?
   (b) What is s_{within}?
   (c) How many degrees of freedom are associated with s_{within}?
   (d) What are the 99% confidence limits on the population mean?

7

Comparing the Means
of an Experimental
and a Control
Group

Until now we have been concerned with methods for characterizing *one* group of subjects by means of descriptive and inferential statistics. These are the necessary steps researchers go through when they start off in a new research area, try to develop a new assay procedure, test out a new piece of apparatus, and so on. Once we know how our baseline or control group is going to perform under a set of standardized conditions, and once we know the probable range of variability in the performance of subsequent samples from the same population, we are then able to move to our prime objective, which is that of carrying out experimental research. In the usual situation this requires introducing one or more experimental variables that we administer to samples from our subject population, and then comparing the experimental mean or means against that of a simultaneously tested control to determine whether our independent variable has a significant influence on performance. (Note that we cannot use the pilot or control group data from our prior study to compare against our experimental group. Both groups must be run at the same time.)

RESEARCH AND RANDOMIZATION

Principles of Randomization

In the simplest research situation we have two groups which we will call a Control Group and an Experimental Group. We shall let n_c represent the number of subjects in the Control Group and n_e will represent the number in the Experimental Group. The total sample consists of N subjects, so that $N = n_c + n_e$. When $n_c = n_e$, the symbol n will be used to designate the number of subjects in any one group. The Control Group is administered Treatment C whereas the Experimental Group is administered Treatment E. A schematic design of our experiment is shown in Figure 7.1.

There are certain obvious similarities between Figure 7.1 and Figure 4.1. In both instances we start off with an arbitrarily defined population which is of interest to the experimental investigator. Let us continue with our example of the ABC strain of rats purchased from the XYZ Laboratory. We telephone in an

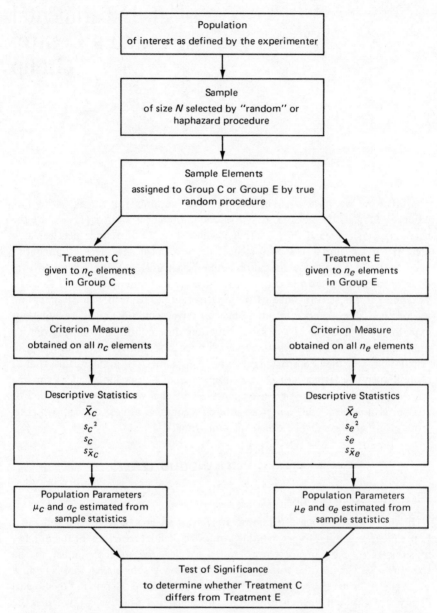

FIGURE 7.1 Schematic outline of an experiment comparing a control and an experimental group.

order for a certain number (N) of animals of specified sex and body weight, and they arrive a few days later by air freight. We may treat these as a "random" sample, but it is much more accurate to say that they are a haphazard sample from the XYZ Production Laboratory. Under no conceivable condition do they come close to meeting the criterion for a random sample as specified by a theoretical statistician. Thus we should be fully aware that this particular sample may have all sorts of biases built into it. but we already know that we have two empirical methods for ascertaining the reproducibility of our results: (1) repeating our experiment, or (2) seeing whether our results are consistent with other data reported in the literature.

Thus we start off with a sample of animals that is admittedly not a true random sample of the population. Our next step is crucial. *It is absolutely necessary that the elements in our sample be assigned to the experimental or control groups by an appropriate random procedure.* This principle is the key to competent research. Even if our sample is biased because of the way it was drawn from the population, if the elements of the sample are assigned at random to Treatment C or Treatment E, then the *difference* between Treatment E and Treatment C is unbiased. This is true because whatever biasing factors are present in our initial sample will also be present in the two subsamples that are assigned to Treatment E and Treatment C. Thus, when we obtain the difference in means between these two treatments, these biases are cancelled out, leaving us with an unbiased evaluation of the effects of our experimental treatment relative to our control condition.

The importance of this procedure cannot be overemphasized. Tables of random numbers or coin tosses should routinely be used in assigning animals to treatment conditions. For example, if there are several animals in a cage and they are to be sorted into experimental and control groups, then the flip of a coin determines whether the first animal taken from the cage goes into Group C or Group E, another flip of the coin determines the assignment for the second animal, and so on. If the animals are housed individually, then each cage can be given an arbitrary number from 1 to N, and a table of random numbers can be used to determine which animals are assigned to the control group and which to the experimental group. The one restriction that may be imposed on this randomization procedure is that half the animals are to be assigned to Treatment C and the other half to Treatment E. If at all possible, there should be the same number of subjects in each group. The reasons for this will be discussed later in this chapter.

The experimental moral of this is plain: If appropriate randomization procedures have been followed in the assignment of animals to experimental and control conditions, then there is an unbiased test of the difference between the experimental and control mean; but if we obtain a significant difference we still have not answered the question concerning the replicability of the finding. It is for this reason that the recommendation is made that the experiment be repeated a second time before publishing it, unless the results are consistent with other data already in the literature. If the latter condition obtains, then an independent replication is not necessary because the experiment itself serves as a confirmation of previous results.

"Semirandom" Procedures

At times it may be more convenient for the experimenter to use a "semirandom" procedure. For example, animals in one rack may be assigned to the control condition while animals in another rack are assigned to the experimental treatment; or animals in one colony room may be designated as controls while animals in a different room are designated as experimentals. If the researcher has two technicians available, as well as two pieces of test apparatus, he or she may find it logistically convenient to have one technician test control animals in one apparatus, while the other technician tests experimental animals in the second apparatus. Examples of these kinds of "semirandom" procedures can be multiplied at great length. *These procedures should always be avoided because they do not meet the specifications of random assignment of subjects to treatment conditions.* There is already enough difficulty in obtaining findings in empirical research that have any degree of generality, to a large extent because our sample is not randomly drawn from a population; if we compound that difficulty by failing to use appropriate randomization procedures when we assign our sample elements to treatment conditions, then there is very little likelihood that we shall ever come up with findings that have importance or generality.

Randomization, Generalization, and Inferences

A note of caution must be added here. Even though the random assignment of sample elements to our treatment conditions ensures us that we have an unbiased test when we compare Treatment E with Treatment C, the degree of generality of this comparison is unknown because we can only crudely specify how our total sample was drawn from our population. For example, suppose that we ordered from the XYZ Production Laboratory male rats that weighed more than 300 grams, and, unknown to us, all the animals that we were sent were over 18 months of age. Let us further suppose that our experimental treatment is a pharmacological preparation that affects learning performance and that we found our experimental group to perform significantly better than our control group. However, the reason for this better performance was that our drug was able to stimulate the central nervous system of an aging rat (it is quite unlikely that rats raised under routine laboratory conditions will live more than two years), so our generalization applies only to aged rats. Having obtained a significant finding, we may order another batch of 300-gram male rats, this time receive animals that are between 200 and 250 days of age, and find that we are unable to repeat our prior findings. Thus we see that the definition of our population as "male rats weighing more than 300 grams from the XYZ Laboratory" may not be sufficient to ensure reproducibility of results. A common problem in biological and behavioral research is to try to isolate the "unknown variable" that appears to be running through the experiments in a random fashion, as indeed may occur.

Returning to Figure 7.1, we see that once the N sample elements have been randomly assigned to Group C (n_c) or Group E (n_e), they are administered Treatment C or Treatment E; then measurements are taken of criterion performance. From these data we compute the usual descriptive statistics, which include the mean of each of our treatment groups, the variance, standard deviation, and standard error of the mean. Descriptive statistics are of relevance to us because they permit us to make certain inferences concerning the population parameters, μ and σ. Note that the descriptive statistics from Treatment C allow us to make inferences concerning a population of animals that receive Treatment C, whereas the descriptive statistics from Treatment E allow us to draw inferences concerning the population parameters for this treatment.

Our major experimental (and statistical) question now is: Does Treatment E bring about a change in the population mean, relative to Treatment C? It is important to keep in mind that this question refers to our hypothetical populations: Our samples are important only insofar as they allow us to make inferences about those populations. The fact that there is a numerical difference between the means of Treatment C and Treatment E is, in and of itself, a trivial fact. Of what importance is it to conclude that Treatment E is better than (or poorer than) Treatment C *for this particular sample?* Such a conclusion has absolutely no degree of generality. The difference between our sample means is important because it allows us to draw inferences concerning the difference between our hypothetical population means. It is for this reason that it is absolutely necessary that we assign our sample elements to our treatment conditions by a random procedure. If we do not do this, then we cannot draw valid conclusions concerning the difference between our experimental and control conditions. If we have assigned subjects randomly, then our next question concerns the evaluation of our sample means to determine whether it is reasonable to infer that these means differ in our hypothetical populations. We now turn to that question.

EXPERIMENTAL AND STATISTICAL HYPOTHESES

The purpose of doing an experiment is to determine whether our independent variable (i.e., experimental treatment) brings about a change in performance relative to that of the control group. The fact that our sample means may differ numerically does not necessarily indicate that our treatment was the cause of this difference, because we know that two samples drawn from the same population and given the *same* experimental conditions will probably have numerically different mean scores because of sampling error. Thus we have to determine whether the difference between the two sample means is greater than we could expect by random sampling from a single population; if the difference is greater, then we may conclude that our experimental treatment significantly influenced the mean score. The purpose of this section is to discuss the logic involved in evaluating the sample means to show how we can draw inferences about the population. But before examining this question from the statistician's perspective, let us look at our problem from the viewpoint of an experimentalist.

Experimental Hypotheses

When we set out to compare an experimental treatment to a control condition, we have one of several experimental hypotheses in mind. The simplest hypothesis is that our independent variable will cause a change in performance of our experimental subjects relative to the control condition. That is, if Treatment E either increases or decreases the experimental mean relative to the control, we want to know about this. This is known formally in statistics as a *two-tailed* test of our hypothesis, because a deviation in either direction from the control group mean (i.e., toward either tail of the distribution) is considered equally important. If we let the expression H_1 designate our experimental hypothesis, remembering that we are always drawing inferences to our hypothetical population, we can express this experimental hypothesis in the following fashion:

$$H_1 : \mu_c \neq \mu_e$$

Either of two conditions will satisfy H_1 : (1) The mean of the experimental population is greater then the mean of the control population, or (2) the mean of the experimental population is less than the mean of the population. This is illustrated in Figure 7.2.

As an example of this kind of hypothesis, we may be interested in studying the relationship between odors and aggression in mice, because there is good evidence that odorous substances, called pheromones, can act to enhance or decrease fighting. Because incoming odors are processed through the olfactory bulbs, we could implant electrodes in this region and stimulate our animals when (1) two of them are together in a cage but are not fighting or (2) the two are fighting. In this kind of situation we are interested in whether electrical stimulation of the olfactory bulbs had *any* effect on aggression, and thus a change in either direction (i.e., either an increase or decrease in fighting) is considered important to us.

The two-tailed experimental hypothesis is the one most commonly used in behavioral and biological research because we often do not know enough about the mechanisms underlying our experimental treatments to be able to specify with a high degree of certitude the direction of a difference that will be brought

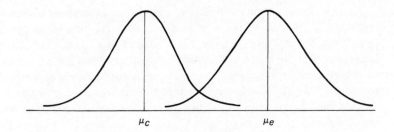

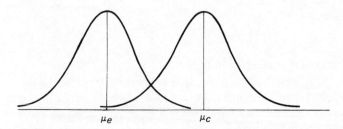

FIGURE 7.2   Either of the two hypothetical conditions shown above will satisfy the experimental hypothesis $H_1 : \mu_c \neq \mu_e$.

about by our intervention. If we are able to specify that our experimental treatment will either increase or decrease the mean, then we have a directional or *one-tailed* hypothesis. For example, if we have reason to believe that our experimental treatment will bring about an increase in mean performance, then our experimental hypothesis is of the following form:

$$H_1 : \mu_e > \mu_c$$

This is depicted graphically in Figure 7.3.

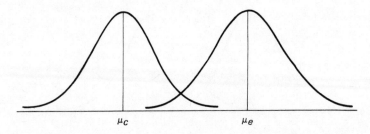

FIGURE 7.3   Illustrating a hypothetical situation that will satisfy our experimental hypothesis $H_1 : \mu_e > \mu_c$.

As an example of this hypothesis, consider an experiment in which we are interested in the question of memory transfer via brain homogenates. Our hypothesis is that learning causes a structural change in the brain. To test that hypothesis we train a group of animals on a particular learning task, then remove their brains, homogenize the tissue, and inject this homogenate into a group of recipients. Our hypothesis specifies that the experimental group of recipients should learn better than an appropriate control group and thus we have a one-tailed or directional hypothesis.

There is another directional experimental hypothesis that we can entertain, namely, that our experimental group's mean will be less than that of the control group. Formally this is expressed as follows:

$$H_1 : \mu_e < \mu_c$$

This experimental hypothesis is depicted in Figure 7.4.

As an example of this hypothesis, assume we are behavioral pharmacologists investigating a new drug that is supposed to act as a central depressant. We believe that one consequence of the drug action will be to reduce the animal's activity, and so we test an experimental group and an appropriate control group of rats in an open field under the hypothesis that the experimental animals should be less active than the controls.

The Choice of Experimental Hypothesis: One-Tailed or Two-Tailed

If we are able to specify the direction of difference of our experimental group relative to our control, then the one-tailed hypothesis is more powerful: That is, it is more likely to pick up a significant difference if the experimental hypothesis is true. The difficulty with this hypothesis is that we generally do not have enough information about our experimental variables to know how they influence the underlying behavioral and biological processes; thus, it is difficult to be certain that the mean difference will always (or almost always) be in the predicted direction. Note that when we specify a one-tailed hypothesis, we are stating that we *know* the direction of difference of the experimental mean, and

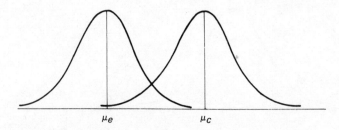

FIGURE 7.4 Illustrating a hypothetical situation that will satisfy our experimental hypothesis $H_1 : \mu_e < \mu_c$.

that if we were to obtain a very large difference in the opposite direction from that which we predicted, this would be experimentally meaningless to us. Consider, for example, the memory transfer study. If we choose a one-tailed hypothesis, predicting that our experimental animals will learn better than our controls, and if we find, in fact, that our experimental animals are incapable of learning, logically that observation is no more relevant to us than the observation that the experimental group and the control group had essentially identical mean scores. Now it is quite obvious that no self-respecting experimentalist is going to act that way. After shaking our heads in wonderment over the data, we would immediately set up a replication of the experiment to see whether we could verify this finding of a failure to learn.

And now consider our behavioral pharmacology problem in which we expect our central depressant drug to reduce the activity of our animals in the open field. If we use a one-tailed hypothesis, we are stating logically that, "If our experimental animals are significantly more active than our controls, we would have no experimental interest in such a finding." It is difficult to imagine competent researchers making such a statement.

In choosing between one-tailed and two-tailed experimental hypotheses, we can use the following rule as a guide; if we obtain a significant difference in the direction *opposite* to our prediction, are we willing to consider this to be an experimentally meaningless piece of information not worth following up? If so, then we can justify the one-tailed hypothesis. If not, then it would be wiser to use a two-tailed hypothesis even though we have a good idea as to where our experimental mean is going to fall relative to the control mean.

Statistical Hypotheses

The hypotheses or H_1's discussed in the previous section are the experimental hypotheses in which we are interested as researchers. However, we cannot test these hypotheses directly because in the realm of inferential logic we are not able to "prove" that a hypothesis is true. However, it is possible to reject or "disprove" a hypothesis in the sense that we can show that it is not reasonable to expect the hypothesis to be true, given the set of data we have in hand. Thus, the logic of our approach is as follows: We shall specify a *statistical* hypothesis that is exactly opposite to our *experimental* hypothesis. We shall then make an evaluation of our statistical hypothesis and if we are able to reject it, we shall accept our experimental hypothesis as the most reasonable alternative interpretation. For this to be logically appropriate the two hypotheses must be mutually exclusive and must together encompass all possible outcomes. We shall designate our statistical hypothesis by the symbol H_0. The statistical and experimental hypotheses for our two-tailed test are as follows:

$$H_0: \mu_c = \mu_e$$
$$H_1: \mu_c \neq \mu_e$$

If we are able to reject H_0, then logically we must accept H_1.

The same logic applies with the one-tailed hypotheses. For the situation where we expect the experimental mean to be greater than the control mean, the statistical and experimental hypotheses are

$$H_0: \mu_e \leqslant \mu_c$$
$$H_1: \mu_e > \mu_c$$

If we are able to reject H_0, we then accept H_1.

For the situation where the experimental hypothesis is that the experimental mean will be less than the control mean we have the following situation:

$$H_0: \mu_e \geqslant \mu_c$$
$$H_1: \mu_e < \mu_c$$

If we reject H_0, then we accept H_1.

Our statistical hypothesis is generally called the *null hypothesis* in the statistical literature, which means that we assume our experimental variable to be without effect (i.e., to have null effect) relative to our control condition. H_1, which is our experimental hypothesis, is typically called the *alternative hypothesis*. If we are able to reject the null hypothesis, then we accept the alternative hypothesis as being a more reasonable way of describing the state of affairs for our experimental conditions.

Therefore, our statistical objective is to evaluate our null hypothesis and specify a set of rules whereby we can determine whether to reject or not to reject it. We first need to decide on a level of significance for rejection.

LEVELS OF SIGNIFICANCE

Our general rule is that we shall reject our null hypothesis H_0 if it is unreasonable. In that event we accept the alternative hypothesis H_1 in its place. What do we mean by the statement "unreasonable"? Statisticians and experimentalists have both agreed to the convention that if the null hypothesis has no more than a .05 probability value associated with it, then we shall conclude that the null hypothesis is unreasonable. Furthermore, if the probability of this occurrence is .01 (i.e., less than 1 in 100), then we consider the null hypothesis to be very unreasonable. These two values—the .05 and the .01 levels—are the same α values that we used in setting the 95% and 99% confidence boundaries around our means in Chapter 5. These α values are called *levels of significance,* and they specify probability values that are sufficiently low that we may reasonably conclude that our null hypothesis is probably false.

Let us consider the ramifications of this type of logical approach to hypothesis testing. A very common statement in the research literature has the following form: "The difference between the means is significant at the .05 level." This statement means that there is less than 1 chance in 20 that a difference as large or larger than that between the experimental and the control group could have occurred by chance alone, on the assumption that the experimental treatment was without effect and that we were sampling from one population. Because we consider this to be an unlikely or unreasonable event, we are willing to reject that hypothesis and thereby accept the alternative hypothesis, which is that the experimental and control means in the population are not numerically the same. We should recognize that when we set the .05 level of significance we shall be wrong 5% of the time, and we are willing to accept that risk. That is, if we were to draw two random samples from the same population, apply the *same* treatment condition to both groups so that there were no differential experimental treatments imposed, and then collect our criterion measures, in 5 out of every 100 experiments we would obtain a significant difference at the .05 level. Thus, when we reject the null hypothesis we have not "proven" it to be false, although we can state that it is an unlikely hypothesis.

Reducing the Risk of Falsely Rejecting the Null Hypothesis

Reducing the α level Perhaps we are not willing to risk being wrong 5 times in 100 and prefer to be more conservative. For example, you may have a particular research idea that, if true, demands that you embark on a whole series of experiments that will require a large investment of animals and time. You are willing to do this if you feel very confident that you have a true experimental phenomenon worth investigating, but you want to be quite certain of this before initiating a major research program. In this instance you may set your significance level to be equal to .01, or perhaps even .001. In the former instance an α level of .01 means that only 1 time in 100 would you reject the null hypothesis when it is true, whereas an α level of .001 means that you only risk making an error 1 time out of 1000 if you reject the null hypothesis. Thus, by your choice of α for a test of significance you specify the amount of risk you are willing to take that you are rejecting the null hypothesis when it is indeed true.

Experimental replication Another approach to this problem of feeling certain that you have a true experimental phenomenon, and one that seems more sensible from a researcher's point of view, is to stay with the .05 level of significance and decide in advance to repeat your experiment if you get a significant finding. In using this approach, you are committing yourself to repeating your study if the mean difference is at or beyond the .05 level. Thus even if you obtained a mean difference so great that it could have occurred less than 1 time in 1000 by random sampling, because you selected an α level of .05 for rejecting the null hypothesis, you must repeat your experiment. Logically you do not decide after inspecting your data whether you should use the .05 or .01 level of significance. This is to be decided before the experiment.

The purpose of repeating your experiment before embarking on a major research program is one that we have been emphasizing continually. After all, the essence of experimentation is to be able to reproduce your findings at will; and it makes more sense to an experimenter to have two studies, each of them independently carried out over different time periods and each significant at the .05 level, rather than one study that is significant at the .01 level. We know enough now about the difficulties and vagaries of experimentation to appreciate fully that two independent replications, each significant at the .05 level, give us a very firm footing for planning a future research program.

The Use of α Levels Other Than .05 or .01

Why the values of .05 and .01 for levels of significance? There is nothing sacrosant or magical about these particular numbers. We should give them the attention they deserve, but not worship at their feet. For example, we have already pointed out that obtaining a significant difference at the .01 level is not sufficient reason for publishing a paper if your experimental finding is either (1) a new phenomenon that has not been reported upon before, or (2) contradictory to established information in the literature. Look now at the other side of the coin. Suppose that someone reports a new experimental procedure which brings about a significant improvement in the behavior of his experimental subjects, and suppose that you repeat the same procedure in your laboratory because you are interested in using the technique described by the first experimenter. Suppose that you obtain a difference that is "significant" at the .07 level, in contrast to his report of a significant difference at the .01 level. Even though your difference at the .07 level does not meet the arbitrarily specified criterion of .05, this may be considered to be a confirmation of the first investigator's finding.

The basic rationale for the use of .05 and .01 levels in scientific research is that science is quite conservative, and evidence has to be very convincing before we are willing to accept the reliability of any new phenomenon. Thus, by insisting that researchers clearly establish that they have isolated a relevant experimental variable by setting a low α level, we reduce the probability of misinformation accumulating in the literature.

There are many instances in which it is reasonable to set a significance level numerically higher than .05. These are often found in an applied context. For example, suppose that a laboratory animal feed salesperson tells us that there is a new product available which, when fed to pregnant animals, results in heavier body weight of the newborn young. The feed only costs a few more cents per bag, this is a reputable company, and so we are not concerned that the food may be ill prepared or contaminated, and we know that animals heavier at birth tend to be somewhat healthier. Under these conditions, we could carry out an experiment comparing the feed we now use in our laboratory against the new product, and we may well decide to set a significance level of .20 for rejecting the null hypothesis that there are no differences between the two feeds. That is, we are willing to be wrong 20 times in 100 because the one consequence of being wrong is that we shall waste some money on the special feed, whereas if we are right, we end up with a heavier and healthier animal. If we are more concerned with the weight than with the money, then a high α level is appropriate. However, if money is tight and can only be spent if we are quite certain that it is a good investment, then we would want to set a low value for α. Here would be a clear-cut instance where a one-tailed test of significance would be appropriate, because we are interested in the feed only if it improves the young's body weight.

TESTING THE NULL HYPOTHESIS

Rationale

The logical approach that we shall use in testing the null hypothesis is to make the assumption that our experimental treatment is without effect so that, in the population, $\mu_c = \mu_e$. *We shall then give our data the opportunity to reject this assumption.* This point is emphasized because it is the basic principle underlying all tests of statistical inference. The logical sequence is as follows: We make one or more assumptions about population parameters (here the assumption is that $\mu_c = \mu_e$). Based on this assumption we make predictions as to how the data should look. We then test our data to see whether their characteristics are consistent with the assumption we have made. If not, we must reject our assumption.

If the assumption is rejected, this is equivalent to rejecting H_0, and we then accept the alternative hypothesis. If our data do not permit us to reject the null hypothesis, this does not "prove" that the hypothesis is true, but merely indicates that it is the more adequate description of our situation to date. Further experiments may well cause us to reject H_0, and it is because of our inability to predict future experimental events that we are never able to prove the null hypothesis to be true.

The logic underlying the test of the null hypothesis is as follows. The hypothesis states that in the population

$$\mu_e = \mu_c$$

and therefore

$$\mu_e - \mu_c = 0$$

The null hypothesis states that in the population the mean *difference* between those who received Treatment C and those who received Treatment E is zero. If Treatment E did not have any effect on the performance of our subjects, then this statement is necessarily true. However, if Treatment E did add (or subtract) an increment to the performance of the experimental subjects relative to the control group, then the difference between μ_e and μ_c will not be zero. The question now becomes: Given two samples that have received Treatment E and Treatment C, in which the numerical values of the means differ, is that difference greater than we can expect by chance when sampling from a population in which the true mean difference is zero? In order to answer that question, it is necessary to generate a hypothetical *distribution of differences*, ascertain the form of that distribution, determine the standard error of the distribution, and then see where the empirical difference found in our experiment (i.e., $\bar{X}_c - \bar{X}_e$) falls on this hypothetical distribution of differences. If the difference we have obtained between our experimental mean and our control mean is such that it could only have occurred 1 time in 20, or less, by random sampling from a population in which the true mean difference is zero, we conclude that this is a sufficiently unlikely event that we are willing to reject our null hypothesis. Let us now look at how we obtain the distribution of differences.

A Distribution of Differences

The easiest way to describe the concept of a distribution of differences is to consider a hypothetical experiment. Imagine that we drew a sample of 20 rats from a research colony, and by a table of random numbers we assigned 10 animals to Treatment E and the other 10 animals to Treatment C. However, we treat all 20 animals identically. We then give our animals a criterion test, compute $\bar{X}_c$ and $\bar{X}_e$, subtract $\bar{X}_e$ from $\bar{X}_c$, and enter that number on a tally sheet. Next, we repeat this experiment a second time, drawing another sample of 20 animals from our colony and following the identical procedure. If we do this for a sufficient number of samples, we shall find that roughly half the numbers on our tally sheet will be positive, the remaining numbers will be negative, and the mean of all numbers should be very close to zero. Indeed, if you can imagine that we repeated this study for an infinite number of samples, then the mean of our distribution would be exactly zero. Now our sample statistic, $\bar{X}_c - \bar{X}_e$, is an unbiased estimate of the population parameter, $\mu_c - \mu_e$, on the assumption that the animals used in the investigation were randomly assigned to Treatment E and Treatment C (note that it is not necessary to assume that the animals were a random sample from a population; it is only necessary to assume that they were randomly assigned to the treatment conditions after they had been selected to be in the experiment). Therefore, we may plot the distribution of our differences, $\mu_c - \mu_e$, in our hypothetical population. Such a distribution of differences is shown in Figure 7.5.

There are several things to note about Figure 7.5. First of all, the mean of the distribution is zero, as we indicated above. Next, the distribution is normal. This will be true if (1) the raw scores are normally distributed, or (2) the number of

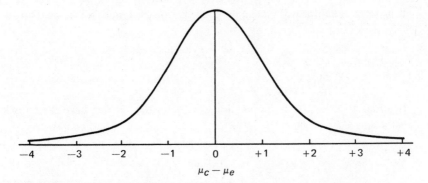

FIGURE 7.5   Distribution of the difference between μ_c and μ_e on the assumption that the population means are identical, thereby resulting in a mean difference of 0.

cases in each of our groups is sufficiently large (approximately 20-30). It is necessary to make this assumption of a normal distribution in order to carry out our test of significance. Now that we know the form of our distribution and its mean, the one remaining value that we have to obtain is the standard deviation of our distribution of differences, called the *standard error of a difference.*

The Standard Error of a Difference

If we have independent observations, we can show that, for the population, the variance of a distribution of differences is equal to the sum of the separate variances. Thus

$$\sigma^2_{\bar{x}_c - \bar{x}_e} = \sigma^2_{\bar{x}_c} + \sigma^2_{\bar{x}_e} = \frac{\sigma^2_c}{n_c} + \frac{\sigma^2_e}{n_e} \tag{7.1}$$

and

$$\sigma_{\bar{x}_c - \bar{x}_e} = \sqrt{\sigma^2_{\bar{x}_c} + \sigma^2_{\bar{x}_e}} = \sqrt{\frac{\sigma^2_c}{n_c} + \frac{\sigma^2_e}{n_e}} \tag{7.2}$$

In order to make a test of the difference between means, we have to assume that the population variances for Treatment C and Treatment E are the same. This is known as the assumption of *homogeneity of variance.* We assume

$$\sigma^2_e = \sigma^2_c = \sigma^2$$

Then

$$\sigma_{\bar{x}_c - \bar{x}_e} = \sqrt{\frac{\sigma^2}{n_c} + \frac{\sigma^2}{n_e}} \tag{7.3}$$

If $n_c = n_e$, then this will yield a minimal value for Formula 7.3. If we let $n_c = n_e = n$, then Formula 7.3 reduces to

$$\sigma_{\bar{x}_c - \bar{x}_e} = \sqrt{\frac{2\sigma^2}{n}} \tag{7.4}$$

In order to estimate the population parameter $\sigma_{\bar{x}_c - \bar{x}_e}$, we have the sample statistic

$$s_{\bar{x}_c - \bar{x}_e} = \sqrt{s_{\bar{x}_c}^2 + s_{\bar{x}_e}^2} = \sqrt{\frac{s_c^2}{n_c} + \frac{s_e^2}{n_e}} \tag{7.5}$$

On the assumption of homogeneity of variance, our best estimate of the population variance σ^2 is obtained by adding together the sums of squares and degrees of freedom from our two independent groups as we discussed in Chapter 3.

$$s_{\text{within}}^2 = \frac{SS_c + SS_e}{(n_c - 1) + (n_e - 1)} = \frac{SS_c + SS_e}{n_c + n_e - 2} \tag{7.6}$$

By substituting s_{within}^2 for s_c^2 and s_e^2 in Formula 7.5, we have the following formula for the standard error of a difference:

$$s_{\bar{x}_c - \bar{x}_e} = \sqrt{\frac{s_{\text{within}}^2}{n_c} + \frac{s_{\text{within}}^2}{n_e}} \tag{7.7}$$

If our sample sizes are the same so that $n_e = n_c = n$, then Formula 7.7 reduces to

$$s_{\bar{x}_c - \bar{x}_e} = \sqrt{\frac{2s_{\text{within}}^2}{n}} \tag{7.8}$$

The computational formula that is algebraically identical with Formula 7.7 is

$$s_{\bar{x}_c - \bar{x}_e} = \sqrt{\left(\frac{SS_c + SS_e}{n_c + n_e - 2}\right)\left(\frac{1}{n_c} + \frac{1}{n_e}\right)} \tag{7.9}$$

If $n_c = n_e = n$, Formula 7.9 reduces to

$$s_{\bar{x}_c - \bar{x}_e} = \sqrt{\frac{SS_c + SS_e}{n(n-1)}} \tag{7.10}$$

In other words, in order to determine the standard error of the difference between means, compute the sums of squares (SS) for the control group and for the experimental group and substitute those values into Formula 7.9 or 7.10 along with the appropriate values for n.

Reasons for Equal n per Group

If at all possible, you should strive to have the same number of subjects in each of your treatment groups. We have already indicated that this will cause your standard error of the difference to be a minimal value. A second statistical advantage to equal n's is that the computational formulas are much simpler, especially when you move beyond the t test and the two-group experiment into experimental designs involving three or more groups (see Chapters 8 and 9). Another reason for having equal numbers of cases per group is that you then give each of your treatment conditions equal weight. Finally, if you have equal numbers per group, this will probably be sufficient to satisfy your assumption of homogeneity of variance (more about this later). In order to ensure equal n's, you should first decide how large an N you need for your experiment, then make an estimate as to how many subjects you expect to lose during the course of the study. That value plus your original N defines the number with which you should start.

THE t TEST FOR A DIFFERENCE BETWEEN MEANS

The statistic used to test the null hypothesis that $\mu_c = \mu_e$ is

$$ t = \frac{(\bar{X}_c - \bar{X}_e) - (\mu_c - \mu_e)}{s_{\bar{x}_c - \bar{x}_e}} \tag{7.11} $$

This t statistic has $n_c + n_e - 2$ degrees of freedom.

Because the null hypothesis is that $\mu_c = \mu_e$, the value $(\mu_c - \mu_e)$ in the numerator of the t test is zero, and often is omitted. It is given here in order to make explicit what hypothesis is being tested. Furthermore, $\mu_c - \mu_e$ does not have to equal zero. Suppose that, in designing an experiment, you decide that your experimental group must differ from the control by at least 4 points in either direction before you are willing to pay serious attention to your experimental variable. Then your null hypothesis and its alternative are as follows:

$$ H_0: \mu_c - \mu_e \leqslant |4| $$
$$ H_1: \mu_c - \mu_e > |4| $$

SUMMARIZING THE STEPS INVOLVED IN TESTING FOR A DIFFERENCE BETWEEN MEANS

1. Specify the null hypothesis, H_0, and its alternative, H_1, which is your experimental hypothesis. The way you state your hypothesis determines whether it is one-tailed or two-tailed.
2. Specify the risk you are willing to take in rejecting the null hypothesis when it is true by choosing a level of significance called an α level. In general this will be .05 or .01.
3. Once you have chosen your level of significance, and knowing how many degrees of freedom you have in your experiment, you can now specify the value of t that will cause you to reject H_0. That value of t defines a *region of rejection*, and you now have the following *decision* rule to guide you:

   Reject H_0 if the observed value of t (designated t_{obs}) falls within the region of rejection.
   Otherwise do not reject H_0.

4. Compute the standard error of the difference from Formula 7.9 or 7.10, then use Formula 7.11 to compute the value of t, and apply the decision rule of Step 3 above.

This procedure can be illustrated by a hypothetical example. Suppose we are behavioral pharmacologists interested in determining whether a new drug has any effect on the open-field activity of rats. We decide to use a sample size of 10 in each group, obtain 20 animals of the same sex and approximately the same weight from our colony room and use a table of random numbers to assign the 20 animals to experimental or control conditions. We then inject the experimental animals with the drug while the controls receive an injection of the vehicle. Next, a technician who does not know which animal belongs to which group tests each animal for three successive days in the open field. Table 7.1 shows the total activity summed over three days for each of the 20 animals as well as the relevant statistics needed to describe the two groups and to test the significance of the difference between the means.

Before initiating the experiment we would have specified our α level for rejecting the null hypothesis. Because we are interested in a difference in either direction, we have a two-tailed hypothesis concerning our experimental drug. With 18 degrees of freedom, the value of t at the .05 level is 2.101, and it is 2.878 at the .01 level.

We have to choose our level of significance prior to conducting the experiment. Our observed value of t is 2.83. If we had selected the .05 level, then we have a significant difference between groups. If we chose the .01 level, on the other hand, then our difference falls just short of being significant.

Our obtained value of t was 2.83. If you look this up in the t table with 18

TABLE 7.1 Hypothetical experiment comparing activity scores of rats given a vehicle injection (control group) with those given a drug (experimental group)

	Control	Experimental
	11	75
	31	52
	43	61
	22	42
	50	50
	58	32
	25	78
	44	55
	39	37
	10	49
ΣX	333	531
ΣX^2	13,461	30,217
n	10	10
$\bar{X} = \Sigma X/n$	33.3	53.1
$SS = \Sigma X^2 - (\Sigma X)^2/n$	2,372.10	2,020.90
$s^2 = SS/n - 1$	263.57	224.54
$s = \sqrt{s^2}$	16.23	14.98
$s_{\bar{x}} = s/\sqrt{n}$	5.13	4.74

$$s_{\bar{x}_c - \bar{x}_e} = \sqrt{\frac{SS_c + SS_e}{n(n-1)}} = \sqrt{\frac{2372.10 + 2020.90}{10(9)}}$$

$$= 6.99$$

$$t = \frac{\bar{X}_c - \bar{X}_e}{s_{\bar{x}_c - \bar{x}_e}} = \frac{33.3 - 53.1}{6.99} = 2.83$$

df, you will find that this value falls between the .02 and .01 levels. Suppose that we had chosen the .05 level in rejecting the null hypothesis. In theory we should report our value to be significant at the .05 level, regardless of its actual numerical value. In practice this is not what researchers do. In all likelihood we would report this value as being significant at the .02 level. The general practice among biological and behavioral researchers is to use the .05 level for rejecting the null hypothesis and then report the actual probability values associated with a particular value of *t*. Thus, when you read in the literature that someone found a significant difference at the .001 level, this does not mean that he set the .001 level as his value of α to reject the null hypothesis. Instead, he is probably using the .05 level (and if you look through his research paper you will probably find some significant differences at the .05 level), but this particular *t* test gave him a large numerical value which placed it at the .001 level.

Setting Confidence Limits Around a Mean Difference

In Chapter 6 we learned how to set confidence limits around a mean as a way of determining the amount of variability, or degree of precision, of that mean. In the same manner we can also set confidence limits around a difference between means. The formula for doing this is directly parallel to Formula 6.2, and is as follows:

$$\text{Confidence limits} = (\bar{X}_c - \bar{X}_e) \pm [t_\alpha \, (df)] \, s_{\bar{x}_c - \bar{x}_e} \qquad (7.12)$$

The use of confidence limits will be illustrated with the data involving the experiment comparing the activity scores of control animals and those given a drug. From Table 7.1 we find that the mean difference between the groups is 19.8 points with a standard error of the difference equal to 6.99. There are 18 degrees of freedom in the experiment. If we wish to set the 95% confidence limits around the mean difference, the value of t needed to satisfy the condition $t_{.05}(18)$ is 2.101. Substituting these values into Formula 7.12, we have

$$\text{Confidence limits} = 19.8 \pm (2.101)(6.99) = 19.8 \pm 14.69$$

Thus the range of the 95% confidence limits is from a mean difference of 5.11 points up to 34.49 points. Our conclusion is that we are 95% confident that the true mean difference $(\mu_c - \mu_e)$ falls somewhere within this interval. Note that the interval does *not* contain the value 0. It is for this reason that we were able to reject the null hypothesis at the .05 level.

To obtain the 99% confidence limits, the value $t_{.01}(18) = 2.878$, and our limits are

$$19.8 \pm (2.878)(6.99) = 19.8 \pm 20.12$$

In this situation the confidence interval goes from -0.32 to $+20.12$. Because the value 0 is contained within this interval, this means that one hypothesis concerning the value $\mu_c - \mu_e$ is that the mean difference is 0. For this reason it is not possible to reject the null hypothesis at the .01 level of significance.

Assumptions Underlying the t Test

The derivation of the formula for the t test has two underlying assumptions: (1) The distribution of the mean differences in the population is normal, and (2) the population standard deviation for Treatment C is equal to the population standard deviation for Treatment E. With moderate-sized samples, the normality assumption is not critical. Rather severe departures from normality can occur without affecting the t distribution very much, that is, without modifying the probability values associated with the particular numerical value of t. A departure from normality is more likely to cause an error when testing a one-tailed hypothesis than a two-tailed hypothesis, depending on the direction of skewness.

With respect to the homogeneity of variance assumption, studies have shown that large deviations from homogeneity can still be tolerated if sample sizes are equal. In older textbooks statistical procedures are described to test the homogeneity assumption, and special formulas are presented to compute t when the variances are found to be heterogeneous. These tests are quite conservative in the sense that the value of t required for significance will be numerically larger than needed for the α level specified by the researcher. In practice these special tests can be safely ignored, especially if one has equal n's in the treatment groups.

From these observations come some general guideliness to aid you in designing your experiments. First, if at all possible have equal numbers of cases per treatment condition. Next, if you have 20-30 cases in each group, you need have no concern about your assumption of normality. Even though 30 sets an upper limit to satisfy this assumption, the lower limit is not known, though it probably falls somewhere between 10 and 20 cases.

A third consideration in designing experiments is whether you want to use independent subjects or use subjects that are *matched* or correlated in some fashion. We shall now turn to that topic.

THE t TEST FOR MATCHED PAIRS

Rationale

One major objective in designing experiments is to minimize the inherent variability in our data. This variability is reflected in the standard error of the mean. We know that one major source of variability is that involved in individual differences. That is, the individuals that will be used in our experiment will differ on many variables including genetic factors, types of rearing histories, and environmental conditions including temperature fluctuations, husbandry procedures, noise level, and so on. This list can be multiplied almost indefinitely. If we randomly assign our subjects to experimental and control conditions, then these factors will be randomly distributed and will not have any biasing effect on our experimental and control means. However, it will increase the variability *within* our experimental and control groups. That is, it will increase the standard deviations, and these standard deviations are the basis for our standard errors.

Quite often we can reduce this variability considerably if we can find a way to use each animal as its own control. For example, in many experiments we can pretest an animal to determine its baseline performance, then interpose an experimental treatment and retest the animal. A comparison of the scores before and after treatment gives us an indication of the effect of our experimental intervention. However, there are two serious problems with this kind of procedure. For one thing, the pretest may influence the animal's behavior so that its performance on a second test would be markedly different whether or not an experimental intervention had been interposed. Thus, it is necessary to use a more complicated design to isolate or control for this possibility. Second, it may not be possible to use this kind of design because the measurement procedure results in the elimination of the animal from the experiment, for example, when one is studying pathology and it is necessary to determine the effects of an experimental treatment.

The ideal experimental design is one in which you have identical twins, one receiving the control condition and the other getting the experimental treatment. Because twins have the same genetic background and have generally been reared under very similar environmental conditions all their lives, they will be much less variable than unrelated pairs would be. Thus, this reduction in variability allows you to have a more precise experiment so that you are more likely to pick up a significant difference between your experimental and control groups if one indeed exists.

An Experimental Example

When working with animal preparations, we can approximate the model of identical twins if we work with a species that gives multiple births, that is, one for which there are littermates. (Indeed, we can get true identical twins when working with inbred mice or rats.) As an example of the use of this procedure, consider the following experimental question: What effect does the kind of environment in which an animal is raised have on that animal's brain weight? The researchers involved with this question decide to raise control animals in an impoverished environment during early life whereas the experimental animals are raised in a complex environment. They take two animals of the same sex from each of 10 litters and by flipping a coin assign one animal of each pair to the impoverished condition and its littermate to the enriched environment. At the termination of the experiment all animals are removed from the two environments, their brains are dissected out, and the total cortex is weighed to the nearest milligram. Table 7.2 shows the results of such an experiment.

Statistical Analysis

The data in Table 7.2 differ from that in Table 7.1 because we now have matched pairs of animals. Therefore, we are able to obtain the *difference* between each littermate pair, and these values are entered in the last column in Table 7.2. It is this distribution of differences that we evaluate by means of our t test. If we let D represent any difference score in the last column of Table 7.2, then $\bar{D}$ represents the mean of that set of scores. The variance of our difference scores is given by the formula

$$s_d^2 = \frac{\Sigma(D - \bar{D})^2}{n - 1} \tag{7.13}$$

The standard error of our mean difference is given by the following formula:

$$s_{\bar{d}} = \sqrt{\frac{\Sigma(D - \bar{D})^2}{n(n - 1)}} \tag{7.14}$$

Formula 7.14 is our definitional formula for the standard error of our difference column. The computational formula that is algebraically equivalent is

$$s_{\bar{d}} = \sqrt{\frac{n \Sigma D^2 - (\Sigma D)^2}{n^2(n - 1)}} \tag{7.15}$$

The t test for matched pairs is now given by the formula

$$t = \frac{\bar{D}}{s_{\bar{d}}} \tag{7.16}$$

TABLE 7.2 Hypothetical experiment comparing the cortical brain weights (in milligrams) of littermate pairs. The control was reared in an impoverished environment while its experimental littermate was reared in an enriched environment

Pair number	Control	Experimental	Difference = E − C
1	620	681	61
2	835	862	27
3	427	447	20
4	770	765	−5
5	536	574	38
6	617	617	0
7	698	740	42
8	811	866	55
9	473	485	12
10	755	802	47
			297

$\Sigma D = 297$

$\Sigma D^2 = 13,461$

$\bar{D} = \Sigma D/n = 297/10 = 29.7$

$s_{\bar{d}}^2 = \dfrac{n \Sigma D^2 - (\Sigma D)^2}{n^2(n - 1)} = \dfrac{(10)(13,461) - (297)^2}{100(9)} = 51.5567$

$s_{\bar{d}} = \sqrt{s_{\bar{d}}^2} = 7.18$

$t = \dfrac{\bar{D}}{s_{\bar{d}}} = \dfrac{29.7}{7.18} = 4.14$

This t test has associated with it $n - 1$ degrees of freedom, where n refers to the number of *pairs* rather than the number of cases. Thus, in the example in Table 7.2, even though we have 20 animals involved in the experiment, these represent 10 pairs so there are 9 df associated with our t test.

At the bottom of Table 7.2 is the information needed to compute t, which is found to equal 4.14. In Table II in the Appendix we see that $t_{.05}(9)$ equals 2.262 and $t_{.01}(9)$ equals 3.250. Thus, our observed value of t falls into the critical region for rejection of the null hypothesis using either the .05 or the .01 level of significance. Therefore, we reject the null hypothesis that environmental rearing conditions have no effect on cortical brain weight.

In order to illustrate the principles involved in using matched pairs, let us analyze the data in Table 7.2 as though we had randomly assigned 20 independent rats to our experimental conditions. That is, we are going to ignore the fact that these animals represent littermate pairs and treat them as though they came from 20 different litters. Table 7.3 shows the pertinent statistics involved in evaluating the data and parallels the analysis in Table 7.1. On the assumption that these are independent subjects, we have $n_c + n_e - 2$ or 18 degrees of freedom. Our value of t is .45, which is clearly insignificant even with twice the number of df as we had in the analysis in Table 7.2.

Principles Involved in Using the Matched Pair Design

There are several important points to be made about the use of the matched pair design. First of all, when you match you end up with half as many degrees

TABLE 7.3 Analysis of the data in Table 7.3 on the assumption that the subjects in the control and experimental groups are independent of each other

	Control	Experimental
ΣX	6,542	6,839
ΣX^2	4,463,518	4,877,989
n	10	10
$\bar{X} = \Sigma X/n$	654.2	683.9
$SS = \Sigma X^2 - (\Sigma X)^2/n$	183,741.6	200,796.9

$$s_{\bar{x}_c - \bar{x}_e} = \sqrt{\frac{SS_c + SS_e}{n(n-1)}} = \sqrt{\frac{183,741.6 + 200,796.9}{10(9)}}$$

$$= \sqrt{4,272.65}$$

$$= 65.36$$

$$t = \frac{\bar{X}_c - \bar{X}_e}{s_{\bar{x}_c - \bar{x}_e}} = \frac{683.9 - 654.2}{65.36} = \frac{29.7}{65.36} = .45$$

of freedom as you have if you use independent samples. If you have a large number of pairs, this does not make much difference; however, if you are working with small numbers, then halving your degrees of freedom may markedly increase the numerical value of t needed for significance.

Next, a matched pair design will be useful only if the variable on which you are matching is related to your criterion measure. In this particular instance we know that there are large individual differences in brain weights among unrelated animals, and we expect the variation in brain weights among littermates to be much less. Therefore, there is a relationship between litter characteristics and brain weight, and so it is reasonable to use matched pairs. On the other hand, suppose that we knew nothing about the lineage of our animals and decided to match them on body weight alone. There is very little relationship between body weight and brain weight, and so we are matching the animals on a variable that is not related to our criterion measure. If our matching variable bears absolutely no relationship to our criterion measure, then the value of t obtained by Formula 7.16 will be numerically identical with the value of t obtained by Formula 7.11, so we have gained nothing by our matching. Indeed, we have lost because we now have half as many degrees of freedom to evaluate t. Thus, it is important to be reasonably certain than your matching variable is a relevant one before using this kind of experimental design.

Another value of using a matched pair procedure is that there is no homogeneity of variance assumption. Because you have *one* distribution of differences, you are not pooling data from your experimental and control groups, so this assumption does not apply.

The power of the matching procedure in this experiment can be seen by comparing the standard errors of the difference in Tables 7.2 and 7.3. Our matching procedure has reduced the random variability from 65.36 units to 7.18 units, an increase in precision of approximately ninefold.

INTERPRETATION OF STANDARD ERRORS

Matched Pairs

We have indicated that the usual way to report data in the literature is in the form of $\bar{X} \pm SE$. Thus for the data in Table 7.2 we would make the following statistical statement in our publication: "The control mean was 654.20 ± 45.18 milligrams whereas the experimental mean was 683.90 ± 47.24 milligrams." These means and standard errors were obtained by taking the set of control data in Table 7.2 and determining its mean and standard error and doing the same thing for the experimental data. This is an accurate way to describe the variability about our means, and yet an experienced researcher would be aghast if you now reported that the difference of 29.7 milligrams was significant, because this value is so much smaller than either standard error. The reason for this seeming paradox is that you did not use the standard errors associated with the separate means to make your t test, but, instead, used the standard error associated with the difference score. This kind of misinterpretation can be avoided if you include the following statement in your research report: "Using matched pairs, the mean difference and its associated standard error was 29.7 ± 7.18 milligrams."

Independent Groups

If a researcher does have independent groups, then the standard errors associated with his means may be used directly to compute the t statistic. For visual inspection of data, a rough rule-of-thumb is as follows: The difference between two means must be approximately 50% greater than the sum of the two standard errors in order for that difference to be significant at the .05 level. For example, in Table 7.1 we see that the control mean and its associated standard error is 33.3 ± 5.13; for the experimental group these values are 53.1 ± 4.74. The mean difference between the groups is 19.8 points. The sum of the two SE's plus 50% of that sum is 14.8 points [5.13 + 4.74 + ½(5.13 + 4.74)]. Because the mean difference is larger than this value, we know that the difference is significant at the .05 level.

DETERMINING THE POWER OF AN EXPERIMENT

Errors Associated with Decisions About Experiments

As researchers our objective is to design our experiments to maximize the chances of rejecting the null hypothesis whenever the null hypothesis is false. Indeed, that is the essence of experimental design. However, we must realize that when we make a decision concerning the outcome of an experiment, we can make one of two kinds of errors. First is the situation where the null hypothesis is true but our experimental data cause us to reject it. This is called a *Type 1 error*. We have control over this error by the numerical value we select for α. Thus, if you choose $\alpha = .05$, this means that you are willing to falsely reject the null hypothesis 1 time out of 20. If you use the .01 level for α, then you are being more conservative with respect to the Type 1 error, and you will only be wrong 1 time in 100 when you reject the null hypothesis.

A Type 1 error can occur only when the null hypothesis is true, and you falsely reject it. Because science is conservative, we wish to minimize this kind of error, and it is for this reason that numerically small values of α are chosen. But suppose that the null hypothesis is false, which means that our experimental hypothesis is true. If, under these conditions, we reject the null hypothesis, then everything is well and good. However, if we do not reject the null hypothesis when it is false, then we are committing what is known as a *Type 2 error*. The probability associated with this kind of error is called β (read *beta*). Table 7.4 shows the conditions under which each kind of error occurs.

TABLE 7.4   The two types of errors that can occur when making a decision about an experiment

Researcher's decision	True state of affairs in the population	
	H_0 is true	H_0 is false (H_1 is true)
Reject H_0	Type 1 error (α)	No error
Do not reject H_0	No error	Type 2 error (β)

In the ideal experimental design we would strive to minimize both the Type 1 and the Type 2 errors. However, this is not easy to do because several factors influence these errors, and some of these factors are interdependent. For example, as we choose numerically smaller values for α, thereby reducing the Type 1 error, this brings about a numerical increase in β, our Type 2 error. Thus, the *value of* α that we select is one factor that determines the value of β. A second factor affecting β is the *variability in our data* as measured by the standard deviations of our experimental and control groups. The greater the variability the greater will be the standard errors of our means, and the less likely we are to pick up a significant difference if one exists. This brings up the third factor that influences β, the *sample size.* Even if our standard deviations are large, we know that by increasing n we can decrease the standard errors of our mean, and so decrease the value of β. Finally, the fourth factor that influences β is the *true mean difference* in the population between our experimental and our control conditions. If this difference is very small, then it is much more difficult to demonstrate a significant effect than if this difference is very large.

Power of the Test of Significance

In Table 7.4 the Type 2 or β error is defined as that error which occurs when we do not reject a false null hypothesis. Rather than express this in negative terms, we can place this into a positive form by stating that the expression $1 - \beta$ is the probability that we shall reject the null hypothesis if it is false. This is called the *power* of the test of significance. The concept of power is depicted in Figure 7.6.

The power concept applies only to the situation where our experimental hypothesis is true and we wish to know the likelihood that we can "prove" it to be true by being able to reject the null hypothesis. Figure 7.6 shows that if H_1 is true, then the true mean difference in the population is not zero, as postulated by the null hypothesis, but is some value greater than zero called d (see the top two figures in Figure 7.6). The bottom set of curves in Figure 7.6 shows the situation when we assume that the null hypothesis is true (the left-hand curve) when in reality H_1 is true (the right-hand curve). Our objective in designing an experiment is to maximize the white area in the right-hand curve, which represents $1 - \beta$ or power.

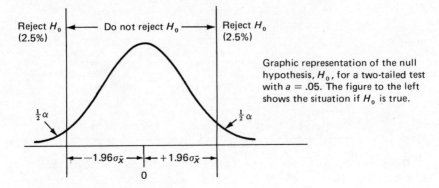

Reject H_0
(2.5%) ◄── Do not reject H_0 ──► Reject H_0
(2.5%)

Graphic representation of the null
hypothesis, H_0, for a two-tailed test
with $a = .05$. The figure to the left
shows the situation if H_0 is true.

$\frac{1}{2}\alpha$ $\frac{1}{2}\alpha$

◄── $-1.96\sigma_{\bar{x}}$ ──► ◄── $+1.96\sigma_{\bar{x}}$ ──►

0

But suppose that our experimental
hypothesis, H_1, is true and that our
experimental mean is greater than
our control mean by some amount
which we shall call d. Then the mean
of our distribution of differences is
not zero, but is the value d as shown
here.

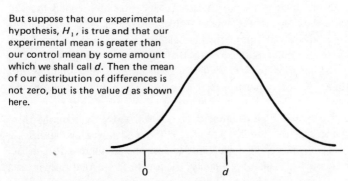

0   d

If we combine these two figures, we see the state of affairs when we
assume H_0 to be true when, in reality, H_1 is true. The probability of
being able to make the correct decision—which is to reject H_0 and
accept H_1—is known as the power of the test of significance and is
defined as $1 - B$. Power represents all those instances in which the
departure from the assumption of zero difference is so great that we
would reject H_0. This is represented by the area to the right of the
"Reject H_0" line. In this example the power is somewhat more than
.50.

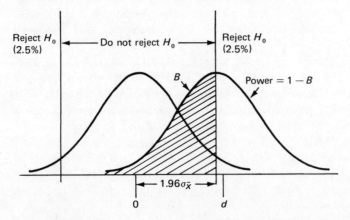

Reject H_0
(2.5%) ◄── Do not reject H_0 ──► Reject H_0
(2.5%)

B Power $= 1 - B$

◄── $1.96\sigma_{\bar{x}}$ ──►

0   d

FIGURE 7.6 The power concept in the test of significance.

Increasing the Power of the Test of Significance

We have noted that four factors influence our Type 2 error, and thus they also influence the power of our test of significance. By inspecting Figure 7.6 we can see how each of them can be varied to increase power.

Level of α If we increase the numerical value of α from, say, .05 to .10, the cutting point for our region of rejection moves closer to the mean of our distribution of differences. This will reduce β and increase the power of our test as shown in Figure 7.7. In Figure 7.7 the z value cutting off all but the extreme 5% in each tail is 1.65. Because a two-tailed test of the null hypothesis at the .10 level is numerically identical with a one-tailed test of H_0 at the .05 level, we find that one way to increase power is to use a one-tailed test. But remember the admonitions we discussed earlier concerning the use of the one-tailed tests.

Within group variability   The variability within our experimental and control subjects, as reflected by our standard deviation, may be able to be reduced (1) by more rigorous controls over environmental events such as standardized husbandry conditions, tighter controls over illumination and noise levels, following protocol exactly in testing animals, and so on; (2) by better control over our subjects through use of known genetic stocks and maintenance of strict rearing conditions from birth onward; and (3) by better experimental designs as, for example, in the use of littermates for control and experimental treatments.

Expected mean difference   The difference between our experimental and control treatments may be made larger as we develop greater understanding of the parameters affecting our experimental variable and get to know the kinds of manipulations that will act to enhance the mean difference. From Figure 7.6 we see that an increase in the value of d causes a greater separation in the two curves at the bottom of the figure, and this results in an increased power.

Sample size   Finally, we know that as sample size increases, the variability around the means decreases, resulting in a decrease in the standard error of the difference. Thus with an increase in n, a constant difference, d, is now more likely to be significant. The variable of sample size is one over which the experimenter has considerable control. In fact, a common question that researchers often ask is, "How many subjects should I use in each of my groups?" Because this is such a critical determiner of the power of the test of significance, we shall give this matter separate consideration.

What Size Sample Should We Use in Our Experiment?

In order to decide on the appropriate n per group in our experiment, we have to make a number of specific statements concerning our experimental design. First, we have to specify a value for α and also specify whether this is for a one-tailed or a two-tailed test. Next, we have to specify the difference we expect to get between our control and experimental means, and express this difference in standard deviation units. Finally, we must specify the power we desire our

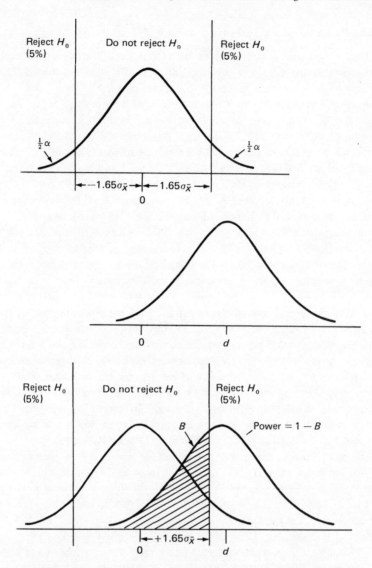

FIGURE 7.7 Showing how an increase in the numerical value of α increases power. α here $= .10$; compare with Figure 7.6, where $\alpha = .05$.

test to have in terms of a probability figure. Once these factors are known, we can use the power table (Table III in the Appendix) to find the desired n per treatment group. This table is adapted from Cohen (1969) and the reader is referred to his book for an extended and lucid discussion of the concept of power.

Use of the Power Tables

The following discussion is based on the assumptions that the n's are equal in both groups ($n_c = n_e = n$) and that the population standard deviations are the same.

Table III in the Appendix is in five sections and lists the following two-tailed α levels: .01, .02, .05, .10, and .20. These values may be halved and used for one-tailed significance tests yielding the following α levels: .005, .01, .025, .05, .10. The two-tailed test is designated as α_2 whereas α_1 indicates the parallel one-tailed test.

In these tables the d value is defined as

$$d = \frac{\text{expected mean difference between experimental and control groups}}{s} \quad (7.17)$$

The d values in Table III are obtained by dividing the expected difference between the experimental and control group by your best estimate of the standard deviation. Quite often that standard deviation is obtained from your pilot data, as in the shuttlebox experiment. However, perhaps you have run some other experiments involving various groups. If so, then we know from Formula 3.8 that the best estimate of the standard deviation is obtained by pooling the sums of squares and degrees of freedom. This formula is repeated below:

$$s_{\text{within}} = \sqrt{\frac{SS_a + SS_b + SS_c + \cdots}{(n_a - 1) + (n_b - 1) + (n_c - 1) + \cdots}} \quad (3.8)$$

We can then define the value d in the following fashion:

$$d = \frac{\text{expected difference between experimental and control groups}}{s_{\text{within}}} \quad (7.18)$$

The power values in Table II range from .25 to .99. The .25 value would rarely if ever be used, because this means the experimenter has only 1 chance in 4 of detecting a significance if one exists. The value $\frac{2}{3}$ is listed in the table because this is the point at which the experimenter has odds of 2 to 1 in his favor of detecting a significance difference when it exists.

The use of this table is best illustrated by an example. Let us return to the experiment in which we found that a control group of rats has a mean shuttlebox learning score of 40.02 correct avoidance responses with a standard deviation of 8.97. Suppose that we wish to investigate a new pharmacological agent which we believe will improve learning. However, we are not fully convinced of this and so we choose to use a two-tailed test. We wish to be quite certain that we do not falsely reject the null hypothesis, and therefore we choose an α level of .01 for our experiment. However, we also wish to be fairly certain that our experiment can detect a significant effect from the drug, if one is present, and so we choose a power of .80. This means that if a difference does exist between the experimental and control groups, we want to be 80% certain of finding that difference. We now have to decide what mean difference we expect to get between our experimental and our control animals. In our judgment, if the change brought about by the new drug is not at least half a standard deviation unit, or 4.5 avoidance responses, which is roughly 10% of the mean value of the control group, then it is not worth pursuing this line of research. In summary, we have the following factors specified:

$$\alpha = .01 \quad \text{(two-tailed)}$$
$$\text{power} = .80$$
$$d = .5 \text{ standard deviation units}$$

When these values are entered into Table III in the Appendix, we find that it is necessary to have 95 animals per group if we wish to be 80% certain of picking up a significant difference of half a standard deviation unit with an α value of .01.

Suppose that we decide we do not want to run a total of 190 animals in this experiment. If so, we can change any of the three factors listed above and then determine the resulting n. For example, we might be willing to settle for odds that are 2 to 1 in our favor rather than demanding .80 power as we did initially. If so, then we would set our power equal to $\frac{2}{3}$, and from Table III we now find that this reduces our n from 95 animals per group to 74 animals per group. That many still be more than we are willing to invest, and so we consider changing our α level from .01 to .05, and staying with our power value of .67. For these conditions we find that we needs 47 animals per group. Considering our other experimental involvements we believe that to be a reasonable number and so we decide to conduct the experiment with $\alpha = .05$ (two-tailed), power = $\frac{2}{3}$, and $d = .5$. Because the values in Table III represent the ideal condition, it is best to carry a few additional subjects along in the experiment, and so we decide that we shall use a total of 100 animals, randomly assigning 50 to each condition.

n for Intermediate Values of d

In Table III only certain values of d are given, and it will often be necessary to find n for a value of d not listed in the table. Cohen (1969) states that a good approximation to n is obtained by making the appropriate substitutions in the following formula:

$$n = \frac{n_{.10}}{100 \, d^2} + 1 \tag{7.19}$$

where $n_{.10}$ is the sample size for a given value of α and specified power for the condition $d = .10$, and d in the above formula is defined as before.

This can be illustrated by an example. In our study of memory transfer via brain homogenates we found that control animals can learn a black–white discrimination in a T-maze with food reward on an average of 40 trials. The standard deviation has been found over repeated experiments to be approximately 8.0. Some preliminary data from a pilot study indicated that recipients of brain homogenates from animals who had learned took an average of 3 fewer trials to learn the discrimination task than did recipients from naive donors. We decide on a one-tailed test with $\alpha = .05$; we also decide that we want to have 3 chances out of 4 of finding the phenomenon if it exists, and so we choose a power value of .75. Our value of $d\left(\frac{3}{8}\right)$ equals .375, and we see in Table III that n will be somewhere between 68 and 120.

We then have the following factors specified:

$$\alpha = .05 \quad \text{(one-tailed)}$$
$$\text{power} = .75$$
$$d = .375 \text{ standard deviation units}$$
$$n_{.10} = 1076$$

We can determine the value of n by substituting in Formula 7.19; we obtain

$$n = \frac{1076}{(100)(.375)^2} + 1 = 76.52 + 1 = 77.52$$

Thus, rounded to the nearest integer, we need to have 78 animals per group to test our hypothesis. To play safe we should carry along a couple of extra animals, so that we end up with 80 animals in each group.

Use of Power Tables for the Matched Pair Case

In the situation where subjects are paired, or the same subject is used twice (as with a pretest and posttest), we use the difference formula to determine our standard deviation:

$$s = \sqrt{\frac{\Sigma(D - \bar{D})^2}{n - 1}}$$

Our value of d now becomes

$$d = \left(\frac{\text{expected mean difference}}{s}\right)\sqrt{2} \tag{7.20}$$

We must use the factor, $\sqrt{2}$, for the matched pair case because the power tables are based on the assumption that we have two error variances which we have pooled, but here we only have one. The n we obtain from the tables is the number of *pairs* needed.

As an example of this case we shall return to our investigation of the effects of a complex environment in early life upon brain weight. We found that the total cortex of enriched animals weighed significantly more than their paired littermate controls who had been reared in an impoverished environment. We now wish to compare our complex environment group with a group reared under standard laboratory conditions, rather than in an impoverished environment, to see whether the enriched group is significantly better than the usual control condition.

In the prior experiment (Table 7.2) the mean difference between groups was 29.7 milligrams and the standard deviation of the difference was 22.71. We again plan to use littermate pairs in our experiment, and so it is reasonable to expect to obtain approximately the same standard deviation. However, because the environmental conditions of standard laboratory rearing are between the conditions of impoverished rearing and the enriched condition, we expect the mean difference between groups to be approximately half of what we obtained in our prior experiment. We decide that we wish to detect a difference of 15 milligrams, and we want to be quite certain that our experiment is powerful enough to pick up this difference, if it exists, and so we specify power of .90. We decide to test our hypothesis at the .05 level using a two-tailed test. We, therefore, have the following factors specified:

$$\alpha = .05 \quad \text{(two-tailed)}$$
$$\text{power} = .90$$
$$s = 22.71$$
$$\text{expected mean difference} = 15 \text{ milligrams}$$

Applying Formula 7.20 for the matched pair case yields the following:

$$d = \left(\frac{15}{22.71} \right) \sqrt{2} = .93$$

When we look this value up in Table III in the Appendix for $\alpha_2 = .05$ and power of .90, we find that n is between 22 and 34 pairs. To get the best value for n we apply Formula 7.19,

$$n = \frac{2102}{(100)(.93)^2} + 1 = 24.30 + 1 = 25.30$$

Therefore, we need 26 pairs of littermates to satisfy the conditions we have stipulated for our experiment.

Post Hoc Evaluation of an Experiment's Power

We often read of experiments in the literature that fail to reject the null hypothesis, and in some instances the researchers have concluded that they have "proven" the null hypothesis. The latter conclusion is meaningful only if the experiment was sufficiently powerful that the null hypothesis had a very good chance of being disproven, if it indeed were false. We can determine *post hoc* the power of an experiment to see whether the experimenter had a sufficiently adequate design to give an opportunity to evaluate the experimental hypothesis critically.

We can also use Table III in the Appendix for this purpose by entering with information concerning the α level used, the number of subjects per group, and d. From this we can determine the power of the experiment. Remember that we are only using this table for the case of equal n's per group and on the assumption that the two population standard deviations are identical. For a more extensive discussion of this issue and for additional tables, see Cohen's text.

As an example, let us consider a common case of an experimenter using a two-tailed test with $\alpha = .05$ and with 20 subjects per group. He reports a mean difference between his experimental and control subjects of 10 units on his criterion measure, and we find from an examination of his data that the average standard deviation is 20 units. We then have the following factors specified:

$$\alpha = .05 \quad \text{(two-tailed)}$$
$$\text{obtained mean difference} = 10$$
$$s = 20$$
$$n = 20$$

The determination of power for the *post hoc* case is dependent on what value we wish to insert for the expected mean difference between the experimental and control groups. In this example the *obtained* mean difference was 10 units. If we are willing to assume that these 10 points also represent the *expected* mean difference, then our value of d is

$$\frac{\text{expected mean difference}}{s} = \frac{10}{20} = .5$$

If we now enter Table III with $\alpha_2 = .05, d = .5$, and $n = 20$, we find that the test has a power between .25 and .50. By linear interpolation we estimate the power to be .33. That is, the experimenter only had one chance in three of disproving H_0 if H_1 were really true and the expected mean difference in the population was equal to 10 units. Under these conditions this is such a weak experimental design that we cannot make any conclusive statements concerning the evaluation of the experimental hypothesis.

However, it may be that the researcher had used a very powerful experimental variable that, in other situations, had resulted in very large differences between the experimental and control groups. Indeed, from his extensive knowledge of the research literature plus his own research experience, he fully expected the mean difference between groups to be 20 points if H_1 were true. If this is the more accurate description of the research situation, then the value of d now becomes $20/20 = 1.0$. When we now enter Table III with $\alpha_2 = .05$, $d = 1$, and $n = 20$, we find by linear interpolation that the power of the test is .88. This is very high power and thus the researcher may be willing to conclude that his experimental variable does not have any significant effect in this research setting.

SUMMARY OF DEFINITIONS

α level   The same as the level of significance.

Alternative hypothesis   The same as the experimental hypothesis.

β error   The same as a Type 2 error.

Distribution of differences   A hypothetical population frequency distribution of the differences between μ_c and μ_e.

Experimental hypothesis   The hypothesis that the researcher wishes to evaluate. This is the alternative hypothesis to the null hypothesis and is generally designated as H_1.

Homogeneity of variance   The assumption that the variances of the various groups in the experiment are homogeneous so that they can be pooled to yield one common value.

Level of significance   The probability level set by the experimenter to reject the null hypothesis. This is also called the α level. The two most commonly used α levels are .05 and .01.

Null hypothesis   The hypothesis that assumes that there are no differences in the population between the control and experimental groups. This is also called the statistical hypothesis and is generally designated as H_0.

One-tailed hypothesis   The situation in which a significant difference in only one direction is considered to be of importance to the researcher.

Power   The probability that a false null hypothesis will be rejected. This is defined as $1 - \beta$.

Standard error of a difference   The standard deviation of the sampling distribution of differences between μ_c and μ_e.

Statistical hypothesis   The same as the null hypothesis.

Two-tailed hypothesis   The situation in which a significant difference in either direction is considered to be of importance to the researcher.

Type 1 error   Rejecting the null hypothesis when it is true. This is also called the α error.

Type 2 error   Not rejecting the null hypothesis when it is false. This is also called the β error.

PROBLEMS

1. For the following set of data compute the value of t. Assume that a two-tailed test is used. If the researcher had selected an α level of .05, what conclusion would he draw from the experiment? If $\alpha = .01$, what is his conclusion?

	Control	Experimental
ΣX	627	433
ΣX^2	67,921	33,909
n	6	6

2. A social psychologist tested a group of students to determine their attitude toward religion (control group). Another group of students saw a movie that portrayed religion in a favorable manner (experimental group). After the movie the students were given the same attitude test. The psychologist was interested in testing the experimental hypothesis that the movie would make for a more favorable attitude toward religion. The following data were collected.

	Control	Experimental
$\bar{X}$	4.83	6.11
s^2	3.27	3.64
n	16	14

(a) What is H_0?
(b) What is the value of t?
(c) With $\alpha = .05$, what interpretation can be made of the data?

3. A behavioral geneticist investigated maze learning in two strains of inbred mice. The learning score was the number of trials needed before an animal could make two successive errorless trials. The following data were obtained.

	Strain A	Strain B
ΣX	150	196
ΣX^2	1987	3449
n	12	12

Obtain the means, sum of square, variances, standard error of the difference, and t. Get the 95% confidence limits for the population mean difference. What conclusion can the researcher draw from the study?

4. For the following data, fill in the blanks beneath the numbers

	Control	Experimental
	39	28
	64	90
	49	111
	63	71
	3	83
	25	50
	47	68
	81	59
	91	69
	45	100

	Control	Experimental
ΣX	_____	_____
ΣX^2	_____	_____
n	_____	_____
$\bar{X}$	_____	_____
SS	_____	_____
s^2	_____	_____
s	_____	_____
$s_{\bar{x}_c - \bar{x}_e}$	_____	
t	_____	
95% confidence limits	_____	

If you computed the confidence limits correctly, the lower limit is negative whereas the upper limit is positive. What relationship does this have to the test of significance when $\alpha = .05$?

5. Even though the mean difference in the previous problem was not significant, the researcher felt that the experimental variable was a strong one and that an increase in number of subjects would result in a significant finding. In order to decide on the appropriate number, she had to make a number of decisions. The two standard deviations were very similar, and so she decided to use s_{within} as her best estimate of the standard deviation in the population. She had obtained a mean difference of 22.2 points but did not feel confident that such a large difference would be obtained in a replication. Based on her experimental knowledge, she decided that 15 points was a more accurate representation of the expected mean difference. Also, she would be interested in pursuing this research area only if the mean of the experimental group was higher than the mean of the control group. Because she wanted to be quite certain that the experiment would detect a significant difference, if one existed, she set her power at .90. To determine the n needed per group to satisfy these conditions, fill in the blanks in the following table.

s within = _____
expected mean difference = _____
d = _____
α = _____ (____-tailed)
power = _____
n per group = _____

6. After determining that $n = 74$ in Problem 5, the researcher realized that her laboratory was not able to house and test that many animals. The most she could accommodate for a research program in this area would be 35 animals per group. She felt it necessary to stay with her d value of .60 and the one-tailed test. Thus, with her n kept constant, she could vary only the value of α or the power of her test. To see whether she could obtain sufficient power to conduct a meaningful experiment, she set up the following table. Fill in the blanks in the table.

Given: $d = .60$ with one-tailed test
$n = 35$ per group

If $\alpha_1 = .01$, with $n = 35$, power = _____
If $\alpha_1 = .025$, with $n = 35$, power = _____
If $\alpha_1 = .05$, with $n = 35$, power = __ _____
If $\alpha_1 = .10$, with $n = 35$, power = _____

Do you think she can do a meaningful experiment? If so, what combination of α and power would you suggest using?

7. An experimental psychologist interested in the effects of verbal instruction on psychomotor performance gave a group of subjects a pretest to determine their level of psychomotor skills. Based on these data he set up matched pairs and assigned the members of the pair to an experimental or a control group by the toss of a coin. The experimental subjects were given a lecture-demonstration on the principles of psychomotor learning while the control subjects were given a lecture on an unrelated subject. Both groups were then given 50 trials on the criterion task and the score was the total number of errors made throughout the 50 trials. The following data were obtained.

Pair no. Control Experimental

Pair no.	Control	Experimental
1	17	10
2	57	31
3	84	42
4	95	81
5	30	21
6	45	35
7	58	50
8	31	33

(a) Use the t test for matched pairs to test the mean difference for significance, using $\alpha = .05$.
(b) Set the 95% confidence limits on the population mean.
(c) Assume that there were two groups of independent subjects (i.e., that there was no matching). Compute the value of t and test H_0 at $\alpha = .05$. How do the values compare? Why?

CHAPTER 8 The Analysis of Variance: Single Classification

The t test is the appropriate statistic to use when there are two groups to compare in an experiment. However, it is inappropriate if there are more than two groups. An example will show why this is so. Consider a situation in which there are 10 treatment groups that we wish to compare with each other. (We shall use the expression "treatment groups" to refer to all of the experimental and control groups within an experiment.) There are 45 possible t tests that can be made with 10 groups. (The formula for determining the number of possible t tests is $N(N - 1)/2$, where $N =$ number of groups in the experiment.) With 45 t tests, we would expect to find two or three "significant" at the .05 level just by chance alone. Suppose that we found four significant differences at the .05 level. Is one or more of these "truly" significant, or is this within chance expectation? There is no way to answer this question by means of the t test.

Thus, when we run multiple t tests on a set of data, we see that our α level changes so that we are testing at a numerically higher level of significance than we have set for our experiment. That is, we may believe we are testing our null hypothesis at the .05 level when, in reality, we may be testing it at the .10 or .15 level, or even higher. Therefore, we need a procedure that allows us to evaluate three or more means at one time so that we can determine whether the variation among these means actually exceeds the α level we have chosen for rejecting H_0. If so, we conclude that the variability is greater than can be expected by chance alone, and so we reject H_0. The statistical procedure called the analysis of variance allows us to do this.

The principle involved in the analysis of variance is to compare the variability among the means of the experimental groups with the average variability found within the experimental treatments. If the former variability is sufficiently larger than the latter, we shall reject our null hypothesis and accept the alternative hypothesis. The logic underlying this principle is probably most easily illustrated by using the t test as the basis for developing the concepts involved in the analysis of variance. Thus, we now return to the t test formula, but this time we shall view it from a slightly different perspective.

THE t TEST, t^2, AND VARIANCES

The formula for the t test is given below.

$$t = \frac{\bar{X}_c - \bar{X}_e}{\sqrt{\{(SS_c + SS_e)/[(n_c - 1) + (n_e - 1)]\}(1/n_c + 1/n_e)}}$$

Without losing any generality, we can let $n_c = n_e = n$. Our formula now becomes

$$t = \frac{\bar{X}_c - \bar{X}_e}{\sqrt{[(SS_c + SS_e)/2(n-1)](2/n)}}$$

If we now square our value of t, this will result in squaring the numerator and getting rid of the square root sign in the denominator:

$$t^2 = \frac{(\bar{X}_c - \bar{X}_e)^2}{[(SS_c + SS_e)/2(n-1)](2/n)}$$

We can take the value $(2/n)$ in the denominator of the above formula and by appropriate algebraic manipulations move it into the numerator to give us

$$t^2 = \frac{n[(\bar{X}_c - \bar{X}_e)^2/2]}{(SS_c + SS_e)/2(n-1)}$$

We can demonstrate by some more algebra that the expression $(\bar{X}_c - \bar{X}_e)^2/2$ in the numerator is another way of computing the variance of our two means, or $s_{\bar{x}}^2$. We also recognize that the denominator of the above formula is s_{within}^2 or MS_{within} from Formula 3.9 when $n_c = n_e = n$. By making the appropriate substitutions we now have the following:

$$t^2 = \frac{n s_{\bar{x}}^2}{s_{within}^2} = \frac{n s_{\bar{x}}^2}{MS_{within}} \qquad (8.1)$$

Let us now consider what the statistics in the numerator and denominator of the above formula tell us about population parameters. The denominator is our best estimate of the squared standard deviation in the population obtained by pooling the sums of squares and degrees of freedom from our experimental and control groups. Up until now we have used the symbol s^2 to designate the variance in our sample, although we also indicated that the expression MS is the same as s^2. In the context of the analysis of variance it is conventional to use the terminology of mean squares to signify variances rather than to use s^2, and we shall also follow that convention.

We know that MS_{within} is an unbiased estimator of the population variance σ^2. And now we come to another statistical convention concerning the analysis of variance: The population variance is designated as σ^2_{error} or σ^2_e. Thus when one talks about *error variance* in the analysis of variance this refers either to the squared standard deviation obtained from the sample or to the variance in the population.

When we state that our sample statistic is an unbiased estimator of a population parameter, this means that the average value of the statistic over an infinite set of samples will be the same value as the population parameter. We use the term *expected value* to refer to this average value of the statistic. We can put this into a simple formula as follows:

$$E(MS_{\text{within}}) = \sigma^2_e \qquad (8.2)$$

This states that the expected value of MS_{within} in the population is σ^2_e.

Now let us turn to the numerator of the t^2 formula. You will remember that the standard error of the mean is defined as

$$s_{\bar{x}} = \frac{s}{\sqrt{n}}$$

If we now square that equation we obtain

$$s^2_{\bar{x}} = \frac{s^2}{n}$$

By transposing we see that

$$ns^2_{\bar{x}} = s^2 = MS_{\text{between}} \qquad (8.3)$$

The term of the right of the formula above, MS_{between}, is introduced here for the first time and is the third convention followed by statisticians in discussing the analysis of variance. The expression MS_{between} signifies that this is the mean square, or variance, between the means of our experimental and control group. The expression MS_{between} is not restricted to the two-group situation and, in more general terms, it refers to the variance among the means of all of our experimental and control groups.

If you look at Formula 8.3 you will see that the expression at the left-hand side of the equation is an estimate of the sample variance, because it is equivalent to s^2, and thus estimates σ^2_{error} in the population if the null hypothesis is true. We can put this into an equation as well.

$$\text{If } H_0 \text{ is true, } E(MS_{\text{between}}) = \sigma^2_e \qquad (8.4)$$

Thus, if H_0 is true, then both the numerator and the denominator of our formula for t^2 are estimating the same value, and the expected value in the population for our t^2 statistic is approximately 1.

However, what happens if our experimental hypothesis is true, namely that our experimental treatments have acted to enhance the mean difference among our groups? In this situation the variability among our means as reflected in the numerator will be greater than the variance among our subjects. Thus,

$$\text{If } H_1 \text{ is true, } E(MS_{\text{between}}) > \sigma_e^2 \qquad (8.5)$$

In this situation the expectation of our t^2 statistic will be greater than 1. If t^2 is sufficiently greater than 1, we shall conclude that the mean difference is greater than can be expected by chance alone, and thus we shall reject the null hypothesis H_0 and accept the alternative H_1.

The statistic t^2 is more commonly called the F statistic. In the same way that the t statistic allows us to evaluate the situation where we have two groups, the F statistic is used to evaluate our data when we have three or more groups. The statistic t^2 is a special case of the F test in the analysis of variance when we have two groups.

THE F TEST FOR TWO OR MORE GROUPS

We can generalize from the two-group situation to the condition where we have any number of treatment means simply by realizing that, regardless of the number of treatment groups in our experiment, we can take the means of those groups and determine the variance of those means. We can then multiply that variance by the number of cases in each group, and obtain the general expression $ns_{\bar{x}}^2$ which is equal to MS_{between}. We have restricted this discussion to the case where the n's are equal for all treatment groups. The same logic applies when the n's vary from one group to another; however, a number of statistical complications arise in that situation, one of which is that the computational formulas become more complicated.

As in the two-group situation, the expected value of MS_{between} is σ_e^2 if the null hypothesis is true. The concept of error variance, or MS_{within}, is carried over directly from the two-group situation. We may now define the F statistic as follows:

$$F = \frac{MS_{\text{between}}}{MS_{\text{within}}} \qquad (8.6)$$

If the null hypothesis is true, the expected value in the population of both the numerator and denominator is σ_e^2, and thus we expect our F value to approximate 1. If our various experimental treatments act to increase the differences among the means of our treatment groups, then the expected value of MS_{between} will be greater than σ_e^2 and our F ratio will be greater than 1. If it is significantly greater than 1, using an α level of .05 or .01, then we shall reject H_0 and accept H_1 as a better way of describing our experimental findings.

THE NULL HYPOTHESIS

We can now formally state the nature of the null hypothesis and its alternative. This hypothesis is also a simple generalization from the t test situation. The null hypothesis is that our experimental treatments are without effect so that the means in the population are identical. Because we now move beyond two groups to an indefinite number of groups, we can no longer talk about an experimmental and a control group using μ_c and μ_e to designate the population means. In this instance we shall use the terms $\mu_1, \mu_2, \mu_3; \ldots$ Therefore,

$$H_0: \mu_1 = \mu_2 = \mu_3 = \ldots$$

Our experimental hypothesis, of course, is that the treatments we have introduced do have an effect so that the means in the population are not identical. Therefore,

$$H_1: \text{not } H_0$$

The null hypothesis and its alternative can be expressed in another way as well. If μ = population mean, and μ_j = any treatment mean (i.e., $\mu_j = \mu_1, \mu_2$, etc.), then $(\mu_j - \mu)^2 = 0$ when H_0 is true. That is, there will be no variance among the population treatment means if H_0 is true. From this approach we can specify H_0 and H_1 as follows:

H_0 : The variance among the population treatment means is 0.
H_1 : the variance among the population treatment means is greater than 0.

Variance and Means

It should now be apparent that the expression "the analysis of variance" refers to the variance of the means as well as to the square of the standard deviation within a group of subjects. This nomenclature tends to confuse the student. To review our logic: The variance among our means will be $1/n$th as large as the average pooled variance within groups if our null hypothesis is true. If, however, our experimental hypothesis is true, then the means will be much farther apart, because of our experimental intervention, than would be the case if we were sampling from the same population for all our groups. Therefore an increase in differences among means is equivalent to an increase in the variance of our mean, and it is this variance that we evaluate relative to our "error" variance by the procedure called "the analysis of variance."

Degrees of Freedom

We have degrees of freedom to evaluate our F test as we did for the t test. Here we need two sets of degrees of freedom: one for the numerator and the other for the denominator. For the general case the degrees of freedom for the numerator is defined as

$$df_{numerator} = number\ of\ treatment\ means - 1 \qquad (8.7)$$

The degrees of freedom for the denominator is the degrees of freedom associated with the pooled variance within groups and is defined as

$$df_{denominator} = (n_1 - 1) + (n_2 - 1) + (n_3 - 1) + \ldots \qquad (8.8)$$

EVALUATING THE F STATISTIC: THE F TABLE

Table IV in the Appendix is used to determine the level of significance of our F ratio. We must enter this table with both sets of degrees of freedom. To use the table, locate the numerator degrees of freedom across the columns. Then locate the denominator degrees of freedom down the rows. At the intersection of the respective row and column will be found two numbers. The upper one signifies the value of F needed for significance at $\alpha = .05$; the lower is the critical value of F for significance at the .01 level.

We have shown that $t^2 = F$ when there are two groups. In this situation there are 1 and $df_{denominator}$ degrees of freedom to evaluate the t^2 statistic. The leftmost column in Table IV represents this situation. We can compare the values in that column with the information in Table II, the t table. Consider the case where there are 10 df to evaluate t. At the .05 level $t = 2.228$; at the .01 level $t = 3.169$. In the F table we see that for 1, 10 df (this is the conventional way of expressing the two sets of degrees of freedom for an F ratio) the .05 value is 4.96, whereas the value is 10.04 for the .01 level. It can be verified that these F values are the squares of the comparable t values.

To illustrate how to use the F table, consider the following set of data.

Source	df	MS	F
Between	3	25	3.12
Within	27	8	

The table above is set up in the standard fashion to report the analysis of variance. The column headed "Source" refers to the source of variability, which in this instance is either the variance between the means of our experimental conditions, or the variance within the experimental treatment groups. The headings of the rest of the table are all familiar to you. The F ratio is found to be 3.12. In Table IV the values needed for significance with 3 and 27 df are 2.96

(for the .05 level) and 4.60 (for the .01 level). The observed value of F falls between the .05 and .01 α levels. If the researcher had specified the .05 α level for rejecting the null hypothesis, he would conclude that he had a significant effect. If he selected the .01 level, he would conclude that his data are not sufficient to enable him to reject the null hypothesis.

FIXED EFFECTS AND RANDOM EFFECTS MODELS

There are a considerable number of analysis of variance models available to the researcher. They differ with respect to the nature of the experimental design, the restrictions placed on the data, and the types of assumptions made. The purpose of this book is to introduce you to the analysis of variance, and therefore we shall discuss only a few of the simpler designs, although these will certainly suffice for much of your research activity.

Fixed Effects Model

In the analysis of variance designs that we discuss, we shall assume that we have a *fixed effects model*. This model is basic to all experimental research. The model assumes that the treatment conditions we have set up in our investigation are the treatments of specified interest to us, and that we do not wish to extrapolate or generalize beyond these particular treatment conditions to *other treament conditions*. Another way of expressing this is to state that the experimental manipulations that we use in our study define the *population of experimental treatments* in which we are interested. Let us consider some experimental examples of the preceding statement.

Examples

A neuroendocrinologist is interested in the effects of neonatal androgen on adult sexual behavior. To investigate this problem he castrates male mice at birth. In adulthood he observes their sexual behavior in the presence of a male and a female mouse. If he obtains a significant effect, this indicates that castration in infancy, which is one method of removing the source of endogeneous androgen, does influence adult sexual behavior. However, he is not able to conclude that other methods of eliminating or reducing androgen in infancy (for example, injection of an antiandrogen or of estrogen) would have the same effects as castration. In order to determine that, he must carry out those particular experiments.

A psychopharmacologist is interested in determining the effects upon learning of drugs that act to stimulate or depress the central nervous system. She selects two stimulants, which belong to chemically different classes, and one depressant. The control group receives a placebo injection. If she finds that the two stimulants improve learning whereas the depressant acts to impede learning, her conclusions are limited to these specific drugs. She is not able to conclude that all drugs that stimulate the central nervous system will act in the same fashion as these do, nor can she conclude that other drugs that depress the central nervous system will also act to depress learning.

An experimental geneticist is interested in the effect of the short ear gene (*se*) on the mouse's ability to resist the lethal effects of radiation. He finds that the length of time before the irradiated animals succumb varies as a function of the number of short ear recessive genes the animal has. He cannot generalize this finding to any other genes.

An experimental psychologist interested in studying aggression with laboratory rats places two males together in a small cage with a grid floor. When the grid is electrified, she finds that the animals act aggressively toward each other. She cannot generalize from this situation to other treatment conditions that induce aggression.

We see from the examples above that the particular experimental treatments we employ in our study have been deliberately selected because they are the ones of direct relevance to us. It is in this sense that these treatments are *fixed effects*. The major statistic of interest in the fixed effects model is the mean of each treatment group. The objective of an experimenter in introducing a particular experimental manipulation is to determine whether he or she can change the mean of that treatment group relative to that of the appropriate control group.

To summarize the fixed effects model: In this model the researcher deliberately selects certain experimental operation (for example, neonatal castration, drugs which have certain effects on the central nervous system, certain genes, a method of studying aggression), and the conclusions which he or she draws must be restricted to those particular experimental operations. His or her interest is in determining how these experimental operations modify the means of the experimental groups.

Random Effects Model

But suppose that as experimenters we wished to determine whether a particular class of treatment had a general effect? How would we do that? Let's go back to our psychopharmacology example. Suppose we were interested in determining whether drugs that stimulate the central nervous system act to improve learning. What we would have to do is to define a population of "drugs that stimulate the central nervous system." We would then list all the drugs that fit within that definition. Using a table of random numbers, we would select from this population a certain number of these drugs (say, half a dozen), administer these to our experimental subjects, and then compare the effects of these drugs against those of appropriate controls. If we found that these drugs acted to improve learning, we could then draw the general conclusion that "drugs that stimulate the central nervous system act to enhance learning." The reason we can make such a broad generalization is that we have defined the population of drugs that stimulate the central nervous system and have randomly selected a sample from this population. In the same way as we can generalize to a subject population when we have randomly sampled subjects from that population, we can also generalize to a treatment population if we randomly sample treatments from that population.

This model is called the *random effects model* (some authors also call it the *variance-component model*). As you can see, the procedures and assumptions underlying the random effects model and the inferences possible are quite different from those of the fixed effects model, and this causes a considerable difference in the test of significance in the analysis of variance.

As experimentalists it is very rarely, if ever, that we are going to define a population of treatment conditions and randomly select several treatments from that population for use in our research. Instead, we are going to deliberately choose our particular experimental manipulation using all the information that we can obtain from our own research, from published research of others, and from information that we gather in conversations with colleagues—all aimed at maximizing the power of our experimental treatment. It is for this reason that discussion in this text will be limited to the fixed effects model.

Generalizations Made from Experimental Procedures

We have discussed the logical considerations that cause us to restrict our generalizations to a particular set of treatments as compared to the situation when we can make a broad generalization to a class of treatment conditions. However, as experimentalists we often write and interpret our data as though we have a random effects model rather than a fixed effects model, although such interpretations are usually cautiously expressed. A psychopharmacologist who had found that the two particular central nervous system stimulants acted to enhance learning might write a summary sentence in his paper such as the following: "These data suggest the possibility that CNS stimulants may have a general effect on improving learning performance." That is a properly cautious statement and may be taken as a working hypothesis to be tested in further research.

As experimenters we often make generalizations beyond our particular experimental conditions to wider populations. For example, from a study with rats we may draw a tentative generalization about humans. We may do this because we have evidence that the animal model is sufficiently similar to the human to allow us to generalize (e.g., endocrine functions in man and rat appear to be quite similar in many instances; electrical stimulation of the brain in man and rat appear to have similar functional properties), or because we are willing to make the assumption that an animal model, especially if it is a mammal, is enough like the human that general laws obtained with the animal may be carried over to the human situation. (This latter position is the basis for most of the research in the area of learning.) In the same way as we generalize beyond our subject population to other subject populations, we also generalize beyond our particular treatment conditions to a broader class of treatment conditions. So a psychopharmacologist may draw a tentative conclusion concerning the effects of CNS stimulants upon learning, and an experimental psychologist may draw tentative conclusions concerning modification of aggression that go beyond his particular way of measuring aggression (e.g., in a small chamber using electrical shock as a motivator) and that also go beyond his particular species (e.g., the rat). These types of tentative generalizations are valid *as long as they are taken as working hypotheses that have to be investigated experimentally.* However, it is *not* possible to make a firm generalization across species or across treatment conditions unless species or treatment conditions have been randomly sampled from a defined population. That is the difference between the fixed effects model and the random effects model.

SINGLE-CLASSIFICATION ANALYSIS OF VARIANCE WITH INDEPENDENT OBSERVATIONS

We are now ready to examine the simplest type of analysis of variance design which is called a *single-classification design* because our various treatments are grouped or classified along a single dimension. (This design is also called a *single-factor experiment* or a *one-way analysis of variance*.) Subjects are assigned at random to one of the treatment conditions. At the end of the experiment we have one numerical score that characterizes each subject's performance on our criterion measure. This one score may be based on a number of observations (e.g., rats may be tested repeatedly in an open field), but eventually one score will be used to describe each subject. Each score is assumed to be an *independent observation* unrelated to any other score. This will be true when we have one score per subject (in contrast to the situation where we have a set of scores per subject—as in a learning task), and when the subjects are unrelated (in contrast to a situation where some of the subjects are related to each other, as with littermates).

Let us consider an abstract example. Assume that we have four experimental treatments that we wish to evaluate. We shall designate these as Treatments 1, 2, 3, and 4. The layout for such an experiment is shown in Table 8.1.

Each subject's score is indicated by an X in Table 8.1. The first subscript beside each X signifies the group of which the subject is a member, and the second subscript indicates an arbitrary number for each subject ranging from 1 to n. There are n subjects per group, and a total of $4n$ subjects in the complete experiment. We shall use the notation N to signify the total number of subjects within an experiment in this design and all further analysis of variance designs. In this particular example $N = 4n$. The sum of the scores within each treatment group is symbolized at the bottom of Table 8.1 by the expressions $\Sigma X_1, \Sigma X_2, \Sigma X_3$, and ΣX_4. If we divided each of these totals by n, this would give us the means for our treatments. However, it is computationally much more convenient to work with totals that with means, and so the means have beeen omitted in Table 8.1.

TABLE 8.1   General layout for the single-classification analysis of variance with independent observations

Treatment 1	Treatment 2	Treatment 3	Treatment 4
$X_{1 \cdot 1}$	$X_{2 \cdot 1}$	$X_{3 \cdot 1}$	$X_{4 \cdot 1}$
.	.	.	.
.	.	.	.
.	.	.	.
.	.	.	.
$X_{1 \cdot n}$	$X_{2 \cdot n}$	$X_{3 \cdot n}$	$X_{4 \cdot n}$
ΣX_1	ΣX_2	ΣX_3	ΣX_4

We now need to determine $MS_{between}$ and MS_{within} in order to compute our F ratio. To do this we have to obtain the appropriate SS and divide by their respective degrees of freedom.

Table 8.2 shows us where these sums of squares come from. This table contains the same information as in Table 8.1 with several additions. We see that we can take the scores of the subjects within Treatment 1 and can compute the sum of squares among those subjects. This is designated in Table 8.2 as $SS_{within\ 1}$. The same can be done for the subjects within Treatments 2, 3, and 4. These four sums of squares are added together to provide the value of SS_{within}.

In order to obtain the sum of squares between treatments we work with the treatment totals at the bottom of Table 8.2. By means of a computational formula, which will be presented shortly, we are able to obtain $SS_{between}$.

Assumptions in the Analysis of Variance Model

We can use Table 8.2 to develop some points about the assumptions underlying the analysis of variance. In Table 8.2 you see that the sums of squares

TABLE 8.2   Showing how SS_{within} and $SS_{between}$ are obtained

$$SS_{within\ 1} \quad + \quad SS_{within\ 2} \quad + \quad SS_{within\ 3} \quad + \quad SS_{within\ 4} = SS_{within}$$

Trt 1	Trt 2	Trt 3	Trt 4
$X_{1 \cdot 1}$	$X_{2 \cdot 1}$	$X_{3 \cdot 1}$	$X_{4 \cdot 1}$
.	.	.	.
.	.	.	.
.	.	.	.
$X_{1 \cdot n}$	$X_{2 \cdot n}$	$X_{3 \cdot n}$	$X_{4 \cdot n}$
ΣX_1	ΣX_2	ΣX_3	ΣX_4

The variation among the treatment totals $= SS_{between}$

within each of our treatment groups are added together to give SS_{within}. This can only be done on the assumption that there is *homogeneity of variance* within treatment groups. That is, the expected value of each sample variance is σ^2_{error} as shown below.

$$E(s^2_{within\ 1}) = \sigma^2_e$$
$$E(s^2_{within\ 2}) = \sigma^2_e$$
$$E(s^2_{within\ 3}) = \sigma^2_e$$
$$E(s^2_{within\ 4}) = \sigma^2_e$$

This assumption is the same one that applies to the *t* test, and the same general rule holds here as with the *t* test: If you have equal numbers of subjects in each treatment group, the homogeneity of variance assumption can be violated without appreciably affecting the *F* test.

The second assumption in the analysis of variance is that our *treatment means are normally distributed.* That assumption is satisfied when (1) the raw scores within the treatment groups are from a normally distributed population, or (2) there are approximately 20–30 cases within each group. Again, the *F* test has been shown to be insensitive to rather considerable departures from the normality assumption. In statistical parlance the *F* test is said to be a *robust* statistic in the sense that many of the assumptions underlying its derivation may be violated without affecting the distribution of this statistic and the corresponding probability values.

The two assumptions of homogeneity of variance and normality pertain to the distribution of scores within each of the treatment groups in Table 8.2. The third assumption, that of *independence between treatment and error*, is concerned with the treatment totals (or means) at the bottom of Table 8.2. These totals are assumed to represent the summation of two factors: (1) the effect of the experimental treatment itself, and (2) the effect of all other conditions in the experiment. Factor 2 represents the sum total of all sources of error and this summation is assumed to go to 0 so that the totals are unbiased estimates of the treatment effects reflected in Factor 1. This assumption is one that cannot be violated. It states that the error associated with any particular raw score must be the same for all treatment conditions. If the errors in measurement differ from one treatment condition to another, then the difference in means may well reflect error differences rather than a difference due to the experimental variable we are studying. In addition, the lack of such independence may cause heterogeneity of the within-group variance. For these various reasons we are not able to interpret our findings.

In order for this assumption to be valid, it is *necessary* to assign the sample elements at random to the treatment conditions. Why is this? Each subject has its own unique effect which will interact with the particular treatment condition to which it is assigned. If all subjects are randomly distributed, in the long run these unique effects will be distributed evenly among all experimental treatments and thus will be independent of the treatment effects.

Randomization is a *necessary* condition to ensure the validity of the independence assumption, but it is not *sufficient*. It is also absolutely necessary that we maintain the same level of experimental controls for all treatment groups. This is why, for example, it is inappropriate to have one research assistant test the animals in treatment groups 1 and 2 while another assistant tests the animals in groups 3 and 4. The effects of the assistants are not balanced across the treatment groups, and a significant effect between either of the first two groups and either of the last two groups may reflect differences in the manner in which the assistants test the animals rather than differences in the treatments themselves. As another example, there is an ample body of literature showing that "experimenter bias" will affect one's observations of animals and humans alike. Even if one technician tests all animals, if he expects that animals in one particular group will perform better (or worse) than animals in another group, this could induce a perceptual bias such that the scores he records for that particular group will be different from what they would have been if he did not have this expectation. It is for this reason that double-blind studies are run. Whenever there is a lack of independence between the treatment condition and the error associated with an animal's score, your experimental data are worthless, and there is no statistician in the world who can help you salvage any useful information from your results.

We see here an interesting and necessary parallel between experimental methodology and statistical assumptions. If there is faulty experimental procedure, as in the situation where there is a lack of independence between treatment and error, there is a concomitant violation of a statistical assumption which renders invalid the model underlying the experimental design. In those situations where the experimenter exercises careful methodology and uses "experimental common sense" by having large numbers of subjects within each group and equal number of subjects per group, the parallel statistical assumptions are much less stringent, and violations of them will not affect the interpretation of one's data in any important manner. There is an important moral to all of this: If you are a good and careful experimental researcher and design your experiments in a rigorous fashion, there will be an appropriate statistical model to enable you to analyze your data. Remember, though, that if your design is a rather complex one it is always advisable to consult with a statistician *before* you start collecting your data to get the statistician's advice concerning the design. Quite often a trained statistician can show you a more efficient way to obtain the same amount of information or to obtain more information for relatively little additional cost.

Computational Formulas for the Single-Classification Analysis of Variance

The basis for the analytical procedures involved in determining the sums of squares is shown in Table 8.3. That table contains the same information as Table

TABLE 8.3 Showing how the three SS are obtained

Trt 1	Trt 2	Trt 3	Trt 4	
$X_{1 \cdot 1}$	$X_{2 \cdot 1}$	$X_{3 \cdot 1}$	$X_{4 \cdot 1}$	The variation
.	.	.	.	among all the raw
.	.	.	.	scores, treated
.	.	.	.	as a single group,
$X_{1 \cdot n}$	$X_{2 \cdot n}$	$X_{3 \cdot n}$	$X_{4 \cdot n}$	$= SS_{total}$
ΣX_1	ΣX_2	ΣX_3	ΣX_4	

The variation among these treatments totals $= SS_{between}$

The variation remaining after subtracting $SS_{total} - SS_{between} = SS_{within}$

8.1., and some of the information from Table 8.2, but in a slightly different arrangement.

Look within the large rectangular box in Table 8.3. Within the block are contained all the observations of our experiment. If we put all these scores together and treat them as one group, we can determine the sum of squares among all these scores. That sum of squares represents the total variation in our experiment and is designated by the expression SS_{total}. By going through some algebra we can show the following to be true:

$$SS_{total} = SS_{between} + SS_{within} \qquad (8.9)$$

In a parallel fashion the degrees of freedom follow the same arrangement:

$$df_{total} = df_{between} + df_{within} \qquad (8.10)$$

In words, the total sum of squares or variation within an experiment can be broken down into two independent and additive components which represent the variation within our experimental groups and the variation between our experimental groups. This is what we have portrayed in Table 8.2. Similarly, the total degrees of freedom within our experiment is also broken down into two independent and additive components representing the df within groups and df between groups. The usual procedure followed in computing the various sums of squares is to compute directly SS_{total} and $SS_{between}$, and then to obtain SS_{within} by subtraction.

Table 8.4 gives the general rules and computational formulas for analyzing data in the single-classification analysis of variance design with independent observations. All the symbols and terminology are familiar ones with the exception of the expression $(\Sigma X)^2/N$, which is called a *correction term*, or *CT*.

TABLE 8.4   General rules and formulas to obtain SS_{total}, $SS_{between}$, and SS_{within} for a single-classification analysis of variance design with independent observations

---

The data table

	Trt 1	Trt 2	Trt 3	Trt 4
X	.	.	.
	.	.	.	.
	.	.	.	.
	.	.	.	.
	.	.	.	.
	.	.	.	.
	ΣX_1	ΣX_2	ΣX_3	ΣX_4

Definitions

X = any raw score within our data table

n = number of subjects per treatment group. The n's are assumed to be equal so that
$n_1 = n_2 = n_3 = n_4 = n$

ΣX_1, ΣX_2, ΣX_3, and ΣX_4 = totals of the scores within Treatments 1, 2, 3, and 4

N = total number of subjects in the experiment = $n_1 + n_2 + n_3 + n_4 = 4n$

Obtain the following values

ΣX = sum of all the raw scores in the experiment

ΣX^2 = sum of squares of each raw score in the experiment

$\dfrac{(\Sigma X)^2}{N}$ = Correction term needed to adjust the data so that deviations are taken from the overall mean of the experiment = CT

Rules and formulas

SS_{total}:   The total sum of squares of all the data, designated as SS_{total}, is always obtained by subtracting the correction term $(\Sigma X)^2/N$ from the sum of the squares of the raw scores:

$$SS_{total} = \Sigma X^2 - \frac{(\Sigma X)^2}{N} = \Sigma X^2 - CT$$

$SS_{between}$:   To obtain the SS among the treatments, square each treatment total, add these values together, divide this amount by the number of scores making up a treatment total, and subtract the correction term:

$$SS_{between} = \frac{(\Sigma X_1)^2 + (\Sigma X_2)^2 + (\Sigma X_3)^2 + (\Sigma X_4)^2}{n} - CT$$

This general rule applies only for the case of equal n's.

SS_{within}:   The within SS (also called error SS) can be obtained by subtracting the between SS from the total SS:

$$SS_{within} = SS_{total} - SS_{between}$$

TABLE 8.4 *(continued)* General rules and formulas to obtain SS_{total}, $SS_{between}$, and SS_{within} for a single-classification analysis of variance design with independent observations

An alternative method: The sums of squares can be computed directly for each treatment group by the formula

$$SS_{within\ 1} = \Sigma X_1^2 - \frac{(\Sigma X_1)^2}{n_1}$$

where

ΣX_1^2 = sum of the squares of the raw scores within Treatment 1

$(\Sigma X_1)^2$ = square of the sum of the raw scores within Treatment 1

n_1 = number of cases within Treatment 1

This formula can be generalized to the other Treatment conditions by simply changing the numerical subscripts used to designate the treatment groups. These separate components are added together to give SS_{within}:

$$SS_{within} = SS_{within\ 1} + SS_{within\ 2} + SS_{within\ 3} + SS_{within\ 4}$$

The value of SS_{within} obtained by this alternative method should be the same as the value obtained by the subtraction method, within rounding error.

That particular expression first occurred in Chapter 3 in Table 3.5 as part of the computational formula for determining the sum of squares of deviations around the mean. It is called a correction term for the following reason: When you take your raw scores, square and sum them, you are essentially obtaining deviations around an assumed mean of 0. However, it is necessary to take the deviations around the true mean. The value $(\Sigma X)^2/N$ allows us to shift our reference point from an assumed mean of 0 to the true mean of the group, hence the expression correction term.

The rules in Table 8.4 are general ones which apply to any single-classification analysis of variance design where there are equal n's. After the instructions in Table 8.4 are followed, it is conventional to put the information into a summary table in the form shown in Table 8.5. The F ratio at the far right at Table 8.5 is evaluated by entering the F table with the degrees of freedom listed in the second column of Table 8.5. We shall now analyze a set of data to show how the computational formulas work and how this is all put together into a summary table.

An Example

Table 8.6 summarizes a set of data in which there are four treatment conditions with five observations per treatment. The computations to determine the sums of squares parallel the instructions given in Table 8.4. SS_{within} is obtained by subtraction. As an exercise you may wish to compute the sums of

TABLE 8.5 General summary table for single-classification analysis of variance with independent observations

Source of variation	Degrees of freedom	Sums of squares	Mean squares	F
Between treatments	Number of treatments $- 1$	$SS_{between}$	$\dfrac{SS_{between}}{df_{between}} = MS_b$	$\dfrac{MS_b}{MS_w}$
Within treatments	$N -$ number of treatments $=$ $(n_1 - 1) + (n_2 - 1) + (n_3 - 1) + \cdots$	SS_{within}	$\dfrac{SS_{within}}{df_{within}} = MS_w$	
Total	$N - 1$	SS_{total}		

squares within each of the four treatment conditions and verify that their sum will equal SS_{within}.

The analysis of variance summary is given in Table 8.7. Our observed value of F is 17.47, and with 3, 16 df this exceeds the tabled value of F at the .05 level (3.24) as well as the tabled value at the .01 level (5.29). The double asterisk beside the F ratio signifies that the observed value of F exceeded $F_{.01}(3,16)$. This is the usual way significance at the .01 level is indicated in a research publication. A single asterisk would indicate significance at the .05 level.

TABLE 8.6 A computational example of the single-classification analysis of variance with independent observations

Treatment 1	Treatment 2	Treatment 3	Treatment 4
7	21	21	19
16	30	15	12
8	25	14	10
10	28	13	9
7	23	18	9
48	127	81	59

$$\Sigma X = 315$$
$$\Sigma X^2 = 5919$$
$$(\Sigma X)^2/N = 4961.25 = CT$$
$$SS_{total} = \Sigma X^2 - \frac{(\Sigma X)^2}{N} = 5919 - 4961.25 = 957.75$$
$$SS_{between} = \frac{(\Sigma X_1)^2 + (\Sigma X_2)^2 + (\Sigma X_3)^2 + (\Sigma X_4)^2}{n} - CT$$
$$= \frac{(48)^2 + (127)^2 + (81)^2 + (59)^2}{5} - 4961.25 = 733.75$$
$$SS_{within} = SS_{total} - SS_{between} = 957.75 - 733.75 = 224.00$$

TABLE 8.7 Summary table for the data in Table 8.6

Source	df	SS	MS	F
Between	3	733.75	244.58	17.47^{**}
Within	16	224.00	14.00	
Total	19	957.75		

$^{**}p < .01.$

We conclude, therefore, from our significant overall F, that there is at least one significant difference among our treatment conditions. But where within our data is this significant difference? Usually one or more of the treatment means will differ from other treatment means, but this does not have to be so. It may be, instead, that one combination of means differs from another combination. For example, the highest and lowest means will often differ from each other. However, there are instances in which this is not true, but where the average of the two lowest means will differ from the highest mean or from the average of the two highest means. There are actually 25 different comparisons that can be made from the four means in Table 8.6. These are listed in Table 8.8. Note that we can compare any one mean against any other mean, as we did with the t test, or we can compare any combination of means against a single mean or some other combination. The question facing us, therefore, is this: Given a significant overall F, how do we pinpoint the significant difference(s)? Indeed, is it even necessary to compute an overall F test before looking for differences among means? Before these questions can be answered, it is necessary to know the specific experimental hypothesis or hypotheses that are being investigated.

There are at least three major qualitatively different hypotheses and one quantitative hypothesis that a researcher can test. Once we know the hypothesis under study, we can then specify the proper computational procedure for an exact test of that hypothesis. Therefore, we shall now turn to a discussion of the kinds of experimental hypotheses that can be studied in the context of single-classification analysis of variance experiments.

Qualitative Experimental Hypotheses

Three types of qualitative hypotheses are commonly investigated. The first is concerned with exploratory experiments comparing a number of treatments against each other. The second involves comparing a control group against several experimental groups. The third investigates planned comparisons set up before the experiment was run.

Exploratory experiments At times a researcher does exploratory work of the "I wonder what would happen if" nature. A number of experimental treatments may be placed together within one design even though the variables manipulated

TABLE 8.8 Showing the 25 possible comparisons that can be obtained from four treatment means

Treatment mean(s)	vs.	Treatment mean(s)
1		2
1		3
1		4
2		3
2		4
3		4
(1 + 2)/2		3
(1 + 2)/2		4
(1 + 2)/2		(3 + 4)/2
(1 + 3)/2		2
(1 + 3)/2		4
(1 + 3)/2		(2 + 4)/2
(1 + 4)/2		2
(1 + 4)/2		3
(1 + 4)/2		(2 + 3)/2
(2 + 3)/2		1
(2 + 3)/2		4
(2 + 4)/2		1
(2 + 4)/2		3
(3 + 4)/2		1
(3 + 4)/2		2
(1 + 2 + 3)/3		4
(1 + 2 + 4)/3		3
(1 + 3 + 4)/3		2
(2 + 3 + 4)/3		1

may have little or no relationship to one another. The objective here is simply to find out whether these various treatments differ among themselves. For example, suppose that an experimental psychologist was interested in determining whether different forms of stressful experience immediately after learning would have an equal effect on interfering with short-term memory. He selects mice as his experimental subject and has them learn a particular task to a specified criterion. The group is then subdivided and exposed to one of the following types of stress: electric convulsive shock, convulsions induced by a chemical manipulation, convulsions induced by intense sound, rapid temperature loss, being placed into a cage where the experimental animal is attacked by a trained aggressor mouse, or being placed into a drum that spins rapidly. After one of those treatments, all animals are retested on the original task.

This type of design is useful when a new area is being explored, and one wishes to investigate a number of variables quickly. The experimental hypothesis under investigation here is that the treatment means in the population differ. The statistical hypothesis and its alternative can be expressed as follows:

$$H_0 : \mu_1 = \mu_2 = \mu_3 \ldots$$

$$H_1 : \text{not } H_0$$

To evaluate H_0 the overall F test is used. If significant, we know there is significance somewhere in the data. To appreciate what this means, we can return to the data in Table 8.6 and the listing of all possible comparisons in Table 8.8. A significant overall F says that at least one of those 25 differences is beyond chance expectation. The usefulness of this finding will depend on the nature of the experiment and the specific comparison or comparisons found to be significant. It should now be apparent that the "I wonder what would happen if" type of experimental design is not a useful research strategy once knowledge has been developed within a field of investigation. Certainly no one investigating factors that interfere with short-term memory would use the type of design given above. Even if we are still at the crude stage of exploring a new phenomenon, to ask the question, "Do these experimental treatments differ among themselves?" is often not a very useful experimental query. For example, what logical basis can a researcher have for comparing two qualitatively different experimental treatments? Of what value is it to the experimental psychologist to know that mice that receive electroconvulsive shock have poorer short-term memory than mice that are attacked by an aggressor? It makes much more sense, even in preliminary work on a new phenomenon, to define a control condition and then use this control group as the reference point against which to evaluate the various experimental treatments. That brings us to our next type of qualitative experimental design.

Comparing a control against several experimental groups   If there are one control group and a number of qualitatively different experimental groups in a research design, the experimental question of interest may well be: "How does each experimental group differ from the control?" This is essentially the same question we asked when we ran our t test comparing an experimental group with a control, but here we have added more experimental groups to our design. This can be a very efficient arrangement. For example, suppose that a pharmacologist has four experimental drugs which she wishes to evaluate for their effectiveness on a particular endpoint. She has no interest in comparing the drugs among themselves, but does want to know whether they change performance relative to a control group. One way to do this is to perform four experiments each containing a control group and an experimental group that receives one of the unknown drugs. This requires eight groups of animals. Instead, the pharmacologist can run one large experiment in which there is one control group and four experimental groups each receiving a different drug. Thus, the same information is obtained with five groups as would have been obtained with eight groups, with a saving of 37.5% in time and animals.

It is apparent that the investigator of short-term memory would have had a more efficient experimental design if he had included a control group in his study against which to compare the various experimental treatments. At times, however, it is not possible to specify a control group. As an example, consider a geneticist who is studying the adrenal activity of different strains of inbred mice. There is no such thing as a control group for inbred animals, so all that the geneticist can do is use an overall F test to see whether the groups differ among themselves. If he wished, though, he might select one of the inbred strains as a "standard" against which to compare the other strains.

If there is a control group against which to compare the various experimental groups, then we have a set of experimental hypotheses and concomitant statistical hypotheses, all of which are essentially equivalent to the hypotheses involved in the t test. For example, if we had one control and three experimental groups, the following hypotheses would be evaluated:

Statistical Hypothesis I	$H_0: \mu_c = \mu_1$
Experimental Hypothesis I	$H_1: \mu_c \neq \mu_1$
Statistical Hypothesis II	$H_0: \mu_c = \mu_2$
Experimental Hypothesis II	$H_1: \mu_c \neq \mu_2$
Statistical Hypothesis III	$H_0: \mu_c = \mu_3$
Experimental Hypothesis III	$H_1: \mu_c \neq \mu_3$

The hypotheses specified above are much more exact than the hypothesis tested in our exploratory experiment. With the exploratory study we were evaluating all 25 comparisons at once, whereas our objective in this design is to make a specific test on just three of those 25 possible comparisons. Therefore, if we have a set of qualitatively different groups in an experiment and if we can include an appropriate control group in our design, the test procedure will be more powerful if we test each experimental group against the control than if we merely test to determine whether some significant difference exists among the treatment groups. That is, the more specific we can make our hypothesis, the more likely we are to obtain a significant difference if one exists. If we just ask, "Are there any differences among these groups?" one significant effect among many means may be hidden because of the random variation among the other means.

As we gain more knowledge concerning our field of experimental investigation, we are able to move beyond the comparison of a control group with an experimental group to explore different combinations of variables in an experiment. Once knowledge has advanced to a certain point we design our experiments with certain specific comparisons in mind that we wish to test. These comparisons are planned during the design phase of our experimental research. This leads us to the third of our qualitative experimental designs.

Planned comparisons The more knowledge we have concerning our experimental variable, the more specifically we can pinpoint hypotheses to be investigated. Thus, within one experiment we may be able to test several hypotheses related to a particular research area. Consider the following example.

A psychobiologist is interested in the effects of infantile stimulation on the rat's activity in the open field in adulthood. The two major techniques used by researchers in this field to stimulate the newborn animal directly are handling and electric shock. The researcher is interested in knowing whether these different procedures have the same effect on activity. The researcher also knows that whenever rats are stimulated in infancy, one of the consequences is that the mother is disturbed, and thus it may be that the differences obtained are a function of the mother's reaction to the disturbance rather than being a function of the stimulation the pups receive. To investigate this hypothesis the researcher has a third group in the experiment in which the mother is removed from the cage as a way of disturbing her while the pups are left untouched. The fourth group in the experiment consists of nondisturbed control animals. The four groups may be summarized as follows.

Group C: undisturbed controls
Group H: animals handled in infancy
Group S: animals shocked in infancy
Group M: mother removed from the nest box during infancy

We shall let μ_c, μ_h, μ_s, and μ_m stand for the population means estimated by the four sample means. The psychobiologist has three hypotheses that she wishes to test.

1. She wants to know whether animals handled in infancy differ from those that received shock in infancy as measured by open-field activity in adulthood. The null hypothesis and its alternative are

$$H_0: \mu_h = \mu_s$$
$$H_1: \mu_h \neq \mu_s$$

2. The researcher wishes to know whether disturbing the mother has the same effect on activity as disturbing the young. She therefore compares the mean of Group M with the average of Group S and Group H, as follows:

$$H_0: \mu_m = \frac{\mu_h + \mu_s}{2}$$

$$H_1: \mu_m \neq \frac{\mu_h + \mu_s}{2}$$

3. Groups H, S, and M were all disturbed during infancy. The psychobiologist wishes to determine whether disturbance of any sort affects activity relative to the nondisturbed control group. To do this she compares the mean of

Group C with the average of Groups H, S, and M. The null hypotheis and its alternative are

$$H_0 : \mu_c = \frac{\mu_h + \mu_s + \mu_m}{3}$$

$$H_1 : \mu_c \neq \frac{\mu_h + \mu_s + \mu_m}{3}$$

The structure of this experiment as reflected by the three hypotheses under investigation is quite different from the structure seen in either of the other two types of situations. When carrying out an exploratory experiment we had no knowledge as to whether any of our variables was going to be effective, and the purpose of the experiment was really to seek information on that question. In the second situation we had a control group against which a number of experimental groups could be assessed, but we were not able to compare various experimental groups among themselves. In the planned comparisons type of experimental design we have sufficient knowledge concerning the nature of our independent variables to be able to set up experimental conditions aimed at elucidating the effects of these variables. The means of the various experimental groups can be arranged in different patterns to test our hypotheses. Thus, Groups H and S are present in all three statistical hypotheses, Group M is present in two of the planned comparisons, and Group C is used in only one comparison.

Whenever it is possible to set up planned comparisons in advance, this will be the most powerful kind of experimental design. There are several reasons for this. First, using the data in Tables 8.6 and 8.8 to illustrate the principle, we see that of the 25 possible comparisons the researcher selects only those that are experimentally relevant, and ignores the others. Thus, because we are evaluating only a relatively few hypotheses out of all those possible, we do not have to be as conservative in our statistical assessment. Next, each planned comparison tests a specific hypothesis, which makes it much easier to interpret any significant differences that are found.

The obvious question at this point is: How many planned comparisons can one have in an experiment? The answer is: as many as are experimentally relevant, but the number should not approach the total number of possible comparisons. There are statements made in some statistical texts that you can only have as many planned comparisons as you have degrees of freedom for your between mean square, and that these comparisons must be orthogonal, or independent, of each other. However, it is not necessary to limit your experiments to orthogonal comparisons, This does have a slight advantage in terms of keeping the Type 1 error constant, but this advantage does not outweigh the utility of exploring all relevant facets of the data. If the experiment was designed to test several specific hypotheses, then these hypotheses may be tested, whether they are orthogonal or not, and whether their number exceeds the degrees of freedom for between groups.

Quantitative Experimental Hypotheses: Regression Experiments

A major advance in scientific investigation occurs when we are able to move from a comparison of qualitatively different variables to a study of the quantitative relationship between an independent variable and our endpoint or dependent variable. We shall use the term *regression experiment* to characterize an experiment in which we systematically vary a quantitative variable as our independent variable. Many examples abound in biology and behavior: An endocrinologist is interested in the dose-response relationship between the number of International Units of ACTH he injects and an animal's corticosterone response. A biochemist studies the time-response properties of a particular compound by injecting it and then taking measurements every half-hour until all activity disappears. An experimental psychologist determines the learning curve for a form discrimination task as a function of number of reinforced trials. A biologist measures different groups of animals every 3 hours over the 24-hour day to determine whether the endpoint has a circadian rhythm.

There are many forms of curves for relationships which can be investigated. We are going to limit ourselves in this discussion to two of the simplest and most common ones. The first is a *linear* relationship between our independent variable and our criterion measure. A hypothetical example is shown in Figure 8.1 for the situation where the experimental endocrinologist varied the number of International Units of the hormone and found that the response measured increased linearly. In Figure 8.1 the dots represent the data points and the line is the best-fitting straight line as determined from a mathematical equation.

A second situation with which we shall be concerned involves *quadratic* relationships. In its pure form a quadratic relationship is a parabola represented by a U or inverted-U curve. As an example, a biologist interested in circadian rhythms could measure a particular endpoint starting at 8 a.m. and taking measurements every 3 hours. Figure 8.2 shows the results of this hypothetical experiment. The endpoint shows little activity at 8 a.m., just after the lights go on in the laboratory. The activity continues to build up during the day, reaching

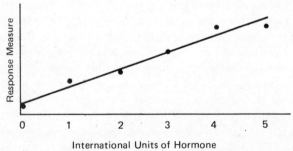

FIGURE 8.1   Example of a linear relationship.

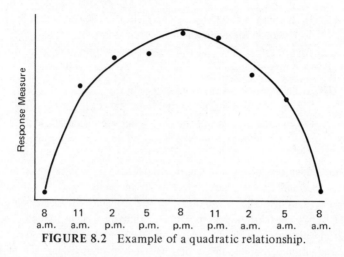

FIGURE 8.2   Example of a quadratic relationship.

a maximum at 8 o'clock that evening just before the lights go out, and then the response drops steadily over the next 12 hours.

Experimental and statistical hypotheses   The experimental hypothesis that is under study when one is investigating quantitative functional relationships is of a different nature than the hypotheses involved in qualitative comparisons. The experimental endocrinologist's hypothesis is that there is a significant linear relationship between the response measure and the amount of hormone injected. If we have no interest in any trend beyond the linear, we can designate the statistical hypothesis and its alternative as follows:

H_0: in the population there is no linear
      relationship between the number of
      International Units of the hormone
      and the response measured.

H_1: In the population there is a linear
      relationship between the number
      of International Units of hormone
      injected and the response measured

The hypotheses that the biological rhythms researcher wishes to test differ from that of the experimental endocrinologist in that the expected form of the curve differs. The biological rhythms researcher expects a major quadratic component with a possible lesser linear trend, and feels that these two will be sufficient to describe the data. The statistical hypotheses and their alternatives are

Hypothesis I	Hypothesis II
H_0: In the population there is no linear relationship between time of day and the response measured	H_0: In the population there is no quadratic relationship between time of day and the response measured
H_1: In the population there is a linear relationship between time of day and the response measured	H_1: In the population there is a quadratic relationship between time of day and the response measured

This concludes our discussion of the kinds of experimental hypotheses that can be investigated. It is apparent that even the simplest type of analysis of variance design allows the researcher considerable latitude for testing a wide variety of hypotheses. When one uses more complex designs (which are beyond the scope of this book), an even greater variety of experimental hypotheses can be evaluated. In all instances the experimental hypothesis is arrived at by first testing the statistical hypothesis. If that can be rejected, then the experimental hypothesis is accepted as the most reasonable alternative. We now turn to a discussion of the computational procedures involved in testing these various statistical hypotheses.

EVALUATING QUALITATIVE HYPOTHESES

Exploratory Experiments

When we carry out an exploratory experiment, it is necessary to have a significant overall F ratio before making any comparisons among the treatment groups. In comparing pairs of means we know that we cannot use a conventional t test because the α level at which we would be testing would be numerically higher than the α level we specify because we shall be capitalizing upon chance variation. That is, some of the "significant differences" we would obtain would be due to chance alone. A number of different techniques have been developed to evaluate the differences among the means. Each technique has slightly diffferent objectives and different sets of assumptions underlying its use. These are discussed in detail in Winer (1971, pp. 185-201). The procedure we shall recommend is one developed by Tukey called the *honestly significant difference* procedure. This technique evaluates all possible pairs of means, comparing the differences obtained with a critical difference derived such that the value of *joint* α is not exceeded. In contrast to the usual or individual α, a *joint* α measures the Type 1 error rate associated with a collection of tests. That is, when all the hypotheses in the collection being tested are true, unless all of the individual decisions are "do not reject," then a Type 1 error is present. The honestly significant difference procedure ensures that the Type 1 error will be no greater than α.

In order to determine the critical difference to reject H_0, we introduce a new statistic called the q statistic in Table V in the Appendix. We enter the q table with the number of treatment groups across our columns and the degrees of freedom associated with MS_{within} down the rows. At the intersection of the appropriate column and row are two numerical values, one listing q at the .05 level and the other listing q at the .01 level. We use the same α level in comparing the differences between means as we used in our overall F test to evaluate the null hypothesis. Once we have obtained our value of q, that value is multiplied by the value obtained from the expression $\sqrt{n\,MS_{\text{within}}}$. The resulting number is the critical difference that must exist between treatment totals for any two treatment groups to differ significantly. We use totals rather than means because it is computationally more convenient, and the formulas have been set up in that fashion.

Procedures for determining which groups differ significantly after an overall F test has been found to be significant are variously called *post hoc comparisons, post mortem tests,* or *a posteriori tests*. All three expressions indicate that these comparisons are made after the fact. Table 8.9 gives the general rules to follow in determining significant differences between pairs of means for any *post hoc* experiment. It should be noted that this procedure works only when there are equal n's per treatment group. Table 8.10 is a computational example based on the data from Tables 8.6 and 8.7.

TABLE 8.9 General rules for determining significant differences between pairs of means on a *post hoc* basis after a significant overall F has been obtained in an exploratory experiment

---

1. Obtain the numerical value for $\sqrt{nMS_{\text{within}}}$, where $n =$ number of observations within a treatment group (it is assumed that n is the same for all treatment groups).

2. From Table V in the Appendix determine the numerical value for the statistic q_α (k, df_{within}), where

$$\alpha = \text{same level of significance as used in evaluating the overall } F \text{ test}$$
$$k = \text{number of treatment groups in the experiment}$$
$$df_{\text{within}} = \text{number of degrees of freedom associated with } MS_{\text{within}}$$

3. Multiply the values obtained in Steps 1 and 2 to get

$$\text{critical difference} = [q_\alpha(k, df_{\text{within}})]\ \sqrt{nMS_{\text{within}}}$$

4. Any two treatment totals (not means) that differ by the amount specified in Step 3 are significantly different at the specified α level.

TABLE 8.10   A computational example using the instructions in Table 8.9 showing how to do a *post hoc* comparison between means. The data are from Tables 8.6 and 8.7

---

1. $MS_{within} = 14.00$

    $n = 5$

    $\sqrt{nMS_{within}} = \sqrt{(5)(14.00)} = \sqrt{70} = 8.367$

2. $\quad\quad\quad k = 4$
    $df_{within} = 16$
    $q_{.05}(4, 16) = 4.05$

3. Critical difference $= (8.367)(4.05) = 33.89$

4. Treatment totals, in order:

    Treatment 1 =   48
    Treatment 4 =   59
    Treatment 3 =   81
    Treatment 2 = 127

    Treatment 2 $-$ Treatment 1 $= 127 - 48 = 79$. This is greater than the critical difference of 33.89, and so Treatments 2 and 1 differ at the .05 level of significance.

    Treatment 2 $-$ Treatment 4 $= 127 - 59 = 68$. This is greater than the critical difference of 33.89, and so Treatments 2 and 4 differ at the .05 level of significance.

    Treatment 2 $-$ Treatment 3 $= 127 - 81 = 46$. This is greater than the critical difference of 33.89, and so Treatments 2 and 3 differ at the .05 level of significance.

    No other difference is significant.

Comparing a Control Against Several Treatment Groups

If the experiment has been designed to compare a control group against a number of experimental groups, it is not necessary to have a significant overall F before proceeding to make the special comparisons. The overall F test is used only for exploratory experiments where the researcher has no specific hypothesis under investigation.

When comparing one control group against a number of experimental groups, the various tests are not independent because the same control group is present in all comparisons. This lack of independence causes a change in the true α level at which one is making the test. Dunnett has determined the numerical value of t needed for significance at different α levels as a function of the number of treatment groups in an experiment. Table VI in the Appendix lists Dunnett's t values. You enter the table with the number of degrees of freedom associated with MS_{within} and the number of treatment groups (including the control). Four α levels are listed: .10, .05, .02, .01, thus giving a wide range of selection for evaluating your statistical hypotheses. (If you were testing a one-tailed hypothesis, you would halve these α levels.) Table 8.11 gives the general rules to follow in comparing the control mean against a number of experimental means. The instructions in Table 8.11 assume that there are equal n's in each treatment group. It is more convenient to work with treatment totals than with the means, and the instructions and formulas in Table 8.11 are based on treatment totals. Table 8.12 gives a computational example of the use of Dunnett's procedure based on the data of Tables 8.6 and 8.7. In this example we assume that Treatment 1 from Table 8.6 is the control group against which each of the experimental groups will be compared.

It is instructive to compare the critical difference obtained in Table 8.12 with that obtained in 8.10. Both tables use the same set of data and so they are directly comparable. We see that the critical difference needed for significance when comparing a control mean against a set of experimental means is less than the critical difference needed for significance in a *post hoc* test. Tukey's honestly significant difference test was designed to be appropriate for all possible differences between pairs, whereas the Dunnett test is concerned with a subset from this group, namely the comparison of the control with a number of experimental groups. Thus, it is reasonable to expect the Tukey test to require the larger critical difference for significance.

Planned Comparisons

The computational procedures involved in planned comparisons are best developed through a specific example. We shall go back to the situation of the psychobiologist interested in studying the effects of infantile stimulation. She had

TABLE 8.11   General rules for comparing experimental means against one control mean

---

1. Obtain the numerical value for $\sqrt{2n\ MS_{\text{within}}}$, where n = number of observations within a treatment group (it is assumed that n is the same for all treatment groups).

2. From Table VI in the Appendix, determine the numerical value of the statistic $t_\alpha(k, df_{\text{within}})$, where

$$\alpha = \text{level of significance}$$
$$k = \text{number of treatment groups (including control) in the experiment}$$
$$df_{\text{within}} = \text{number of degrees of freedom associated with } MS_{\text{within}}$$

3. Multiply the values obtained in Steps 1 and 2 to get

$$\text{critical difference} = [t_\alpha(k, df_{\text{within}})]\ \sqrt{2n\ MS_{\text{within}}}$$

4. Any experimental group whose treatment total (not mean) differs from the control group total by the amount specified in Step 3 is significantly different at the specified α level.

TABLE 8.12 A computational example using the instructions in Table 8.11 showing how to compare a control mean to a number of experimental means. The data are from Tables 8.6 and 8.7, where treatment 1 is assumed to be the control group

---

1. $MS_{within} = 14.00$

   $n = 5$

   $\sqrt{2n\, MS_{within}} = \sqrt{(2)(5)(14.00)} = 11.832$

2. $k = 4$
   $df_{within} = 16$
   $\alpha = .05$
   $t_{.05}(4, 16) = 2.59$

3. Critical difference $= (11.832)(2.59) = 30.64$

4. Treatment total 4 — Treatment total 1 (control) $= 59 - 48 = 11$
   Treatment total 3 — Treatment total 1 (control) $= 81 - 48 = 31$
   Treatment total 2 — Treatment total 1 (control) $= 127 - 48 = 79$

   Treatments 2 and 3 both exceed the critical difference, and so we conclude that these experimental treatments differ from the control group at the .05 level.

---

four groups in her experiment—Controls, Handled, Shocked, and Mother removed—and we used the terminology μ_c, μ_h, μ_s, and μ_m to indicate the population means to which we make inferences. The experimenter had three experimental hypotheses that she wished to investigate, and these were expressed as statistical hypotheses. Those three hypotheses are reproduced below in their original form and in a slightly altered form.

$$\text{Statistical Hypothesis I:} \quad \mu_h = \mu_s \quad \text{or}$$
$$(1)\mu_h - (1)\mu_s = 0$$

$$\text{Statistical Hypothesis II:} \quad \mu_m = \frac{\mu_h + \mu_s}{2} \quad \text{or}$$
$$(1)\mu_m - \frac{1}{2}\mu_h - \frac{1}{2}\mu_s = 0$$

$$\text{Statistical Hypothesis III:} \quad \mu_c = \frac{\mu_h + \mu_s + \mu_m}{3} \quad \text{or}$$
$$(1)\mu_c - \frac{1}{3}\mu_h - \frac{1}{3}\mu_s - \frac{1}{3}\mu_m = 0$$

What we have done above is to rewrite our null hypotheses so that the difference between our two population means (or sets of means) is zero, and we have also made explicit the coefficients involved in our comparisons. We now prepare a table listing our treatment groups and our statistical hypotheses, and we enter into that table the coefficients involved, in the following form.

Statistical hypothesis	Treatment group			
	Cont	H	S	M
I	0	1	−1	0
II	0	$-\frac{1}{2}$	$-\frac{1}{2}$	1
III	1	$-\frac{1}{3}$	$-\frac{1}{3}$	$-\frac{1}{3}$

Whenever a blank cell appears in the table, a zero is entered.

Several features of this table deserve emphasis. First, the sum across any row (i.e., for any of the three statistical hypotheses) is 0. This has to be the case because our null hypothesis stated that the difference in the population is 0. Whenever the sum of the coefficients equals zero, this is called a *comparison* or a *contrast* by statisticians.

The second feature of this table is seen when the coefficients in any two rows are cross-multiplied and summed. Consider Statistical Hypotheses I and II. If we cross-multiply the coefficients, we get

$$(0)(0) + (1)\left(-\frac{1}{2}\right) + (-1)\left(-\frac{1}{2}\right) + (0)(1) = 0$$

For Statistical Hypotheses I and III we get

$$(0)(1) + (1)\left(-\frac{1}{3}\right) + (-1)\left(-\frac{1}{3}\right) + (0)\left(-\frac{1}{3}\right) = 0$$

For Statistical Hypotheses II and III we get

$$(0)(1) + \left(-\frac{1}{2}\right)\left(-\frac{1}{3}\right) + \left(-\frac{1}{2}\right)\left(-\frac{1}{3}\right) + (1)\left(-\frac{1}{3}\right) = 0$$

Whenever the sum of the products of two sets of coefficients equals 0, these two components are said to be *orthogonal comparisons,* or comparisons that are independent of each other. There are as many orthogonal comparisons within a set as there are degrees of freedom for $MS_{between}$. With four groups there are 3 *df*, and one set of orthogonal comparisons is given above. However, it should be apparent that there are many other ways of getting a set of orthogonal comparisons.

We shall use the data of Tables 8.6 and 8.7 to show how to carry out the computations. Let Treatment 1 = Control, Treatment 2 = Handled, Treatment 3 = Shocked, and Treatment 4 = Mother removed. Assume that the data in Table 8.6 are the activity scores of the rats in the open field. We now place in our table of coefficients the treatment totals (not means) from our experiment as shown in Table 8.13.

TABLE 8.13 A computational example of the procedure involved in planned comparisons. The data are from Tables 8.6 and 8.7

Treatment ΣX		Cont 48	H 127	S 81	M 59	C	Σc^2	$n\,\Sigma c^2$	$\dfrac{C^2}{n\Sigma c^2}$
	I	0	1	-1	0	46	2	10	211.60
Statistical									
	II	0	$-\frac{1}{2}$	$-\frac{1}{2}$	1	-45	$\frac{3}{2}$	$\frac{15}{2}$	270.00
hypothesis									
	III	1	$-\frac{1}{3}$	$-\frac{1}{3}$	$-\frac{1}{3}$	-41	$\frac{4}{3}$	$\frac{20}{3}$	252.15
									733.75

$$\text{Test of Hypothesis I:}\quad F = \frac{211.60}{14} = 15.11^{**}$$

$$\text{Test of Hypothesis II:}\quad F = \frac{270}{14} = 19.29^{**}$$

$$\text{Test of Hypothesis III:}\; F = \frac{252.15}{14} = 18.01^{**}$$

We next multiply the coefficients by each of the treatment totals for each statistical hypothesis and enter the resulting value at the right of our table under column C, which stands for comparison. For Statistical Hypothesis I we have

$$(0)\,(48) + (1)\,(127) - (1)\,(81) + (0)\,(59) = 46$$

For Statistical Hypothesis II we have

$$(0)\,(48) - \left(\tfrac{1}{2}\right)(127) - \left(\tfrac{1}{2}\right)(81) + (1)(59) = -45$$

For Statistical Hypothesis III we have

$$(1)\,(48) - \left(\tfrac{1}{3}\right)(127) - \left(\tfrac{1}{3}\right)(81) - \left(\tfrac{1}{3}\right)(59) = -41$$

If we let c stand for any coefficient in our table, we can now obtain the SS for a comparison as follows:

$$SS_{\text{comparison}} = \frac{C^2}{n\Sigma c^2}$$

In words this says that the numerator is the square of the C value obtained in the table above. This is divided by the square of the coefficients involved in assessing a particular hypothesis and by the number of cases per treatment group (n).

For Statistical Hypothesis I we have

$$SS_{comparison} = \frac{(46)^2}{5[(1)^2 + (-1)^2]} = \frac{2116}{10} = 211.60$$

For Statistical Hypothesis II we have

$$SS_{comparison} = \frac{(-45)^2}{5[(-1/2)^2 + (-1/2)^2 + (1)^2]} = \frac{2025}{7.5} = 270$$

For Statistical Hypothesis III we have

$$SS_{comparison} = \frac{(-41)^2}{5[(1)^2 + (-1/3)^2 + (-1/3)^2 + (-1/3)^2]} = \frac{1681}{6.6667} = 252.15$$

Because each comparison is based on 1 df, these SS's are also mean squares. These data are now put into an analysis of variance summary table similar to Table 8.7 as shown below.

Source	df	SS	MS	F
[Between]	[3]	[733.75]	–	
I. H vs. S	1	211.60	211.60	15.11[**]
II. H + S vs. M	1	270.00	270.00	19.29[**]
III. Cont vs. H + S + M	1	252.15	252.15	18.01[**]
Within	16	224.00	14.00	

Each of our three hypotheses is evaluated by dividing $MS_{comparison}$ by MS_{within}. Each F test has $\frac{1}{16}$ df and all three F ratios are significant beyond the .01 level $[F_{.01}(1, 16) = 8.53]$.

If the three $SS_{comparisons}$ are added up, their total is found to be equal to $SS_{between}$. This *must* be true, within rounding error, if the comparisons are orthogonal and if all the df are used in individual comparisons. The between row is in brackets in the table above because one generally does not carry out an overall F test on treatment groups. Instead one goes immediately into the planned comparisons.

Let us look at the interpretation the psychobiologist would make of her data. Her first significant F means that handling has greater effect on increasing activity than does shock. The second finding is that animals that are handled or shocked in infancy are more active in adulthood than animals whose mothers have been removed from the home cage. The third result is that nondisturbed controls are less active in the open field than animals disturbed in infancy.

We stated that a number of sets of orthogonal comparisons are available for any one experiment. To demonstrate this, here is another set of orthogonal comparisons which the researcher could investigate:

1. Does handling have the same effect as shock?
2. Does disturbing the mother have an effect relative to the control group?
3. Does the situation where we directly manipulate the young (i.e., via Handling or Shock) have an effect relative to the situations where we do not manipulate the young (i.e., Control and Mother removed)?

Those three questions are orthogonal to each other (as an exercise, verify this), and the statistical evaluation of these hypotheses is given below.

Source	df	SS	MS	F
I. H vs. S	1	211.60	211.60	15.11[**]
II. Cont vs. M	1	12.10	12.10	.86
III. H + S vs. Cont + M	1	510.05	510.05	36.43[**]
Within	16	224.00	14.00	

What interpretations would the experimenter make of her data now? Her first conclusion is that handling has a greater effect than electric shock on open-field performance in adulthood (the same conclusion as she drew from the other analysis). The second conclusion is that removing the mother from the nest box has no effect on changing open-field activity relative to a control group (a conclusion she could not draw from the prior analysis). The third conclusion is that directly manipulating the animal in infancy increases open-field activity in adulthood (this conclusion overlaps with Statistical Hypothesis III in the prior analysis). It is apparent that the psychobiologist draws somewhat different conclusions depending on which set of hypotheses she chooses to evluate. Thus, planned comparisons are useful only if you have specific hypotheses you wish to assess and if you are willing to ignore other hypotheses that could be evaluated. If you can formulate exact hypotheses, the procedure of planned comparisons is the most powerful of the three methods we have described in the sense that it is most likely to reject the null hypothesis if it is false.

In the two examples of planned comparisons given above, the comparisons within each set were orthogonal to each other. This was done to illustrate how to test for and obtain orthogonal comparisons, and is not meant to imply that it is necessary to have all comparisons orthogonal to each other. As stated earlier (on page 147) the experimental questions take precedence over statistical considerations, and all hypotheses (i.e., comparisons) that the experiment was designed to test should be evaluated, whether orthogonal or not.

TABLE 8.14 Comparisons of the critical difference needed for significance at the .05 level for the data in Tables 8.6 and 8.7 for exploratory experiments, comparing a control with a set of experimental groups, and planned comparisons

Experimental design	Critical difference
Exploratory experiment	33.89
Control vs. experimental groups	30.64
Planned comparisons	25.08

Comparison Between Procedures

We have indicated that the planned comparison procedure is the most powerful of the three methods used in evaluating mean differences. The critical difference required for significance using planned comparisons is the same difference as is required for significance with a t test. For the data of Table 8.6 this value turns out to be 25.08. The critical differences for the three kinds of experimental designs are summarized in Table 8.14. Notice that the critical difference in our example of an exploratory experiment is approximately 35% greater than the critical difference needed for significance using planned comparisons. The moral of this is plain. If you can pinpoint your experimental hypotheses exactly so that you can set up a planned comparison, you are much more likely to pick up a significant difference if it exists. Remember also that the exploratory experiment demands that the overall F test be significant before examining differences among the mean. Neither of the other two types of experimental design has that requirement.

EVALUATING REGRESSION EXPERIMENTS

Statistical tests of regression experiments are also called *tests for trend*. The primary objective is to determine whether there is a significant functional relationship (that is, curve or trend) between some quantitatively manipulated independent variable (IV) and an endpoint (X). If there is a significant relationship, then we may also be interested in determining the equation describing the best-fitting line for our data. We indicated earlier that linear and quadratic curves are frequently found in biological and behavioral research, and Figures 8.1 and 8.2 gave examples of these curves. In testing statistical hypotheses concerning linear and quadratic (as well as other), trends, we shall use a general polynomial equation which has the following form:

$$X = \alpha + \beta_1(IV) + \beta_2(IV)^2 + \beta_3(IV)^3 + \beta_4(IV)^4 + \cdots \qquad (8.11)$$

where X = our measured variable and IV refers to the numerical value of our indepedent variable.

In the equation given above the coefficient α is a constant designating the value of our criterion measure X when the independent variable is equal to 0 (the control condition). The β's are coefficients associated with the different powers of the IV's. For example, if there is a significant linear relationship between X and the independent variable, then a change in IV must be accompanied by a concomitant change in X which is reflected by the β_1 coefficient or slope of our line. Similarly, if there is a significant quadratic trend, then the β_2 coefficient in the population has a value other than zero.

These coefficients may take the value zero as well as other numerical values. For example, if all the coefficients from β_2 onward are zero, then we have a linear equation as follows:

$$X = \alpha + \beta_1(IV) \qquad (8.12)$$

If β_1 and β_2 have numerical values other than 0, and all other β coefficients are 0, then we have the quadratic equation

$$X = \alpha + \beta_1(IV) + \beta_2(IV)^2 \qquad (8.13)$$

The linear equation above is also called a first-degree equation because the power of the independent variable is 1. The quadratic equation is also called a second-degree equation because the value of IV is squared. The third-degree equation is called a cubic equation, the fourth-degree equation is called quartic, the fifth-degree equation is called quintic, and so on.

The polynomial equation given above refers to population parameters. The sample equation is

$$X = a + b_1(IV) + b_2(IV)^2 + b_3(IV)^3 + b_4(IV)^4 + \cdots \qquad (8.14)$$

where the sample statistic a estimates the population value α, b_1 is an estimate of β_1, and so on.

Experimental and Statistical Hypotheses Revisited

On page 149 we presented statistical and experimental hypotheses concerning examples of linear and quadratic trends. For the example involving the test for a linear relationship between amount of hormone injected and some response measure, the hypotheses were

> H_0: in the population there is no linear relationship
> between the number of International Units of the
> hormone and the response measured

$$H_1: \text{not } H_0$$

That hypothesis is somewhat vague because we have not specified how we can evaluate the linear relationship. However, given our general polynomial equation, we can now restate this hypothesis in exact quantitative form:

$$H_0: \beta_1 = 0$$

$$H_1: \beta_1 \neq 0$$

In words, a statistical test for the linear trend asks whether the slope of the best-fitting straight line is significantly different from zero. If not, we conclude that there is no evidence for a linear trend in the data. If this test is significant, then the statistical hypotheses is rejected, and its alternative is accepted. This means that there is evidence for a significant linear relationship between the independent variable and the criterion measure X.

The same reasoning applies in determining whether we have significant linear and quadratic trends for the example of biological rhythms. The coefficient associated with the quadratic portion of our equation is β_2, and our statistical and experimental hypotheses are

Hypothesis I	Hypothesis II
$H_0: \beta_1 = 0$	$H_0: \beta_2 = 0$
$H_1: \beta_1 \neq 0$	$H_1: \beta_2 \neq 0$

We see from the two sets of hypotheses above that it is possible to test separately the linear and quadratic components of our general polynomial equation. We can extend this and make tests of the cubic, quartic, quintic, and other components as well. However, it is often difficult to make biological or behavioral sense out of curves beyond the third degree. Also, it makes no sense to fit a mathematical equation just because a procedure is available. It is necessary to relate the form of the equation to the phenomenon that is being investigated. Therefore, we now turn to a discussion concerning the forms of the curves that can be derived from the general polynomial equation.

Polynomial Curves

Suppose that we have the equation $X = 3 + 2.5(IV)$. The curve for the equation is shown in Figure 8.3. As we systematically vary our independent variable from 0 units (our control condition) to 5 units, we find a linear increase in our criterion measure X. The basal or resting level for our criterion measure is 3 units because this is the value of X when our independent variable is at 0 amount. For each unit increase in IV we have an increase of 2.5 units in X, thus giving us a slope of 2.5.

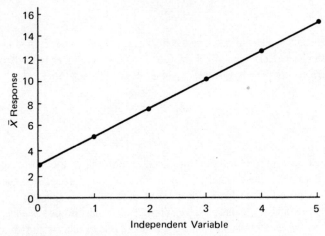

FIGURE 8.3   Plot of the equation $X = 3 + 2.5(IV)$.

Let us now turn our linear equation into a quadratic one by adding the expression $-.5(IV)^2$ to the equation above, yielding

$$X = 3 + 2.5(IV) - .5(IV)^2$$

The curve for that equation is shown in Figure 8.4. The quadratic equation starts at 3 when IV equals 0, rises to a maximum when $IV = 2.5$, and then drops off in a symmetrical fashion.

Examples of pure cubic, quartic, and quintic curves are shown in Figure 8.5.

The Use of Orthogonal Polynomials in the Test for Trend

In our discussion concerning tests of trend, we indicated that statistical tests

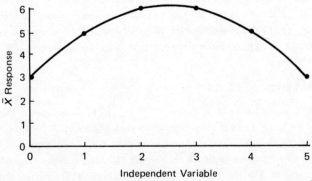

FIGURE 8.4   Plot of the equation $X = 3 + 2.5(IV) - .5(IV)^2$.

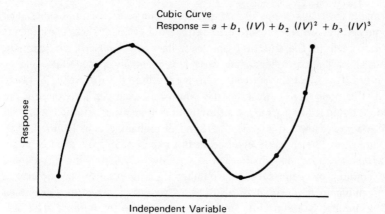

Cubic Curve
Response $= a + b_1\ (IV) + b_2\ (IV)^2 + b_3\ (IV)^3$

Response

Independent Variable

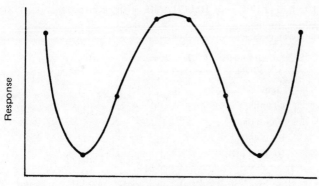

Quartic Curve
Response $= a + b_1\ (IV) + b_2\ (IV)^2 + b_3\ (IV)^3 + b_4\ (IV)^4$

Response

Independent Variable

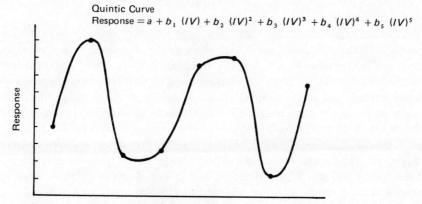

Quintic Curve
Response $= a + b_1\ (IV) + b_2\ (IV)^2 + b_3\ (IV)^3 + b_4\ (IV)^4 + b_5\ (IV)^5$

Response

Independent Variable

FIGURE 8.5   Examples of cubic, quartic, and quintic polynomial curves.

can be made to evaluate the linear, quadratic, cubic, etc., components. The statistic procedures we shall describe to extract and evaluate these components will be restricted to the situation where we have equal n's for all treatment conditions and where our independent variable is in equally spaced units.

The necessity of equally spaced units means that the difference between successive treatment levels must be the same. For example, an experimental psychologist could study learning as a function of intensity of reinforcement and he could use shock levels of .1, .3, .5, .7, and .9 milliamperes as his method of varying intensity. The shock levels used in this experiment provide an example of an *arithmetic series* in which each succeeding number is increased by a constant amount over the preceding number (in this instance the increase is always .2 milliamperes). We may also have a *geometric series* in which each succeeding number is a constant multiplier of the preceding number. This also satisfies our demand for equal intervals, because taking logarithms of a geometric series will convert the data to an arithmetic series. For example, suppose that a biochemist wishes to study a particular compound over a very wide range. He decides to use 1, 10, 100, 1,000, and 10,000 units of the compound. If he takes logarithms of the values, he has

Dose of compound	Log dose
1 unit	0
10 units	1
100 units	2
1000 units	3
10,000 units	4

In plotting his data, the biochemist would use log dose as his independent variable. Therefore, the requirement that amounts of the independent variable must be equally spaced may be satisfied by using either an arithmetic or a geometric series.

If these requirements are met, then there is a very simple computational procedure that will allow us to test the data for significance of trends. In addition, it is possible to use these same computations to derive the best-fitting mathematical equation. Thus, our discussion will be restricted to the situation of equal intervals and equal numbers. If the n's are unequal, or if unequal intervals are used, there are adjustments that can be made in the computations. For a discussion of these see Gaito (1965).

When the n's are equal and the spacing is equal, then it can be shown that the linear, quadratic, cubic, etc., components in the general polynomial equation are independent of each other and the method of *orthogonal polynomials* may be used to test for trend. When we use orthogonal polynomials for trend analysis, we break down our $SS_{between}$ into orthogonal components associated with the linear, quadratic, cubic, etc., trends, and each of these has associated with it one degree of freedom.

Computational Procedures for Trend Analysis

In order to make tests for trends, we use the coefficients of orthogonal polynomials in Table VII in the Appendix. We enter the table with the number of treatment groups in the experiment. Next to that is a listing of the degree of the polynomial from linear to quintic. Within the table are the coefficients needed to do the calculations. These coefficients have certain features worth noting. First of all, in summing across any row the total is 0. This satisfies the definition of a comparison which we discussed in the section on planned comparisons. Next, the sum of the cross-products of any two components also is equal to 0. You will recall that when this condition is satisfied the two components are said to be orthogonal to each other.

Now look at several of the linear trends in Table VII. The coefficients always follow a linear series starting with negative values, going through 0, and ending with positive values. If we were to plot the coefficients against the values of X given at the top of the table they would form a perfect straight line. Next consider the quadratic coefficients. If we were to plot them against the values of X, we would obtain a U-shaped curve. The coefficients associated with the cubic component all give a curve that starts low, then rises, then drops and starts to rise again. That is, all the cubic curves reverse direction twice. We see, therefore, that the coefficients within Table VII represent the actual form of a pure linear, quadratic, cubic, etc., function.

The next to the last column in Table VII is headed Σc^2, and the numbers in that column are the sum of the squares of the coefficients for each row. The numbers in the last column headed by the symbol λ (lambda) will be explained when we get to the section concerning the calculation of the best-fitting equation for our data.

The procedure of orthogonal polynomials will be illustrated using the data from Table 8.6, where we shall now assume that Treatment Groups 1, 2, 3, and 4 represent an arithmetic series for our independent variable. The computations are given in Table 8.15.

We enter the treatment totals (not means) at the top of our table. There are four groups in the experiment and from Table VII we take the coefficients of orthogonal polynomials and enter them into Table 8.15. Once the coefficients have been entered, we proceed in the same fashion as we did in computing planned comparisons in Table 8.13. We multiply the treatment total by the coefficient associated with the particular degree of the polynomial, sum the products, and enter the resulting value at the right of the table under column C. For the linear trend we have

TABLE 8.15   A computational example of the procedures involved in trend analysis with orthogonal polynomials. The data are from Tables 8.6 and 8.7. The treatment groups are assumed to form an arithmetic series. $n = 5$ per group

Treatment	1	2	3	4	C	Σc^2	$n\Sigma c^2$	$\dfrac{C^2}{n\Sigma c^2}$
ΣX	48	127	81	59				
Linear	-3	-1	1	3	-13	20	100	1.69
Quadratic	1	-1	-1	1	-101	4	20	510.05
Cubic	-1	3	-3	1	149	20	100	222.01
								733.75

$$\text{Test for linear trend: } F = \frac{1.69}{14} = .12$$

$$\text{Test for quadratic trend: } F = \frac{510.05}{14} = 36.43^{**}$$

$$\text{Test for cubic trend: } F = \frac{222.01}{14} = 15.86^{**}$$

$$(-3)(48) + (-1)(127) + (1)(81) + (3)(59) = -13$$

For the quadratic component we have

$$(1)(48) + (-1)(127) + (-1)(81) + (1)(59) = -101$$

For the third-degree component we have

$$(-1)(48) + (3)(127) + (-3)(81) + (1)(59) = 149$$

The sum of the squares of the coefficients for the linear component is 20, for the quadratic component 4, and for the cubic component 20. These values can be obtained directly from Table VII or they can be obtained by squaring and summing the coefficients across each row. To obtain the sum of squares of the linear component we employ the following formula:

$$SS_{\text{linear}} = \frac{C_{\text{lin}}^2}{n\Sigma c^2} \qquad \frac{(-13)^2}{(5)(20)} = 1.69$$

The computations for this value are shown in Table 8.15.

The same procedure is followed for the quadratic sum of squares:

$$SS_{\text{quad}} = \frac{(-101)^2}{(5)(20)} = 510.05$$

For the cubic component we have

$$SS_{\text{cubic}} = \frac{(149)^2}{(5)(20)} = 222.01$$

Because each comparison has 1 df, these SS's are also mean squares. These data are now put into an analysis of variance summary table as shown below:

Source	df	SS	MS	F
[Between]	[3]	[733.75]		
Linear	1	1.69	1.69	.12
Quadratic	1	510.05	510.05	36.43**
Cubic	1	222.01	222.01	15.72**
Within	16	224.00	14.00	

We use MS_{within} as the error term to test our orthogonal trend components. The F ratio for the linear trend is .12, which is clearly insignificant. The F for the quadratic trend is 36.43, which, with 1, 16 df, is significant beyond the .01 level; and the F for the cubic component of 15.72 is also significant beyond the .01 level. How do we interpret these data? The first step is to plot the empirical curve, as we have done in Figure 8.6.

An inspection of that figure immediately reveals why the linear component was so small. The best-fitting straight line through these data would be a curve

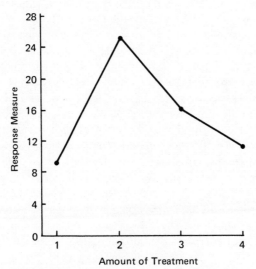

FIGURE 8.6 A plot of the means from the data in Table 8.15.

essentially parallel to the axis of the independent variable (i.e., the slope would be very close to zero). The curve is in the general form of an inverted U, and our significant quadratic component is consistent with that observation. The cubic component is more difficult to interpret. The slope of the curve between segments 2 and 3 is greater than that between segments 3 and 4. If that change is a meaningful one to the researcher, then an interpretation can be made. If not, then the cubic component can be viewed as a "residual" which says that there is still a significant amount of variation left over after we have accounted for the linear and quadratic trends.

The data in Table 8.15 illustrate an important point about curve fitting via orthogonal polynomials. By the use of a polynomial we can always fit any set of data perfectly with an equation that uses $N - 1$ terms of the polynomial, where N refers to the number of groups in the experiment. That is, a cubic equation can be obtained that would perfectly predict the four means in Figure 8.6. One way of demonstrating this is to look in Table 8.15 in the last column where the sums of squares for linear, quadratic, and cubic components are totaled. The value at the bottom of that column is 733.75, which is equal to $SS_{between}$ in Table 8.6. Thus, you can always statistically account for all the variation in a regression experiment. However, it is necessary to interpret the statistical findings within your research context in order to give meaning to the data obtained.

Example: Fitting a Linear Dose-Response Curve

To illustrate the use of trend analysis, a biological and a behavioral example will be given. A common problem in biology is to develop a bioassay that shows a linear increase in the response measured as a function of increasing amounts of the biological material under investigation. We shall take our example from endocrinology. A standard bioassay for estrogen is the increase in uterine weight as a function of the presence of this hormone. Assume that we are endocrinologists interested in relating uterine weight to estradiol (one type of estrogen) in the range from 2.5 micrograms to 80 micrograms. We decide to use a geometric series consisting of 2.5, 5, 10, 20, 40, and 80 micrograms of estradiol, because doubling the dosage will give us an approximately constant increment in uterine weight. We use the mouse as our experimental subject, draw a sample of 60 animals from our colony, randomly assign 10 animals to each of our 6 treatment groups, inject them with the specified dosage of estradiol, and then return them to their home cages. The next day the animals are etherized, and the uteri were removed, trimmed of fat, and weighed on a torsion balance to the nearest milligram. Table 8.16 lists the six treatment groups in terms of micrograms of estradiol and the total uterine weight in milligrams for the 10 animals within each group. At the bottom of Table 8.16 are the computations for determining SS_{total}, $SS_{between}$, and SS_{within}.

Because we are interested in determining whether the curve exhibits a significant linear trend, and whether there are significant components above the

TABLE 8.16 Total uterine weight (milligrams) as a function of amount of estradiol injected; and the breakdown of the sums of squares

Treatment (estradiol)	2.5 μg	5 μg	10 μg	20 μg	40 μg	80 μg
$\Sigma X_{\text{uterine weight}}$	112	242	400	583	756	931

$$\Sigma X = 3024$$

$$\Sigma X^2 = 209{,}337$$

$$n = 10/\text{group}$$

$$N = (6)(10) = 60$$

$$CT = (3024)^2/60 = 152{,}409.60$$

$$SS_{\text{total}} = \Sigma X^2 - CT = 209{,}337 - 152{,}409.60 = 56{,}927.40$$

$$SS_{\text{between}} = \frac{(\Sigma X_{2 \cdot 5})^2 + (\Sigma X_5)^2 + \cdots + (\Sigma X_{80})^2}{n} - CT$$

$$= \frac{(112)^2 + (242)^2 + \cdots + (931)^2}{10} - 152{,}409.60$$

$$= 48{,}519.80$$

$$SS_{\text{within}} = SS_{\text{total}} - SS_{\text{between}} = 56{,}927.40 - 48{,}519.80 = 8{,}407.60$$

linear, we next proceed to carry out a trend analysis using orthogonal polynomials of the treatment totals given in Table 8.16. The computational procedures for doing so are shown in Table 8.17. We are able to use the orthogonal polynomial procedure because we have equal n's at each point of our curve and because our intervals are equally spaced on a logarithmic scale. The first three rows of Table 8.17 show the equal spacing. The first row lists the dosages of estradiol injected and the second row lists the logarithms (to base 10) of the dosages. The log-dose values are equally spaced, with .301 log units between each value. Because they are equally spaced, we can substitute the numbers 1, 2, 3, 4, 5, and 6 for the six dose levels of estradiol, and that is what

TABLE 8.17 Computational example showing how to obtain the linear, quadratic, and cubic sums of squares for the data in Table 8.16

	2.5	5	10	20	40	80				
Treatment (in μg)										
Log of treatment	.398	.699	1.000	1.301	1.602	1.903				
Level of estratiol (k)	1	2	3	4	5	6				$\frac{C^2}{n\Sigma c^2}$
$\Sigma X_{\text{uterine weight}}$	112	242	400	583	756	931	C	Σc^2	$n\Sigma c^2$	
Linear	-5	-3	-1	1	3	5	5,820	70	700	48,389.14
Quadratic	5	-1	-4	-4	-1	5	285	84	840	96.70
Cubic	-5	7	4	-4	-7	5	-235	180	1,800	30.51

we have done in the third row. We designate these numbers by the symbol k and shall return to them when we obtain the best-fitting straight line for the data. First, however, we have to determine whether there is any significant trend in the data, and the necessary computations are shown in Table 8.17.

We have six treatment groups, and we decide that it will be sufficient to obtain the linear, quadratic, and cubic components of the curve. From Table VII we obtain the coefficients for these three components for the condition where there are six groups, and these are listed in Table 8.17. We follow the same computational procedure as described in Table 8.15, and the far right column in Table 8.17 contains the sums of squares of the linear, quadratic, and cubic components. We then collate these data in Table 8.18 which is the summary table for the analysis of variance. From Table 8.16 we obtain the $SS_{between}$ of 48,519.80 and SS_{within} of 8,407.60. The $SS_{between}$ is listed in brackets because we do not make an F test on that component.

With six groups there are 5 degrees of freedom, and we have used up one degree of freedom each in determining our linear, quadratic, and cubic components. The sums of squares for these components are taken from Table 8.17 and entered into Table 8.18. Because we have 5 df among our treatment groups, we have 2 df remaining after listing our linear, quadratic, and cubic components, and we find the sum of squares of this remainder by the following formula:

$$SS_{remainder} = SS_{between} - SS_{lin} - SS_{quad} - SS_{cub} = 3.45$$

MS_{within} of 155.70 is used to evaluate the orthogonal trend components. The linear trend is found to be highly significant with an F of 310.78. All the remaining F ratios are less than 1. We therefore conclude that a linear function adequately describes the functional relationship between the logarithm of the amount of estradiol injected and uterine weight.

TABLE 8.18 Summary analysis of variance table of the data in Tables 8.16 and 8.17

Source	df	SS	MS	F
[Between]	[5]	[48,519.80]		
Linear	1	48,389.14	48,389.14	310.78**
Quadratic	1	96.70	96.70	
Cubic	1	30.51	30.51	
Remainder	2	3.45	1.72	
Within	54	8,407.60	155.70	
Total	59	56,927.40		

Obtaining the best-fitting straight line   In addition to finding a linear trend, it is often very useful to be able to write the equation for the best-fitting straight line. When we use the method of orthogonal polynomials there is a very simple way to get the equation. The equation for a straight line is of the form

predicted response measure $= a + b$(level of independent variable)

We shall let X' = any predicted response measure and rewrite the equation as follows:

$$X' = \bar{X} + \left[\lambda_{\text{lin}} \left(\frac{C_{\text{lin}}}{n \Sigma c_{\text{lin}}^2} \right) (k - \bar{k}) \right] \tag{8.15}$$

where $\bar{X}$ is the mean of our observed scores, λ_{lin} is a constant obtained from the table of orthogonal polynomials, C_{lin} and $n \Sigma c_{\text{lin}}^2$ are values obtained in computing the analysis of variance, k represents the levels of the independent variable in the study, and $\bar{k}$ is the mean level of k.

For this particular example the computations are given in Table 8.19. The mean uterine weight ($\bar{X}$) is obtained from Table 8.16, where $\Sigma X = 3024$. With 60 observations, $\bar{X} = 3024/60 = 50.40$. C_{lin} and $n \Sigma c_{\text{lin}}^2$ are both taken directly from Table 8.17. In Table VII of the Appendix we find that the value for λ_{lin} for the linear component when there are six groups is 2 (see the righmost column of Table VII). The value for $\bar{k}$ is 3.5. When we make the appropriate substitutions we obtain

$$X' = 50.40 + 2 \left(\frac{5820}{700} \right) (k - 3.5)$$

When we simplify this we obtain

$$X' = -7.80 + 16.63k$$

In words, the equation says that the best-fitting straight line for predicting uterine weights (X') from the level of estradiol injected (k) is $-7.80 + 16.63k$. At the bottom of Table 8.19 are given the predicted weights for the six levels used in the experiment. The best-fitting straight line as well as the actual mean uterine weights are plotted in Figure 8.7.

In interpreting Figure 8.7 and the equation, it is necessary to keep in mind that k represents the logarithms of the original dosages of estradiol. That is, biologically we have a log-linear relationship between dose level and the response measure. If we wished to use this equation for some intermediate dosage of estradiol to predict the expected uterine weight, we would get the log of the dosage and then interpolate that value to get the corresponding value of k. We could then insert k in our equation to find the expected weight. As an example, suppose that we are

TABLE 8.19 Procedure to obtain best-fitting straight line by orthogonal polynomials using the data of Tables 8.16 and 8.17

The regression equation is

$$X' = \bar{X} + \left[\lambda_{lin} \left(\frac{C_{lin}}{n\Sigma c_{lin}^2} \right) (k - \bar{k}) \right]$$

where

X' = predicted uterine weight in milligrams

$\bar{X} = \dfrac{3024}{60} = 50.40$

$\lambda_{lin} = 2$

$C_{lin} = 5820$

$n\Sigma c_{lin}^2 = 700$

$\bar{k} = 3.5$

$X' = 50.40 + 2 \left(\dfrac{5820}{700} \right) (k - 3.5)$

$X' = -7.80 + 16.63 \, k$

k = level of estradiol	X' = predicted uterine weight (milligrams)
1	8.83
2	25.46
3	42.09
4	58.72
5	75.35
6	91.98

interested in knowing the predicted uterine weight if we inject 30 micrograms of estradiol. The log of 30 is 1.477. At the top of Table 8.17 we find the data concerning the 20- and 40-microgram levels. We can put all of this together as follows:

Treatment (micrograms)	Log treatment	Level of k
20	1.301	4
30	1.477	
40	1.602	5

By linear interpolation we find that log 30 micrograms (= 1.477) is .585 of the way into the interval between log 20 micrograms and log 30 micrograms (.176/.301 = .585). Therefore the value of k for log 30 micrograms is 4.585, and the predicted uterine weight is $X' = -7.80 + 16.63(4.585) = 68.45$ milligrams.

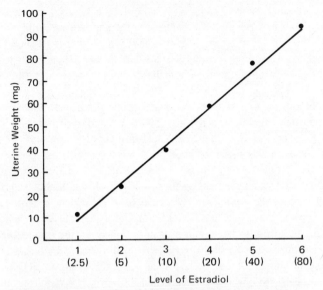

FIGURE 8.7 Functional relationship between level of estradiol injected and uterine weight of the mouse. The dots represent the data points, and the straight line is the best-fitting line determined from the equation

predicted uterine weight $= -7.80 + 16.63$(level of estradiol)

The actual micrograms of estradiol administered are given in parentheses along the X axis.

A Behavioral Example

Assume that we are experimental psychologists interested in determining the form of the function relating retention of a learned response to intensity of the reinforcement. To investigate this problem we decide to study the behavior of rats in a passive learning situation. Our passive learning apparatus is a two-chambered unit. The larger chamber, into which the animal is placed, is painted white and is brightly illuminated. The second chamber is black. A small open doorway connects the two chambers. The floor of the unit is a grid that can be electrified. We have a shock generator that permits us to vary the intensity of the current (in milliamperes) across the grid.

When a rat is placed into the brightly illuminated white chamber, the typical response is to move quickly into the dark chamber. When the animal does this, we close the door behind the rat and administer 3 seconds of electric shock. The animal is then removed, returned to his home cage, and left undisturbed for 24 hours. He is then brought back to the apparatus, placed into the white chamber, and his latency (in seconds) to enter the dark chamber is measured.

We have decided to use .1, .3, .5, .7, .9, 1.1, 1.3, and 1.5 milliamperes of shock. Forty rats are selected from our colony and are randomly assigned in groups of 5 to one of the eight shock conditions. Table 8.20 shows the total response latency in seconds during the retention test for five animals within each group. At the bottom of the table are the computations to determine SS_{total}, $SS_{between}$, and SS_{within}.

The question of experimental interest is to determine the form of the function relating shock intensity to retention, and we therefore decide to evaluate the linear, quadratic, and cubic components of our curve. The analysis of these components is given in Table 8.21 together with the final analysis of variance table. From the analysis of variance table we see that there is an F of 57.76 for the linear component; with 1, 32 df this is highly significant. In addition, the quadratic component has an F of 7.33, which is significant at the .05 level. Neither the cubic component nor the remainder approaches significance. From these results we would conclude that retention of a learned passive avoidance response improves as intensity of reinforcement increases (the linear component), but at a decreasing rate (the quadratic component). That is, for each stepwise increase in shock intensity there is a progressively lesser increase in response latency.

TABLE 8.20   Total response latency (in seconds) to enter a dark chamber as a function of prior shock intensity in the dark chamber; and the breakdown of the sums of squares

	Treatment (milliamperes of shock)							
	.1	.3	.5	.7	.9	1.1	1.3	1.5
$\Sigma X_{latency}$	25	131	197	240	283	298	310	327

$\Sigma X = 1,811$

$\Sigma X^2 = 104,647$

$n = 5/\text{group}$

$N = (8)(5) = 40$

$$CT = \frac{(\Sigma X)^2}{N} = \frac{(1,811)^2}{40} = 81,993.025$$

$$SS_{total} = \Sigma X^2 - CT = 22,653.975$$

$$SS_{between} = \frac{(25)^2 + (131)^2 + \cdots + (327)^2}{5} - CT = \frac{486,117}{5} - CT = 15230.375$$

$$SS_{within} = SS_{total} - SS_{between} = 7,423.600$$

TABLE 8.21 Computational example showing how to obtain the linear, quadratic, and cubic sums of squares for the data in Table 8.20; and the final analysis of variance summary table

Treatment	.1	.3	.5	.7	.9	1.1	1.3	1.5				
Level of treatment (k)	1	2	3	4	5	6	7	8				C^2
$\Sigma X_{latency}$	25	131	197	240	283	298	310	327	C	Σc^2	$n\Sigma c^2$	$\overline{n\Sigma c^2}$
Linear	−7	−5	−3	−1	1	3	5	7	3,355	168	840	13,400.03
Quadratic	7	1	−3	−5	−5	−3	1	7	−1,195	168	840	1,700.03
Cubic	−7	5	7	3	−3	−7	−5	7	383	264	1,320	111.13

Source	df	SS	MS	F
[Between]	[7]	[15,230.375]		
Linear	1	13,400.030	13,400.030	57.76[**]
Quadratic	1	1,700.030	1,700.030	7.33[*]
Cubic	1	111.130	111.130	
Remainder	4	19.185	4.796	
Within	32	7,423.600	231.988	

Obtaining the best-fitting second-degree equation We can extend Equation 8.15 to include the second-degree trend as follows:

$$X' = \bar{X} + \lambda_{lin}\left(\frac{C_{lin}}{n\Sigma c_{lin}^2}\right)(k - \bar{k}) + \lambda_{quad}\left(\frac{C_{quad}}{n\Sigma c_{quad}^2}\right)\left[(k - \bar{k})^2 - \frac{N_k^2 - 1}{12}\right]$$

$$(8.16)$$

where all terms are defined the same as with Equation 8.15. There is one new term in the equation, N_k. This is defined as the number of levels of k. Table 8.22 shows the computational procedure to find the best-fitting second-degree equation, and Figure 8.8 plots the equation together with the actual data points.

Orthogonal Polynomials and Curve Fitting

It is important to reiterate that the procedure of orthogonal polynomials as presented here requires that we have equal numbers of subjects at each treatment point and that the intervals be equally spaced as an arithmetic series or a geometric series. If we have a geometric series, then a logarithmic transformation puts our numbers into an arithmetic series (see Table 8.17). Furthermore, we have converted the arithmetic series into levels of value k, where k is 1, 2, 3, . . . Equations 8.15 and 8.16 apply only when k is used as the values of the independent variable.

TABLE 8.22 Procedure to obtain best-fitting second-degree equation by orthogonal polynomials using the data in Tables 8.20 and 8.21

The regression equation is

$$X' = \bar{X} + \lambda_{\text{lin}} \left(\frac{C_{\text{lin}}}{n \Sigma c_{\text{lin}}^2} \right) (k - \bar{k}) + \lambda_{\text{quad}} \left(\frac{C_{\text{quad}}}{n \Sigma c_{\text{quad}}^2} \right) \left[(k - \bar{k})^2 - \frac{N_k^2 - 1}{12} \right]$$

where

$X' = $ predicted response latency in seconds

$\bar{X} = \dfrac{1,811}{40} = 45.275$

$\lambda_{\text{lin}} = 2$

$C_{\text{lin}} = 3355$

$n \Sigma c_{\text{lin}}^2 = 840$

$\bar{k} = 4.5$

$\lambda_{\text{quad}} = 1$

$C_{\text{quad}} = -1195$

$n \Sigma c_{\text{quad}}^2 = 840$

$N_k = 8$

$$X' = 45.275 + 2 \left(\frac{3355}{840} \right) (k - 4.5) + 1 \left(\frac{-1195}{840} \right) \left[(k - 4.5)^2 - \frac{(8^2 - 1)}{12} \right]$$

$$= 45.275 + (7.988k - 35.946) + \left[-1.423 \left(k^2 - 9k + 20.25 - \frac{63}{12} \right) \right]$$

$$= -12.016 + 20.795k - 1.423k^2$$

$k = $ level of shock	$X' = $ predicted response latency (seconds)
1	7.356
2	23.882
3	37.562
4	48.396
5	56.384
6	61.526
7	63.822
8	63.272

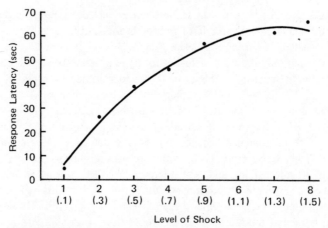

FIGURE 8.8 Functional relationship between level of shock administered and response latency. The dots represent the data points and the curve is the best-fitting line determined from the equation

predicted response latency $= -12.016 + 20.795k - 1.423k^2$

The actual milliamperes of shock admnistered are given in parentheses along the X axis.

The method of orthogonal polynomials has the very valuable feature that each succeeding component is independent of (i.e., orthogonal to) all previous components. Thus, Equations 8.15 and 8.16 are identical up through the linear component, and then the quadratic portion is added to Equation 8.15 to yield Equation 8.16. Similarly, the cubic component can be added to Equation 8.16 to yield a third-degree equation. That component is

$$\lambda_{cub}\left(\frac{C_{cub}}{n\Sigma c_{cub}^2}\right)\left\{(k-\bar{k})^3 - (k-\bar{k})\left[\frac{3(N_k)^2 - 7}{20}\right]\right\} \quad (8.17)$$

Equations up to the third degree are useful as a way of describing many behavioral and biological processes. They can also be used to approximate logarithmic, exponential, and S-shaped (learning) curves, and the orthogonal polynomial method is a simple way to show the general form of a relationship. However, these curves are not the same as logarithmic, exponential, and S-shaped curves, and if you are interested in fitting a specified kind of curve to your data it is probably better to fit that curve by appropriate curve fitting procedures rather than approximate the curve via orthogonal polynomials.

The Use of Curves with the Fixed Effects Model

In studying the formal relationship between an independent quantitative variable and a measured response, we are still working within the context of the fixed effects model. That is, the experimenter has deliberately selected the values of the independent variable to be used in the study. Because these are not randomly selected from a population, in terms of statistical inference the experimenter can only draw a conclusion to the population of values represented by the treatment conditions. Thus, in the endocrinology example we can talk about 2.5, 5, 10, 20, 40, and 80 micrograms of estradiol but technically we cannot draw conclusions about 30 micrograms because we do not have a group that received 30 micrograms of estradiol. Similarly, we cannot talk about the effects of 120 micrograms because no such group was represented in the experiment. The two examples given in the previous sentence—estimating uterine weights for 30 micrograms and 120 micrograms—differ in one very important respect: We can estimate the uterine weight for 30 micrograms by *interpolating* among the points in our experiment because 30 micrograms fits within the limits of our experimental treatments, but we would have to *extrapolate* beyond the most extreme treatment group in our experiment in order to estimate the uterine weight for 120 micrograms of estradiol. Even though the fixed effects model technically limits one to those points specifically selected, as experimenters we are willing to interpolate within the limits of our experimental variable to unknown values not included in the experiment. However, it is very risky to extrapolate to values more extreme than those in the study. Thus, if you need information about an extreme value of your independent variable, this should be built into your experimental design as a specific treatment group. Even interpolation is dangerous if there are insufficient data points to allow you to obtain a good estimate of the form of the curve.

POWER AND SAMPLE SIZE FOR THE SINGLE–CLASSIFICATION ANALYSIS OF VARIANCE EXPERIMENT

Power in Exploratory Experiments

The procedure described below permits us to determine the appropriate sample size to achieve a given power for an exploratory experiment. It is assumed that n is the same in all groups. If we are conducting an exploratory experiment with no specific experimental hypotheses to guide us, then it is necessary to have a significant overall F test (remember, this is not a necessary condition for the three other types of experiments we discussed above). For this type of design, in order to determine n we need the following information: the level of α (with the analysis of variance this is always two-tailed); the desired power; an estimate of s_{within}; and an estimate of the expected value of $s_{\bar{x}}$, the standard deviation of the sample means. We again use Cohen's (1969) tables and procedure. We define a statistic called f, as follows:

$$f = \frac{s_{\bar{x}}}{s_{within}} \tag{8.18}$$

That is, f indicates the percentage of variability expected among the sample means, relative to the variability to be expected within a homogeneous group of subjects.

Tables VIII, IX, and X in the Appendix permit us to determine n with specified power for α levels of .01, .05, and .10. In order to use these tables it is first necessary to select a value of α, then within the table select the subtable corresponding to the numerator df for the F test. Within this subtable locate the value of f and the desired power. The numerical value at the intersection of these two values is the n needed per group.

An example will illustrate the use of there tables. Let us return to the behavioral pharmacology investigation of shuttlebox learning as a function of pharmacological agents. We now have four new compounds we wish to evaluate, and we decide, from past experience, that the variability among the learning means for these compounds (i.e., $s_{\bar{x}}$) must be at least 25% of the variability of our standard deviation (s_{within}). Therefore, $f = .25$. We wish to be fairly certain to find a difference, if one exists, and so we select a high numerical value for α, .10. We also want good power and decide that we wish to be 90% certain of detecting a difference, if there is one. We now have the following conditions specified:

$$\alpha = .10$$
$$df_{numerator} = 3$$
$$f = .25$$
$$power = 90\%$$

We find in Table X that an n of 48 per group satisfies these requirements.

n for Intermediate Values of f

Tables VIII, IX, and X contain only certain values of f, and it will often be necessary to find n for a value of f not listed in the table. Cohen (1969) presents the following formula to find n:

$$n = \frac{n_{.05}}{400f^2} + 1 \tag{8.19}$$

where $n_{.05}$ is the sample size for a given α, $df_{numerator}$, and power; and f is defined as above.

As an example, suppose that we decided that the variability among our sample means must be one-third our within-group standard deviation. Then $f = .33$. Making the appropriate substitutions in Formula 8.19, we get

$$n = \frac{1180}{(400)(.33)^2} + 1 = \frac{1180}{43.56} + 1 = 28.08$$

Therefore, we should have 29 subjects per group.

Power in Designs Other Than Exploratory Experiments

Exploratory experiments are done when we have the least information about our subject matter. We have already seen (Table 3.13) that this type of experiment demands the largest critical difference in order to achieve significance. If we have information that allows us to plan individual comparisons, design a regression experiment, or compare a control group against several experimental groups, then, for a given n, our experiment will be more powerful than if we are forced to resort to a significant overall F test. Thus, if we select n by the method described for exploratory experiments, we know that we have more than sufficient numbers of subjects per group. Another approach is to use the method described in Chapter 7 on the t test to determine the necessary n.

Post Hoc Evaluation of an Experiment's Power

Tables VIII, IX, and X also permit us to evaluate the power of a completed experiment. The f statistic is determined from the data of the experiment. The other needed information are the α level, the numerator df, and the n per group.

To illustrate, suppose that we plan to conduct an experiment comparing five different groups with $n = 40$ per group, and an α of .05. We decide that the standard deviation among our five treatment means must be at least 30% of the within-group standard deviation to justify further research on the topic. Thus $f = .30$. When these values are entered into Table IX, we find that the power of our test is 94%. Suppose that we do the experiment and get no significant finding. We may now feel quite certain that there are not sufficient differences among the five treatments means to warrent continuing this line of research.

MISSING VALUES

Nothing is more aggravating to a researcher than to have an excellently designed experiment go awry because of a missing value in the data. For example, an animal escapes from its cage and has to be discarded from the study; a chemical assay becomes contaminated and has to be thrown away; a research technician misplaces the original data sheet. Because these events do occur in research, it is best to plan for them. It is for this reason that the suggestion was made earlier that extra animals be carried along in an experiment to replace those whose data cannot be used. A second approach is to randomly discard animals from the other groups in the experiment to give you equal numbers of cases per group. However, it is as aggravating to randomly discard good data as it is to have a missing value. There is a third alternative, which is a very workable one if there is only one value missing within any particular treatment group.

Suppose that we are doing an experiment in which there are three treatment groups and six animals per group. The data of one animal from one of the groups has been lost, and yet we wish to analyze our data as though we had equal n's. What we would do is compute the mean of that group based on our five remaining animals, and insert that mean value into the missing cell of this group. We would then go ahead and carry out an analysis of variance as though we had equal numbers of cases per group. The one adjustment we would make would be to reduce our degrees of freedom for our within source of variation by 1 to adjust for this missing value. An example of this is given in Table 8.23.

TABLE 8.23 Computational example showing how to fit a missing value and compute the analysis of variance

	Group 1	Group 2	Group 3
	10	38	24
	12	45	31
	7	51	19
	4	42	23
	15	46	28
	6	Missing (44.4)	23
ΣX	54	266.4	148
ΣX^2	570	11,921.36	3,740

$\Sigma X = 468.4$

$\Sigma X^2 = 16,231.36$

$N = 18$

$CT = \dfrac{(\Sigma X)^2}{N} = 12,188.81$

$SS_{total} = \Sigma X^2 - CT = 4,042.55$

$SS_{between} = \dfrac{(54)^2 + (266.4)^2 + (148)^2}{6} - CT = 3,776.02$

$SS_{within} = SS_{total} - SS_{between} = 266.53$

Source	df	SS	MS	F
Between	2	3,776.02	1,888.01	99.16[**]
Within	14[a]	266.53	19.04	
Total	16	4,042.55		

[a]One df lost in fitting missing value.

This procedure is recommended only for conditions where you have independent observations and where there is only one value missing from any treatment group. There are procedures for fitting missing values under conditions where there are two or more values missing from a treatment group, and where more complex experimental designs are involved. The reader is referred to Cochran and Cox (1957) for a detailed discussion of this matter.

SUMMARY OF DEFINITIONS

A posteriori tests   The same as *post hoc* comparisons.

Arithmetic series   A sequence of numbers in which each succeeding number is increased by a constant amount over the preceeding number.

Error variance   In the analysis of variance, this refers either to the error term in the sample used to evaluate treatment effects or to the variance in the population.

Expected value   The average value of a statistic over an infinite set of samples.

Exploratory experiment   An experimental design consisting of several qualitatively different treatment groups that are compared with each other. This is often called "I wonder what would happen if" kind of experiment.

Fixed effects model   An analysis of variance model in which the specific treatments or experimental procedures used are the only ones of relevance to the researcher. Generalizations cannot be made to other treatment conditions.

Geometric series   A sequence of numbers in which each succeeding number is a constant multiplier of the preceeding number. Taking logarithms of a geometric series converts it to an arithmetic series.

Honestly significant difference procedure   After an overall significant F ratio has been found in an exploratory experiment, the honestly significant difference procedure is applied to determine which pairs of means differ significantly.

Orthogonal comparisons   Statistical comparisons that are independent of each other.

Orthogonal polynomials   A method for determining the linear, quadratic, cubic, etc., trends of a polynomial equation for the situation of equal n per point and equal spacing of the independent variable.

Planned comparisons   An experimental design in which specific hypotheses, which have been designated in advance, are tested.

Post hoc *comparisons*   Any statistical procedure used to determine which means differ from each other after an overall F test has been found to be significant. Also called postmortem tests and *a posteriori* tests.

Post mortem tests   The same as *post hoc* comparisons.

Random effects model   An analysis of variance model in which the treatments or experimental procedures used have been selected by a random procedure from a population of treatment conditions. With this model, generalizations can be made to the treatment population.

Regression experiment An experimental design in which the independent variable is quantitative and is systematically varied by the researcher.

Robust statistic A statistic whose distribution and probability values remain much the same even though the assumptions underlying its derivation have been violated.

Single-classification analysis of variance An experimental design in which the various treatments are classified along a single dimension.

Tests for trend A statistical test of a regression experiment to determine if the linear, quadratic, cubic, etc., trends are significant.

PROBLEMS

1. Given the following data based on independent observations, compute the analysis of variance.

Treatment 1	Treatment 2	Treatment 3
89	37	80
32	42	58
54	53	84
61	60	76
81	22	79
49	52	48
13	34	98

2. For Problem 1, compute the sums of squares within each of the three treatment groups. The summation of these values is equal to what statistic?
3. Assume that the experiment in Problem 1 is an exploratory study. Use Tukey's honestly significant difference procedure to compare the means.
4. Assume that Treatment 1 in Problem 1 is a control group. Use Dunnett's procedure to compare the control group to each of the two experimental groups.
5. A researcher was interested in the effects of drugs on learning. He took a group of rats and put them on a chronic injection schedule of heroin, methadone, or amphetamine. One control group received injections of saline, and another control group was not injected. Three weeks after the start of the drug treatment the researcher began testing the rats on a series of three maze problems. He recorded the number of errors made on each maze until the learning criterion was achieved. His final learning score was the total number of errors made on all three mazes. These error scores are shown below.

No injection	Saline	Heroin	Methadone	Amphetamine
46	25	44	35	46
37	22	21	49	41
36	34	24	18	35
30	38	39	19	26
28	37	32	12	48
26	43	14	17	30
27	19	9	23	33

(a) Do an overall analysis of variance of these data.
(b) The researcher had designed the experiment to test several specific hypotheses. These hypotheses are as follows:

(i) $\mu_{no\ injection} = \mu_{saline}$
(ii) $\mu_{heroin} = \mu_{methadone}$
(iii) $(\mu_{no\ injection} + \mu_{saline})/2 = (\mu_{heroin} + \mu_{methadone})/2$
(iv) $(\mu_{heroin} + \mu_{methadone})/2 = \mu_{amphetamine}$
(v) $(\mu_{no\ injection} + \mu_{saline})/2 = \mu_{amphetamine}$

Using planned comparisons, test these hypotheses.

6. A developmental psychologist was interested in determining whether newborn infants would attend differentially to figures with differing amounts of information. To vary information content she decided to use geometrical figures of 3, 4, 5, 6, and 7 sides because previous work had shown that the amount of information increased with the number of angles present in a figure. Babies were randomly assigned to one of the five geometrical figure groups. Each baby was then placed in a cradle board facing a screen. On the screen was projected a figure of 3, 4, 5, 6, or 7 sides, depending on the group to which the baby had been assigned. The figure was left on for a constant amount of time, and the number of seconds the baby looked at the figure was recorded. This procedure was repeated with four different figures, and the baby's final score was the total time in seconds that he/she had looked at the figure. The data are presented below.

3 sides	4 sides	5 sides	6 sides	7 sides
20	73	81	43	23
28	83	70	58	48
79	26	57	93	73
21	15	72	58	30
42	56	66	23	24
15	33	55	14	13
43	51	83	64	38
15	43	66	27	46
63	63	32	29	56
65	70	31	45	12

(a) The researcher expected that there would be an optimal level of information that the baby would attend to and that figures with lesser or greater information would be looked at for less time. Statistically, this is equivalent to saying that there will be a quadratic relationship between the independent variable, the number of angles, and the dependent variable of time spent looking at the figures. Test this hypothesis.
(b) In addition to a major quadratic function, the researcher thought that a minor linear trend might also be present. Test this hypothesis.
(c) How would the researcher interpret her findings?

(d) If she had done an overall F test of the between MS, and not made the specific test for a quadratic trend, how would she have interpreted her data?

7. A biologist is interested in the functional relationship between litter size at birth and weaning weight. At birth pups are removed from their mothers, mixed together, and then randomly given to mothers in units of 2, 4, 6, 8, 10, 12, or 14. At weaning the pups are weighed and the mean litter weight to the nearest tenth of a gram is obtained. This litter mean is the basic statistic used in the analysis. There were four mothers randomly assigned to each of the seven litter-size units, and the following data were obtained:

2 pups	4 pups	6 pups	8 pups	10 pups	12 pups	14 pups
68.2	75.4	70.0	49.7	54.8	31.8	34.9
71.4	67.7	64.4	55.5	52.1	44.3	26.5
63.6	72.7	57.1	63.4	47.1	41.9	19.4
74.0	52.6	56.7	59.3	42.6	43.2	31.0

(a) Do a trend analysis of these data, extracting the linear and quadratic components.

(b) Obtain the equation for the best-fitting straight line where the independent variable of litter size is coded in units of k.

(c) Obtain the equation for the best-fitting second-degree equation.

(d) From the equations obtained in (b) and (c) above, determine the predicted mean litter weights.

CHAPTER 9

The Analysis of Variance: Nested Designs, Randomized Blocks, and Factorial Experiments

In Chapter 8 we were concerned with developing the rationale for the single-classification analysis of variance experiment and showing how to evaluate qualitative and quantitative experimental hypotheses using this basic design. In this chapter we shall examine two variations in the one-way analysis of variance design—the nested experiment and the randomized blocks design—and we shall also look briefly at an experimental design, called a factorial experiment, which allows us to assess simultaneously two or more independent variables. In contrast to single-factor designs, factorial designs permit the evaluation of what are called interaction effects.

NESTED DESIGNS

Up to now individual subjects have been assigned at random to our experimental treatments, and differences between subjects within the same treatment defined our error variance (usually called MS_{within}) used in our test of significance. Such differences are due to random effects because of the randomization procedure followed. There are times, however, when it is not possible to assign subjects at random to treatment conditions, but where it is necessary to use a larger unit as the basis for randomization. In animal research a very common procedure is to use a complete litter as the unit in assignment to a particular treatment condition. We shall use this example as the basis for developing the principles involved in nested designs.

The use of whole littters is often seen in developmental research when an experimental treatment has to be administered sometime between birth and weaning. Consider the following example. A neuroendocrinologist is interested in the effects of thyroxine upon later body weight and behavior. He chooses two levels of thyroxine: .5 micrograms of thyroxine per gram of body weight and 1 microgram per gram body weight. A third group is to receive an injection of the vehicle, and a fourth group is to remain undisturbed. The injections are administered once daily on days 1, 2, and 3 of life. The researcher decides to administer a particular treatment to a complete litter of animals. He mates a group of female rats and sets them aside to be used in the study. Twelve females become pregnant and he randomly assigns three of them to each of the four treatment conditions. Starting the day after birth, the pups receive the injection of thyroxine, of the vehicle, or are not disturbed. At weaning the endocrinologist selects three males from each of the 12 litters and sets them aside to be tested later on. The remaining animals from the litter are not used in this study. The layout for this experiment is shown in Table 9.1.

General Layout

In Table 9.1 we see that Litters 1, 2, and 3 were assigned to the Nondisturbed Control condition; Litters 4, 5, and 6 were assigned to the Vehicle Injection condition; and so on. The scores of the three males within each litter are designated by the usual symbol X and two subscripts which indicate the animal and the litter to which it belongs.

This design is quite different from the previous ones we have discussed, and we should consider some of the implications. Note that the differences among the four treatment conditions in the experiment are due in part to differences caused by the unique effects associated with individual litters. The unique effects of Litters 1, 2, and 3 are confined to the Nondisturbed Control condition; the unique effects of Litters 4, 5, and 6 are restricted to the Vehicle Injection condition; and so on. When an effect (in this case the litter) is restricted to a single treatment condition, that effect is said to be *nested* within that treatment condition. When this happens we speak of *nested designs* or *hierarchical designs*. In this example we say that litters are nested within the experimental treatments.

Now consider the three male rats within each litter. Their own unique effects are doubly confined to the litter of which they are a member and to the treatment condition to which the litter (not the subject) has been randomly assigned. Thus, we say that subjects are nested within litters and also within experimental treatments. The breakdown of the sources of variation and degrees of freedom for the design of Table 9.1 is shown in Table 9.2.

TABLE 9.1 Experimental layout for endocrinological study using a nested design

Nondisturbed controls			Vehicle injection			.5 microgram of thyroxine injection			1.0 microgram of thyroxine injection		
Lit. 1	Lit. 2	Lit. 3	Lit. 4	Lit. 5	Lit. 6	Lit. 7	Lit. 8	Lit. 9	Lit. 10	Lit. 11	Lit. 12
$X_{1 \cdot 1}$	$X_{2 \cdot 1}$	$X_{3 \cdot 1}$	$X_{4 \cdot 1}$	$X_{5 \cdot 1}$	$X_{6 \cdot 1}$	$X_{7 \cdot 1}$	$X_{8 \cdot 1}$	$X_{9 \cdot 1}$	$X_{10 \cdot 1}$	$X_{11 \cdot 1}$	$X_{12 \cdot 1}$
$X_{1 \cdot 2}$	$X_{2 \cdot 2}$	$X_{3 \cdot 2}$	$X_{4 \cdot 2}$	$X_{5 \cdot 2}$	$X_{6 \cdot 2}$	$X_{7 \cdot 2}$	$X_{8 \cdot 2}$	$X_{9 \cdot 2}$	$X_{10 \cdot 2}$	$X_{11 \cdot 2}$	$X_{12 \cdot 2}$
$X_{1 \cdot 3}$	$X_{2 \cdot 3}$	$X_{3 \cdot 3}$	$X_{4 \cdot 3}$	$X_{5 \cdot 3}$	$X_{6 \cdot 3}$	$X_{7 \cdot 3}$	$X_{8 \cdot 3}$	$X_{9 \cdot 3}$	$X_{10 \cdot 3}$	$X_{11 \cdot 3}$	$X_{12 \cdot 3}$

TABLE 9.2   Sources of variation and degrees of freedom for the design in Table 9.1

Source	df
Between experimental treatments	3
Litters within experimental treatments	8
Subjects within litters within experimental treatments	24
Total	35

The experimental design in Table 9.1, and as expressed in the sources of variation in Table 9.2, differs from our other experimental designs in several ways.

1. This is the first experimental design we have encountered in which there are two sources of random variation— litters, and subjects within litters. As usual, treatments are considered to be a fixed effect.
2. Even though there are two sources of random variation, only *one* of these has been assigned by a random procedure to treatment conditions—the litter variable. Subjects within litters have *not* been randomly assigned because they remain associated with the same litter throughout the experiment.
3. Because subjects are not randomly assigned to treatment conditions, and because litters are randomly assigned, litters is the proper source of error variance to be used in evaluating the experimental treatments. The appropriate F test is

$$F = \frac{MS_{\text{between treatments}}}{MS_{\text{litters within treatments}}}$$

Choice of Error Term

The general principle to follow in determining the proper error term in a nested design where there is a hierarchy of nested random variables is that the nested variable that is highest in this hierarchy is used to evaluate a fixed treatment effect. Litters is above subjects in the nesting hierarchy. This is seen by the physical arrangement of the design in Table 9.1 and also by the fact that the word "within" occurs once with litters and twice with subjects in Table 9.2.

Why should MS_{litters} be used as the error term rather than MS_{subjects}? Our experimental question is whether or not the administration of thyroxine changes the performance of the experimental animals. Because litters are nested within experimental treatments, we know that some of the differences we obtain among our treatment means will be a function of unique effects of the litters. What we wish to know is whether the differences among these treatment means are significantly greater than the variation brought about by litter differences. It follows, therefore, that variation among litters must be the error term against which to evaluate variation among treatment conditions. Thus, the litter becomes the unit of analysis in this experiment.

This point can be made clear by taking the extreme condition. Suppose that only one litter is assigned to each of the four experimental treatments in Table 9.1. If a significant difference were obtained among the treatment groups, the experimenter could not tell whether that difference was due to the experimental treatments that he introduced or whether it reflected unique effects among the litters themselves. In this instance there is complete *confounding* between the litter effect and the treatment effect, and no useful information is obtained from such an experiment. In order to obtain variation among litters, it is necessary to have at least two litters randomly assigned to each treatment condition. This is necessary because that is the variation used as the denominator in the F test to assess the effects of the experimental treatments. If that variation is absent because only one litter is assigned to each treatment condition, no valid denominator is available for the F test.

We have been considering the relationship of the treatment mean square to the litter mean square. What is the relationship of MS_{litter} to $MS_{subjects}$? In most instances the mean square for the litter effect will be larger than the mean square associated with subjects, in many cases significantly so. To understand this, consider the many factors that tend to make animals in a litter similar. Littermates share a common genetic heritage, they all develop in the same uterine environment, they are reared from birth to weaning by one maternal figure who probably behaves in a similar fashion to all her offspring, and they all share the same source of nutrition. Because of these common prenatal and postnatal factors, we expect the variability among littermates to be less than the variability among independent animals. The factors listed above which are common among littermates are absent when different litters are compared with each other, and it is for this reason that we expect the variability within litters to be less than the variability between litters.

The ideas we have developed in the context of litters nested within treatments and subjects nested within litters hold for all nested designs. Whenever subjects exist as part of a randomly assigned larger block, and when all the subjects within that block receive the same experimental treatment, then the block becomes the base for analysis. In such a situation the use of $MS_{subjects}$ as the error term will bias the test in the direction of making the F ratio too large.

Deficiencies in Nested Designs

The major deficiency in a nested design is that there are relatively few degrees of freedom associated with the error mean square. For example, in Table 9.2 there are 8 df for $MS_{litters}$. Because the litter rather than the subject is used as the unit of analysis to determine the error term, a reasonable question to ask is whether it is worthwhile to test more than one subject per litter. If only one male from each litter in Table 9.1 were tested, we would have the following sources of variation and *df* in our experiment.

Source	df
Between experimental treatments	3
Litters within experimental treatments	8

In this situation we have the same number of degrees of freedom to evaluuate the treatment effect as in Table 9.2, and yet we have only used one-third as many subjects as in the experimental design in Table 9.1. There is considerable merit in the suggestion of using only one animal per litter, especially if sufficient litters are available to allow one to have approximately 20 or more degrees of freedom for the error term. When this is not possible, as with this particular example, then there are two reasons why it is of value to sample two or more animals per litter. The first has to do with the stability of the litter mean, and the second is concerned with making a preliminary test on the mathematical model of the experiment. We shall consider each of these in turn.

Stability of the Litter Mean in Nested Designs

The stability of the litter mean is reflected by its standard error, and we know that the standard error decreases as the number of cases in our sample (or litter) increases. Recall that the formula for the standard error is

$$s_{\bar{x}} = \frac{s}{\sqrt{n}}$$

The larger the number of cases used in determining the litter mean, the smaller will be the variability of that mean. This increased stability acts to decrease the variation among litters. Therefore, we would expect $MS_{litters}$ based on three observations within each litter to be numerically less than $MS_{litters}$ based on only one observation per litter.

Making a Preliminary Test on the Model of a Nested Experiment

The model for the nested design as depicted in the sources of variation in Table 9.2 assumes that there is a significant effect due to litters. Although this is often the case, there are still many instances in the literature in which the variation among litters is no greater than the variation within litters. Suppose, for example, that the numerical values of $MS_{litters}$ and $MS_{subjects}$ in Table 9.2 were approximately the same. In such a situation there would not be any evidence for a significant litter effect, and it would be permissible for the experimenter to pool the sums of squares and the degrees of freedom from these two sources of random variation to give MS_{within} with 32 df. This makes a very considerable difference in our evaluation of the treatment effects. From our F table we find that $F_{3,8}(.05) = 4.07$, whereas $F_{3,32}(.05) = 2.90$. The critical value needed for significance with the second F ratio is approximately 29% less than that needed for significance with the first F ratio.

When we are in doubt as to whether a nested variable should be a source of variation in our analysis of variance model, we may carry out what is called a *preliminary test on the model*. The purpose of the preliminary test is to determine whether there is any evidence from the experiment itself that the nested variable should be left in as part of the model. When carrying out a preliminary test on a particular nested random variable, the next nested random variable in the hierarchy is used as the denominator of the F test. Therefore, in this particular situation our preliminary test to determine whether a significant source of variation is associated with litters is

$$F = \frac{MS_{litters}}{MS_{subjects}}$$

When carrying out a preliminary test, it is quite important to avoid Type 2 error, that is, accepting the null hypothesis of no litter effect when it should be rejected. The Type 2 error can be kept numerically small by setting a high α level for the preliminary test. An α of .25 is recommended for this type of preliminary test, and Table IVa in the Appendix lists the .25 level. Table 9.3 gives the general procedures involved in carrying out a preliminary test. A computational example is given in the next section.

Computational Procedures for Nested Experiments

Table 9.4 gives the general layout for the nested design in which there are two levels of nesting (i.e., litters within treatments, and subjects within litters). This is the same design as shown in Table 9.1 except that the symbols have been changed to make it somewhat more general. The summations at the bottom of each column represent the total score for all subjects within a litter. For example,

$$\Sigma X_{L_1} = X_{1 \cdot 1} + \cdots + X_{1 \cdot n}$$

The summation at the bottom of each treatment panel represents the total sum of all scores within that panel. For example,

$$\Sigma X_{T_1} = X_{1 \cdot 1} + \cdots + X_{3 \cdot n} = \Sigma X_{L_1} + \Sigma X_{L_2} + \Sigma X_{L_3}$$

From Table 9.2 we know that there are three sources of variation in the design shown in Table 9.4. It is instructive to see how the SS associated with

TABLE 9.3 General procedure for performing a preliminary test to determine whether two nested random variables can be pooled to give one error variance

1. Obtain

$$F = \frac{MS_{\text{litters within treatments}}}{MS_{\text{subjects within litters}}}$$

2. Use $\alpha = .25$ to reject H_0.

3. If H_0 is rejected, conclude that there is significant litter variation, and use the litter mean square to evaluated the treatment effects, as follows:

$$F = \frac{MS_{\text{between treatments}}}{MS_{\text{litters within treatments}}}$$

with df_{between} and df_{litters}.

4. If H_0 is not rejected, conclude that there is no evidence for a litter effect, and pool the SS's and df's to obtain MS_{within} and df_{within} as follows:

$$MS_{\text{within}} = \frac{SS_{\text{litters within treatments}} + SS_{\text{subjects within litters}}}{df_{\text{litters}} + df_{\text{subjects}}}$$

$$df_{\text{within}} = df_{\text{litter}} + df_{\text{subjects}}$$

Use MS_{within} to evaluate the treatment effects:

$$F = \frac{MS_{\text{treatment}}}{MS_{\text{within}}}$$

with df_{between} and df_{within}.

TABLE 9.4 General layout for the single-classification analysis of variance with independent observations in which litters are nested within treatments and subjects are nested within litters

Treatment 1			Treatment 2			Treatment 3			Treatment 4		
Lit. 1	Lit. 2	Lit. 3	Lit. 4	Lit. 5	Lit. 6	Lit. 7	Lit. 8	Lit. 9	Lit. 10	Lit. 11	Lit. 12
$X_{1 \cdot 1}$	$X_{2 \cdot 1}$	$X_{3 \cdot 1}$	$X_{4 \cdot 1}$	$X_{5 \cdot 1}$	$X_{6 \cdot 1}$	$X_{7 \cdot 1}$	$X_{8 \cdot 1}$	$X_{9 \cdot 1}$	$X_{10 \cdot 1}$	$X_{11 \cdot 1}$	$X_{12 \cdot 1}$
$\cdot$	$\cdot$	$\cdot$	$\cdot$	$\cdot$	$\cdot$	$\cdot$	$\cdot$	$\cdot$	$\cdot$	$\cdot$	$\cdot$
$\cdot$	$\cdot$	$\cdot$	$\cdot$	$\cdot$	$\cdot$	$\cdot$	$\cdot$	$\cdot$	$\cdot$	$\cdot$	$\cdot$
$\cdot$	$\cdot$	$\cdot$	$\cdot$	$\cdot$	$\cdot$	$\cdot$	$\cdot$	$\cdot$	$\cdot$	$\cdot$	$\cdot$
$X_{1 \cdot n}$	$X_{2 \cdot n}$	$X_{3 \cdot n}$	$X_{4 \cdot n}$	$X_{5 \cdot n}$	$X_{6 \cdot n}$	$X_{7 \cdot n}$	$X_{8 \cdot n}$	$X_{9 \cdot n}$	$X_{10 \cdot n}$	$X_{11 \cdot n}$	$X_{12 \cdot n}$
ΣX_{L_1}	ΣX_{L_2}	ΣX_{L_3}	ΣX_{L_4}	ΣX_{L_5}	ΣX_{L_6}	ΣX_{L_7}	ΣX_{L_8}	ΣX_{L_9}	$\Sigma X_{L_{10}}$	$\Sigma X_{L_{11}}$	$\Sigma X_{L_{12}}$
ΣX_{T_1}			ΣX_{T_2}			ΣX_{T_3}			ΣX_{T_4}		

TABLE 9.5 Showing where SS_{subjects} is obtained from the design in Table 9.4

$SS_{\text{within}_{L_1}} + SS_{\text{within}_{L_2}} + SS_{\text{within}_{L_3}} + \cdots = SS_{\text{subjects within litters within experimental treatments}}$

Lit. 1	Lit. 2	Lit. 3	Lit. 4	Lit. 5	Lit. 6	Lit. 7	Lit. 8	Lit. 9	Lit. 10	Lit. 11	Lit. 12
$X_{1 \cdot 1}$	$X_{2 \cdot 1}$	$X_{3 \cdot 1}$	$X_{4 \cdot 1}$	$X_{5 \cdot 1}$	$X_{6 \cdot 1}$	$X_{7 \cdot 1}$	$X_{8 \cdot 1}$	$X_{9 \cdot 1}$	$X_{10 \cdot 1}$	$X_{11 \cdot 1}$	$X_{12 \cdot 1}$
.	.	.	.	.	.	.	.	.	.	.	.
.	.	.	.	.	.	.	.	.	.	.	.
.	.	.	.	.	.	.	.	.	.	.	.
$X_{1 \cdot n}$	$X_{2 \cdot n}$	$X_{3 \cdot n}$	$X_{4 \cdot n}$	$X_{5 \cdot n}$	$X_{6 \cdot n}$	$X_{7 \cdot n}$	$X_{8 \cdot n}$	$X_{9 \cdot n}$	$X_{10 \cdot n}$	$X_{11 \cdot n}$	$X_{12 \cdot n}$
ΣX_{L_1}	ΣX_{L_2}	ΣX_{L_3}	ΣX_{L_4}	ΣX_{L_5}	ΣX_{L_6}	ΣX_{L_7}	ΣX_{L_8}	ΣX_{L_9}	$\Sigma X_{L_{10}}$	$\Sigma X_{L_{11}}$	$\Sigma X_{L_{12}}$

these sources of variation are derived. Table 9.5 is a simplified version of Table 9.4 and shows that the sum of squares for subjects within litters within experimental treatments is obtained by taking the sum of squares within each litter in the experiment and adding them all together.

In Table 9.6 we see that the litters within treatments sum of squares is obtained by taking the variation among the three litters within each treatment condition and summing them together.

Finally, in Table 9.7 we see that the sum of squares for treatment effects is obtained by working with the four treatment totals.

We could write a set of formulas based on the layouts in Tables 9.5, 9.6, and 9.7 to obtain the needed sums of squares. However, that is a fairly tedious procedure, and by a slight rearrangement we can simplify our computation

TABLE 9.6 Showing where SS_{litter} is obtained from the design in Table 9.4

Treatment 1			Treatment 2			Treatment 3			Treatment 4		
Lit. 1	Lit. 2	Lit. 3	Lit. 4	Lit. 5	Lit. 6	Lit. 7	Lit. 8	Lit. 9	Lit. 10	Lit. 11	Lit. 12
$X_{1 \cdot 1}$	$X_{2 \cdot 1}$	$X_{3 \cdot 1}$	$X_{4 \cdot 1}$	$X_{5 \cdot 1}$	$X_{6 \cdot 1}$	$X_{7 \cdot 1}$	$X_{8 \cdot 1}$	$X_{9 \cdot 1}$	$X_{10 \cdot 1}$	$X_{11 \cdot 1}$	$X_{12 \cdot 1}$
.	.	.	.	.	.	.	.	.	.	.	.
.	.	.	.	.	.	.	.	.	.	.	.
.	.	.	.	.	.	.	.	.	.	.	.
$X_{1 \cdot n}$	$X_{2 \cdot n}$	$X_{3 \cdot n}$	$X_{4 \cdot n}$	$X_{5 \cdot n}$	$X_{6 \cdot n}$	$X_{7 \cdot n}$	$X_{8 \cdot n}$	$X_{9 \cdot n}$	$X_{10 \cdot n}$	$X_{11 \cdot n}$	$X_{12 \cdot n}$
ΣX_{L_1}	ΣX_{L_2}	ΣX_{L_3}	ΣX_{L_4}	ΣX_{L_5}	ΣX_{L_6}	ΣX_{L_7}	ΣX_{L_8}	ΣX_{L_9}	$\Sigma X_{L_{10}}$	$\Sigma X_{L_{11}}$	$\Sigma X_{L_{12}}$

$SS_{\text{litters within Trt 1}} + SS_{\text{litters within Trt 2}} + SS_{\text{litters within Trt 3}} + SS_{\text{litters within Trt 4}}$

$= SS_{\text{litters within experimental treatments}}$

TABLE 9.7 Showing where SS_{trt} is obtained from the design in Table 9.4

Treatment 1			Treatment 2			Treatment 3			Treatment 4		
Lit. 1	Lit. 2	Lit. 3	Lit. 4	Lit. 5	Lit. 6	Lit. 7	Lit. 8	Lit. 9	Lit. 10	Lit. 11	Lit. 12
$X_{1.1}$	$X_{2.1}$	$X_{3.1}$	$X_{4.1}$	$X_{5.1}$	$X_{6.1}$	$X_{7.1}$	$X_{8.1}$	$X_{9.1}$	$X_{10.1}$	$X_{11.1}$	$X_{12.1}$
.	.	.	.	.	.	.	.	.	.	.	.
.	.	.	.	.	.	.	.	.	.	.	.
.	.	.	.	.	.	.	.	.	.	.	.
$X_{1.n}$	$X_{2.n}$	$X_{3.n}$	$X_{4.n}$	$X_{5.n}$	$X_{6.n}$	$X_{7.n}$	$X_{8.n}$	$X_{9.n}$	$X_{10.n}$	$X_{11.n}$	$X_{12.1}$
ΣX_{L_1}	ΣX_{L_2}	ΣX_{L_3}	ΣX_{L_4}	ΣX_{L_5}	ΣX_{L_6}	ΣX_{L_7}	ΣX_{L_8}	ΣX_{L_9}	$\Sigma X_{L_{10}}$	$\Sigma X_{L_{11}}$	$\Sigma X_{L_{12}}$
	ΣX_{T_1}			ΣX_{T_2}			ΣX_{T_3}			ΣX_{T_4}	

$$SS_{\text{between trt}}$$

considerably. Table 9.8 contains the same information as in Table 9.4, but Table 9.8 is broken into three zones of data. What we do is compute the sums of squares within each of these three zones and by the appropriate subtractions we are able to obtain the sums of squares needed in our analysis of variance. Table 9.9 gives the general rules and formulas needed to calculated the various sums of squares, and Table 9.10 is a general summary of the final analysis of variance table.

To illustrate the computational procedures, we shall now return to our neuroendocrinologist who administered thyroxine to infant rats during their first three days of life. The three males within each litter were left undisturbed in their cages until 65 days of age. At that time the animals were removed from the cage, weighed to the nearest gram, and then given a form-discrimination learning task using food as a reinforcer. We shall first consider the body weight data. Table 9.11 lists the body weights for all animals in the experiment as well as the computations involved in obtaining the necessary sums of squares for the nested design.

In planning his experiment the endocrinologist had decided that he wished to make three specific comparisons among the four treatment groups. He wished to compare the undisturbed control group with the group that received the injection of the vehicle because there was evidence in the literature that the stress of the injection procedure could influence later behavior and body weight. He also wished to compare the .5 microgram thyroxine group with the 1.0 microgram group to see whether the dose level had an effect. Finally, he wished to compare the two groups that received thyroxine with the two groups that had not received thyroxine. Therefore, after computing SS_{trt} in Table 9.11, the experimenter determined the sums of squares for his three planned comparisons.

TABLE 9.8 Showing how a nested design is broken into zones for purposes of computing the necessary sums of squares

Treatment 1			Treatment 2			Treatment 3			Treatment 4			
Lit. 1	Lit. 2	Lit. 3	Lit. 4	Lit. 5	Lit. 6	Lit. 7	Lit. 8	Lit. 9	Lit. 10	Lit. 11	Lit. 12	
$X_{1\cdot1}$	$X_{2\cdot1}$	$X_{3\cdot1}$	$X_{4\cdot1}$	$X_{5\cdot1}$	$X_{6\cdot1}$	$X_{7\cdot1}$	$X_{8\cdot1}$	$X_{9\cdot1}$	$X_{10\cdot1}$	$X_{11\cdot1}$	$X_{12\cdot1}$	The variation among all the raw scores, treated as a single group, $= SS_{total}$
.	.	.	.	.	.	.	.	.	.	.	.	
.	.	.	.	.	.	.	.	.	.	.	.	
$X_{1\cdot n}$	$X_{2\cdot n}$	$X_{3\cdot n}$	$X_{4\cdot n}$	$X_{5\cdot n}$	$X_{6\cdot n}$	$X_{7\cdot n}$	$X_{8\cdot n}$	$X_{9\cdot n}$	$X_{10\cdot n}$	$X_{11\cdot n}$	$X_{12\cdot n}$	
ΣX_{L_1}	ΣX_{L_2}	ΣX_{L_3}	ΣX_{L_4}	ΣX_{L_5}	ΣX_{L_6}	ΣX_{L_7}	ΣX_{L_8}	ΣX_{L_9}	$\Sigma X_{L_{10}}$	$\Sigma X_{L_{11}}$	$\Sigma X_{L_{12}}$	The variation among all the litter totals, treated as a single group, $= SS_{litters}$ within trt $+ SS_{between}$ trt
ΣX_{T_1}			ΣX_{T_2}			ΣX_{T_3}			ΣX_{T_4}			The variation among all the treatment totals $= SS_{between}$ trt

TABLE 9.9 General rules and formulas to obtain SS_{total}, $SS_{between}$, $SS_{litters}$, and $SS_{subjects}$ in a nested design, using Table 9.4 as the data table

The data table
See Table 9.4 for layout of the data table.

Definitions

X = any raw score within the data table

n = number of subjects within a nested unit (the litter). The n's are assumed to be equal so that $n_1 = n_2 = n_3 = \cdots$

k = number of nested units within a treatment group. The k's are assumed to be equal so that $k_1 = k_2 = k_3 = \cdots$

t = number of treatment groups.

$\Sigma X_{L_1}, \Sigma X_{L_2}, \Sigma X_{L_3}, \cdots$ = totals of the scores within the nested unit.

$\Sigma X_{T_1}, \Sigma X_{T_2}, \cdots$ = totals of the scores within Treatments 1, 2, $\cdots$

N = total number of subjects in the experiment, $= n_1 + n_2 + n_3 + \cdots$

TABLE 9.9 (*continued*) General rules and formulas to obtain SS_{total}, $SS_{between}$, $SS_{litters}$, and $SS_{subjects}$ in a nested design, using Table 9.4 as the data table

---

Obtain the following values

ΣX = sum of all the raw scores in the experiment.

ΣX^2 = sum of the square of each raw score in the experiment.

$(\Sigma X)^2 / N$ = correction term = CT.

Rules and formulas

SS_{total} The total sum of squares of all the data is always obtained by subtracting the correction term from the sum of the squares of the raw scores:

$$SS_{total} = \Sigma X^2 - \frac{(\Sigma X)^2}{N} = \Sigma X^2 - CT$$

$SS_{between\ treatments}$ To obtain the SS among the treatments, square each treatment total, add these values together, divide this amount by the number of scores making up a treatment total, and subtract the correction term:

$$SS_{trt} = \frac{(\Sigma X_{T_1})^2 + (\Sigma X_{T_2})^2 + (\Sigma X_{T_3})^2 + \cdots}{kn} - CT$$

For the example in Table 9.4, $k = 3$.

$SS_{litters\ within}$
$_{treatments}$ To obtain the litters-within-treatments SS, square each litter total, add these values together, divide this amount by the number of scores making up a litter total, and subtract the correction term. This value is the total variation among all litters. From that value is subtracted $SS_{between\ treatments}$, as shown below:

$$SS_{litters\ within\ treatments} = \frac{(\Sigma X_{L_1})^2 + (\Sigma X_{L_2})^2 + \cdots}{n}$$

$$- CT - SS_{between\ treatments}$$

An alternative method: The sum of squares can be computed directly following the rules given at the end of Table 8.4.

$SS_{subjects\ within\ litters}$ The subjects SS is usually obtained by subtracting $SS_{litters}$ and $SS_{between\ treatments}$ from SS_{total}:

$$SS_{subjects\ within\ litters} = SS_{total} - SS_{litters\ within\ treatments} - SS_{between\ treatments}$$

An alternative method: The sum of squares can be computed directly following the rules given at the end ot Table 8.4.

TABLE 9.10 General summary table for single-classification analysis of variance with independent observations in which litters are nested within treatments and subjects are nested within litters

Source of variation	Degrees of freedom	Sums of squares	Mean squares	F
Between treatments	$t-1$	$SS_{\text{treatments}}$	$\dfrac{SS_{\text{trt}}}{t-1}=MS_{\text{trt}}$	$\dfrac{MS_{\text{trt}}}{MS_{\text{litters}}}$
Litters within treatments	$t(k-1)$	SS_{litters}	$\dfrac{SS_{\text{litters}}}{t(k-1)}=MS_{\text{litters}}$	$\dfrac{MS_{\text{litters}}}{MS_{\text{subjects}}}$
Subjects within litters within treatments	$kt(n-1)$	SS_{subjects}	$\dfrac{SS_{\text{subjects}}}{kt(n-1)}=MS_{\text{subjects}}$	
Total	$knt-1=N-1$	SS_{total}		

TABLE 9.11 Body weight in grams at 65 days for the animals in Table 9.1, and computations of the sums of squares

Controls			Vehicle			.5 microgram of thyroxine			1.0 microgram of thyroxine		
Lit. 1	Lit. 2	Lit. 3	Lit. 4	Lit. 5	Lit. 6	Lit. 7	Lit. 8	Lit. 9	Lit. 10	Lit. 11	Lit. 12
185	210	220	240	213	195	249	210	183	205	261	239
176	230	222	231	217	188	232	219	206	219	278	260
189	223	235	221	251	201	270	239	201	224	287	289
550	663	677	692	681	584	751	668	590	648	826	788

$$\Sigma X_c = 1{,}890 \qquad \Sigma X_v = 1{,}957 \qquad \Sigma X_{.5} = 2{,}009 \qquad \Sigma X_{1.0} = 2{,}262$$

$$\bar{X}_c = 210.00 \qquad \bar{X}_v = 217.44 \qquad \bar{X}_{.5} = 223.22 \qquad \bar{X}_{1.0} = 251.33$$

$$\Sigma X = 8{,}118$$

$$\Sigma X^2 = 1{,}860{,}002$$

$$N = 36$$

$$CT = \frac{(8{,}118)^2}{36} = 1{,}830{,}609$$

$$SS_{\text{total}} = 1{,}860{,}002 - CT = 29{,}393$$

$$SS_{\text{trt}} = \frac{(1{,}890)^2 + (1{,}957)^2 + (2{,}009)^2 + (2{,}262)^2}{9} - CT = 8{,}799.22$$

$$SS_{\text{litters within trt}} = \frac{(550)^2 + (663)^2 + \cdots + (788)^2}{3} - CT - SS_{\text{trt}} = 15{,}767.75$$

$$SS_{\text{subjects within litters}} = SS_{\text{total}} - SS_{\text{trt}} - SS_{\text{litters within trt}} = 4{,}826.03$$

Because these comparisons were orthogonal he could check his calculations by adding together the sums of squares of the three comparisons and seeing that they were equal to SS_{trt}. In general the planned comparisons will not necessarily be orthogonal, and so this additivity principle may not hold. The final analysis of variance is in Table 9.12.

The endocrinologist first made a preliminary test on his model by dividing $MS_{litters}$ by $MS_{subjects}$. The resulting value of F was 9.80, which is significant beyond the .01 level. Thus, the experimenter must conclude that there are significant litter differences in body weight, and $MS_{litters}$ now becomes the proper denominator to use in evaluating the three planned comparisons. No significant differences were found even though inspection of the treatment means in Table 9.11 suggests that the group that received 1.0 micrograms of thyroxine weighs more than the other groups in the experiment.

Note what would have happened if the experimenter had ignored the fact that litters were nested within treatment conditions and had analyzed his data as though he had nine randomly selected independent subjects within each of the treatment groups. The sums of squares for litters within treatments and subjects within litters would have been pooled, resulting in $MS_{within} = 643.56$. If this is used as the denominator to evaluate the three planned comparisons, the latter two comparisons are now found to be significant. This would have resulted in a serious misinterpretation of the data.

Now consider the learning scores. The animals were required to learn to discriminate between two geometric forms. When they touched the correct form, they received a food reward. Each animal was tested until it made 10 successive correct responses. The number of trials required to reach this criterion of learning was the animal's score. These data and the calculations of the sums of squares are given in Table 9.13. The analysis of variance summary is given in Table 9.14.

TABLE 9.12 Summary table for the data in Table 9.11

Source	df	SS	MS	F
Between treatments	[3]	[8,799.22]		
Control vs. vehicle	1	249.39	249.39	0.13
.5 microgram vs. 1.0 microgram	1	3,556.06	3,556.06	1.80
Thyroxine vs. no thyroxine	1	4,993.78	4,993.78	2.53
Litters within treatments	8	15,767.75	1,970.97	9.80[**]
Subjects within litters	24	4,826.03	201.08	
Total	35	29,393		

[**]$p < .01$.

TABLE 9.13 Learning scores for the animals in Table 9.1, and computation of the sums of squares

Controls			Vehicle			.5 microgram of thyroxine			1.0 microgram of thyroxine		
Lit. 1	Lit. 2	Lit. 3	Lit. 4	Lit. 5	Lit. 6	Lit. 7	Lit. 8	Lit. 9	Lit. 10	Lit. 11	Lit. 12
37	24	38	33	42	30	20	50	36	22	32	30
34	81	42	31	51	66	18	36	37	13	16	52
39	47	54	25	36	32	40	27	39	26	31	40
110	152	134	89	129	128	78	113	112	61	79	122

$$\Sigma X_c = 396 \qquad \Sigma X_v = 346 \qquad \Sigma X_{.5} = 303 \qquad \Sigma X_{1.0} = 262$$
$$\bar{X}_c = 44.00 \qquad \bar{X}_v = 38.44 \qquad \bar{X}_{.5} = 33.67 \qquad \bar{X}_{1.0} = 29.11$$

$$\Sigma X = 1,307 \qquad N = 36$$
$$\Sigma X^2 = 53,941 \qquad CT = \frac{(1,307)^2}{36} = 47,451.36$$

$$SS_{total} = 53,941 - CT = 6,489.64$$

$$SS_{trt} = \frac{(396)^2 + (346)^2 + (303)^2 + (262)^2}{9} - CT = 1,102.53$$

$$SS_{litters\ within\ treatments} = \frac{(110)^2 + (152)^2 + \cdots + (122)^2}{3} - CT - SS_{trt} = 1,562.44$$

$$SS_{subjects\ within\ litters} = SS_{total} - SS_{trt} - SS_{subjects\ within\ treatments} = 3,824.67$$

TABLE 9.14 Summary table for the data in Table 9.13

Source	df	SS	MS	F
Between treatments	[3]	[1,102.53]		
Control vs. vehicle	1	138.89	138.89	.83
.5 microgram vs. 1.0 microgram	1	93.39	93.39	.55
Thyroxine vs. no thyroxine	1	870.25	870.25	5.17*
Within treatments	32	5,387.11	168.35	
Litters within treatments	8	1,562.44	195.30	1.23
Subjects within litters	24	3,824.67	159.36	
Total	35	6,489.64		

$^*p < .05.$

The preliminary test of the model yielded an F of 1.23, which is not significant at the .25 level. Therefore, the sums of squares and degrees of freedom were pooled to give MS_{within} with 32 df. This was used as the error term to evaluate the three planned comparisons. The researcher found that those groups that received thyroxine in infancy learned the discrimination task significantly faster than the groups that did not receive thyroxine.

RANDOMIZED BLOCKS DESIGNS

We now come to the last of the single-classification analysis of variance designs, called randomized blocks. The terminology comes from agriculture, where this design originated. In agriculture a "block" refers to an area of land that is relatively homogeneous. All experimental treatments are assigned randomly within this area of land or block. This procedure is repeated over a number of blocks that are randomly selected. The principle involved here is that placing all treatments on relatively homogeneous land makes it easier to detect differences among treatments within a block as compared to randomly assigning treatments over a very wide acreage where the land may differ in a number of critical variables (e.g., fertility, moisture, contour). It is apparent that this procedure increases efficiency if each block of land is homogeneous and if there are large differences among the blocks. This kind of design has been carried over into behavioral and biological research in a variety of ways.

Examples

As an example, suppose that a group of ethologists were interested in studying the effects of certain experimental treatments upon zoo animals. Each zoo may be thought of as a random block with the ultimate objective of deriving a generalization that could be applied to the population of zoos from which these samples were drawn. If all treatment conditions can be applied randomly to animals within each of the zoos, then this meets the statistical criteria for a randomized blocks design.

Another use is the common situation in which each subject in the experiment is considered to be a block and receives all experimental treatments, usually in a random sequence. Thus a behavioral pharmacologist who was interested in studying the effects of several drugs on activity could decide to use each subject as its own control by randomly administering each drug (including the vehicle) to each subject. There are problems with this kind of design, however. For one thing, a sufficient time must elapse to allow the effects of the drug to disappear before the next drug is administered, thus causing the experiment to be extended over a relatively long interval of time. Second, there is the assumption that the behavior being measured remains the same over time even with repeated testings. Otherwise it is difficult to know whether the change in behavior is due to the experimental intervention or to the effects of repeated testings. There are experimental designs that can separate and evaluate these effects, but they are beyond the scope of this text. Whether it is correct to use each subject as its own control in a randomized blocks design is ultimately an experimental question that can only be answered by your research knowledge of the effects of such a procedure.

A good example of the use of a randomized blocks design in biological and behavioral research is with litters of animals. In the same way that an agriculturist can assign different treatments to the plots within a block of land, a biologist or behaviorist can assign different experimental treatments to animals within a litter. We have already encountered this kind of design in Chapter 7 in the discussion concerning the t test for matched pairs (see page 105). In that example rat littermates were reared in either an impoverished environment or an enriched environment.

General Layout

The general layout for a randomized blocks design with litters representing "blocks" is shown in Table 9.15. In order to make the randomized blocks design as general as possible, we have introduced two new symbols in Table 9.15, the letters R and C, which refer to the number of rows and columns, respectively. Thus R litters of animals are used, and there are C treatments. One animal within each litter is randomly assigned to each of the treatment conditions. Therefore, there are as many animals per litter as there are treatments.

There are three sources of variation present in Table 9.15: an effect due to litters, one due to the experimental treatments, and a residual effect called an interaction. These are summarized in Table 9.16.

Before getting involved in the computational procedures, it is helpful to see where the several sources of variation listed in Table 9.16 come from. These are shown in Table 9.17. The variation among the treatment totals at the bottom of

TABLE 9.15 General layout for a single-classification randomized blocks design where litters represent blocks

	Treatment 1	Treatment 2 $\cdots$	Treatment C	Total
Litter 1	$X_{1 \cdot 1}$	$X_{2 \cdot 1}$	$X_{C \cdot 1}$	ΣX_{L_1}
Litter 2	$X_{1 \cdot 2}$	$X_{2 \cdot 2}$	$X_{C \cdot 2}$	ΣX_{L_2}
.	.	.	.	.
.	.	.	.	.
.	.	.	.	.
Litter R	$X_{1 \cdot R}$	$X_{2 \cdot R}$	$X_{C \cdot R}$	ΣX_{L_R}
Total	ΣX_{T_1}	ΣX_{T_2} $\quad \cdots$	ΣX_{T_C}	

TABLE 9.16   General summary tables for a randomized blocks single-classification analysis of variance where litters represent blocks

Source of variation	Degrees of freedom	Sums of squares	Mean square	F
Between litters	Number of litters -1 $= R-1$	SS_{litters}	$\dfrac{SS_{\text{litters}}}{R-1} = MS_{\text{lit}}$	$-^{a}$
Between treatments	Number of treatments -1 $= C-1$	$SS_{\text{treatments}}$	$\dfrac{SS_{\text{treatments}}}{C-1} = MS_{\text{trt}}$	$\dfrac{MS_{\text{trt}}}{MS_{\text{int}}}$
Interaction	$(R-1)(C-1)$	$SS_{\text{interaction}}$	$\dfrac{SS_{\text{interaction}}}{(R-1)(C-1)} = MS_{\text{int}}$	
Total	$RC-1$	SS_{total}		

[a]A valid F test to assess MS_{litters} is not available in this design because treatments are considered to be fixed.

the table allows us to obtain $SS_{\text{treatments}}$. Similarly, variation among the litter totals to the right of the table permits us to compute SS_{litters}. As usual, the variation among all of the raw scores within the table gives us SS_{total}, and when we subtract from this total the sums of squares due to litters and to treatments we end up with a sum of squares variously called "residual," "error," or "interaction." The preference here is to use the term $SS_{\text{interaction}}$. The general rules and formulas needed to compute the various sums of squares for the randomized blocks design are given in Table 9.18.

TABLE 9.17   Showing where the various sums of squares are obtained in a randomized blocks design

The variation among the raw scores $= SS_{\text{total}}$

	Treatment 1	Treatment 2 $\cdot\ \cdot$	Treatment C	Total	
Litter 1	$X_{1\cdot 1}$	$X_{2\cdot 1}$	$X_{C\cdot 1}$	ΣX_{L_1}	
Litter 2	$X_{1\cdot 2}$	$X_{2\cdot 2}$	$X_{C\cdot 2}$	ΣX_{L_2}	The variation among the litter totals $= SS_{\text{litters}}$
.	.	.	.	.	
.	.	.	.	.	
.	.	.	.	.	
Litter R	$X_{1\cdot R}$	$X_{2\cdot R}$	$X_{C\cdot R}$	ΣX_{L_R}	
Total	ΣX_{T_1}	ΣX_{T_2} $\cdot\ \cdot\ \cdot$	ΣX_{T_C}		

The variation among the treatment totals $= SS_{\text{treatments}}$

$$SS_{\text{total}} - SS_{\text{litters}} - SS_{\text{treatments}} = SS_{\text{interaction}}$$

TABLE 9.18 General rules and formulas to obtain the sums of squares in a randomized blocks design

The data table

	Treatment 1	Treatment 2 $\cdots$	Treatment C	total
Litter 1	$X_{1 \cdot 1}$	$X_{2 \cdot 1}$	$X_{C \cdot 1}$	ΣX_{L_1}
Litter 2	$X_{1 \cdot 2}$	$X_{2 \cdot 2}$	$X_{C \cdot 2}$	ΣX_{L_2}
.				
.				
.				
Litter R	$X_{1 \cdot R}$	$X_{2 \cdot R}$	$X_{C \cdot R}$	ΣX_{L_R}
Total	ΣX_{T_1}	ΣX_{T_2} $\cdots$	ΣX_{T_C}	ΣX

Definitions

X = any raw score within the data table

$\Sigma X_{T_1}, \Sigma X_{T_2}, \cdots, \Sigma X_{T_C}$ = totals of the scores within Treatments 1, 2, $\cdots$, C

$\Sigma X_{L_1}, \Sigma X_{L_2}, \cdots, \Sigma X_{L_R}$ = totals of the scores within Litters 1, 2, $\cdots$, R

N = total number of subjects in the experiment = RC

Obtain the following values

ΣX = sum of all the raw scores in the experiment

ΣX^2 = sum of the square of all raw scores in the experiment

$(\Sigma X)^2 / N$ = correction term = CT

Rules and formulas

SS_{total} — The total sum of squares of all the data is always obtained by subtracting the correction term from the sum of the squares of the raw scores:

$$SS_{\text{total}} = \Sigma X^2 - CT$$

$SS_{\text{between treatments}}$ — To obtain the treatment SS, square each treatment total, add these values together, divide this amount by the number of observations making up a treatment total, and subtract the correction term:

$$SS_{\text{treatment}} = \frac{(\Sigma X_{T_1})^2 + (\Sigma X_{T_2})^2 + \cdots + (\Sigma X_{T_C})^2}{R} - CT$$

$SS_{\text{between litters}}$ — To obtain the litter SS, square each litter total, add these values together, divide this amount by the number of subjects within a litter, and subtract the correction term:

TABLE 9.18 (*continued*) General rules and formulas to obtain the sums of squares in a randomized blocks design

$$SS_{litter} = \frac{(\Sigma X_{L_1})^2 + (\Sigma X_{L_2})^2 + \cdots + (\Sigma X_{L_R})^2}{C} - CT$$

$SS_{interaction}$ To obtain the interaction SS—which is used as the error term in this design—subtract $SS_{treatment}$ and SS_{litter} from the total sum of squares:

$$SS_{interaction} = SS_{total} - SS_{treatment} - SS_{litter}$$

Computational Examples

Our first computational example of this design will be our study from Chapter 7 comparing the brain weights of littermates reared under impoverished or enriched conditions. Table 9.19 contains the same scores as in Table 7.2, but in this instance the data are analyzed by means of the randomized blocks design rather than via the matched t test from Chapter 7. The computations at the bottom of Table 9.19 are done according to the instructions given in Table 9.18. The analysis of variance summary of these data is given in Table 9.20.

Recall that we have shown earlier that $t^2 = F$ when there are only two treatment conditions. The value of t from Table 7.2 is 4.14, and the square of this value is 17.14. This is within rounding error of the F value of 17.11 given in Table 9.20.

For a more detailed example of this use of the randomized blocks design we shall return to our neuroendocrinologist who is interested in the effects of thyroxine in infancy upon the later learning behavior of rats. He had previously used a nested design (see Table 9.13) and had found that the two groups receiving thyroxine in infancy learned more rapidly than the two groups that had not received thyroxine. Because he knew of no report in the literature that thyroxine given in infancy would influence discrimination learning, the researcher wished to replicate his experiment before publishing the results. He was also interested in determining whether there was a dose-response relationship between amount of thyroxine and later performance on the learning task. Even though the comparison of his .5 microgram group with his 1.0 microgram group was not significant (see Table 9.14), inspection of the means in Table 9.13 shows the following:

Dose	Number of trials to learning criterion
Vehicle	38.44
.5 microgram of thyroxine	33.67
1.0 microgram of thyroxine	29.11

TABLE 9.19 Analysis of the data of table 7.2 in a randomized blocks design

Litter	Treatment C	Treatment E	Total
1	620	681	1,301
2	835	862	1,697
3	427	447	874
4	770	765	1,535
5	536	574	1,110
6	617	617	1,234
7	698	740	1,438
8	811	866	1,677
9	473	485	958
10	755	802	1,557
Total	6542	6839	13,381

$\Sigma X = 13,381$

$\Sigma X^2 = 9,341,507$

$CT = 8,925,558.05$

$SS_{total} = \Sigma X^2 - CT = 388,948.95$

$$SS_{treatments} = \frac{(6,542)^2 + (6,839)^2}{10} - CT = 8,956,968.5 - CT = 4,410.45$$

$$SS_{litters} = \frac{(1,301)^2 + \cdots + (1,557)^2}{4} - CT = 9,334,776.5 - CT = 382,218.45$$

$$SS_{interaction} = SS_{total} - SS_{treatments} - SS_{litters} = 2,320.05$$

These data suggest a linear trend, and therefore the endocrinologist decided it was worthwhile to see whether a dose-response relationship could be obtained in the same experiment planned as a replication of his original finding.

Rather than use a nested design, as in his previous experiment, the researcher decided that he could increase the sensitivity of his experiment by administering

TABLE 9.20 Summary of the data analysis of Table 9.19

Source	df	SS	MS	F
Litters	9	382,218.45	42,468.72	–
Treatments	1	4,410.45	4,410.45	17.11**
Interaction	9	2,320.05	257.78	
Total	19	388,948.95		

$**p < .01$.

the different treatments to animals within a litter. Therefore he chose to use a randomized blocks design. He decided to have four treatment conditions: injection of the vehicle, injection of .5 microgram of thyroxine, injection of 1.0 microgram of thyroxine, and injection of 1.5 micrograms of thyroxine. The injections were to be administered on days 1, 2, and 3 as had been done previously. He decided to use 10 litters in this study and to restrict the experiment to male rats. When a litter was born, it was immediately sexed and the litter size reduced to four males. Because it was necessary to identify each animal individually, the researcher injected india ink underneath one of the paws of each pup.

By means of a table of random numbers, the neuroendocrinologist assigned each animal within the litters to one of the four treatment conditions, and the injections were administered once daily during the first three days of life. The animals were group-housed after weaning until 65 days of life, at which time they were given the same form-discrimination learning task as had been used in the prior experiment. Table 9.21 presents the number of trials that each animal took to reach the learning criterion. The computations of the sums of square are given at the bottom of that table.

The endocrinologist had designed his experiment for two purposes: to try to replicate his prior finding that neonatal animals given thyroxine learned a form discrimination faster than controls, and to determine whether there was a dose-response relationship between amount of thyroxine administered in infancy and later learning scores. To make a test of the dose-response hypothesis he decide to extract the linear and quadratic components via orthogonal polynomials. In testing to see whether he could replicate his prior finding, the endocrinologist decided to compare his control group against the mean of the three thyroxine groups. Table 9.22 shows the computations involved in deriving the sums of squares to test these hypotheses.

Several comments should be made about this table. First of all, both a quantitative hypothesis (test of a dose-response relationship) and a qualitative hypothesis (comparing the control group against the thyroxine groups) are being tested. Next, the qualitative hypothesis is not orthogonal to the quantitative hypothesis. However, in this instance experimental considerations take precedence over statistical theory. The researcher has legitimate experimental questions to ask, and the lack of statistical niceties should not deter him from testing both hypotheses even though they are not orthogonal. Also, as the experimenter is restricting himself to three comparisons, each with one degree of freedom, we would not expect the α level to change very much in this instance. The final analysis of variance summary of these data is given in Table 9.23.

The endocrinologist found no evidence of a significant effect that could be attributed to the thyroxine treatment. In trying to determine the reason for his failure to repeat his prior finding, he compared his present data with the results from his previous experiment. The control mean of this study was 36.40 trials to learn as compared to 38.44 trials for the control $\bar{X}$ in the prior experiment.

TABLE 9.21 Learning scores for animals receiving injections of a vehicle or of doses of thyroxine in infancy

Litter number	0	.5 microgram	1.0 microgram	1.5 micrograms	Total
1	37	45	42	39	163
2	37	32	31	30	130
3	33	38	37	42	150
4	49	44	38	44	175
5	31	36	34	30	131
6	47	41	43	44	175
7	40	37	37	39	153
8	25	15	30	25	95
9	31	29	21	24	105
10	34	38	37	38	147
Total	364	355	350	355	1,424

$\Sigma X = 1,424$

$\Sigma X^2 = 52,770$

$N = 40$

$CT = \dfrac{(1,424)^2}{40} = 50,694.40$

$SS_{total} = \Sigma X^2 - CT = 2,075.60$

$SS_{treatments} = \dfrac{(364)^2 + \cdots + (355)^2}{10} - CT = 10.20$

$SS_{litters} = \dfrac{(163)^2 + \cdots + (147)^2}{4} - CT = 1,667.60$

$SS_{interaction} = SS_{total} - SS_{treatments} - SS_{litters} = 397.80$

TABLE 9.22 Breakdown of the treatment totals in Table 9.21 to test one quantitative hypothesis and one qualitative hypothesis

Treatment ΣX	0 364	.5 microgram 355	1.0 microgram 350	1.5 micrograms 355	C	Σc^2	$n\Sigma c^2$	$\dfrac{C^2}{n\Sigma c^2}$
Linear	−3	−1	−1	3	−32	20	200	5.12
Quadratic	1	−1	−1	1	14	4	40	4.90
Control vs. thyroxine	1	$-\frac{1}{3}$	$-\frac{1}{3}$	$-\frac{1}{3}$	10.67	1.33	13.33	8.54

TABLE 9.23 Analysis of variance summary of the data of Tables 9.21 and 9.22

Source of variation	Degrees of freedom	Sums of squares	Mean squares	F
Litters	9	1,667.60	185.29	—
Treatments	[3]	[10.20]		
Linear	1	5.12	5.12	.35
Quadratic	1	4.90	4.90	.33
Control vs. thyroxine	1	8.54	8.54	.58
Interaction	27	397.80	14.73	

Certainly these are comparable. The mean square for litters in this experiment was 185.29 units as compared to 195.30 units in the prior experiment (see Table 9.14). These are also highly comparable and indicate that his method of sampling litters from his colony is consistent from one experiment to another. He has gained precision by using randomized blocks over using the nested design, because $MS_{interaction} = 14.73$ units for the randomized blocks design whereas $MS_{subjects} = 159.36$ units for the nested design (see Table 9.14).

Therefore, in terms of various statistical considerations that he can bring to bear upon his data, the two experiments appear to be highly comparable, and the randomized blocks experiment is more precise than the nested design experiment. There are two possibilities that can account for the discrepant results between the two studies. First of all, the significant finding obtained from the first experiment could have been due to chance alone. That difference was at the .05 level, which means that 5 times out of 100 one would expect to get a significant difference by sampling error alone. The second possibility is that the shift from a nested design to a randomized blocks design changed the structure of the experiment sufficiently to eliminate the effect of the thyroxine treatment. In the nested design all four males within a litter received the same treatment, whereas in the randomized blocks design the four males within a litter received the vehicle or differing doses of thyroxine. This procedural difference may have brought about subtle changes in the behavior of the littermates which influenced their interactions with each other and with their mother, and these changes in social dynamics may have modified their later learning scores. In order to test this hypothesis the neuroendocrinologist would have to repeat his experiment yet another time using a nested design to see whether he could verify the results of his first study.

It is important for the student to realize that there is a complex interplay between statistical designs and experimental procedures. When a particular design is chosen to be used in an experiment, that design specifies a set of experimental operations, and it is necessary for you as an experimenter to decide whether those operations are appropriate ones to be administered to your subjects. These are experimental considerations, not statistical ones, and the final decision as to the appropriate design to use must be made by the experimentalist, though often it is wise to make the decision with the advice of a statistician.

FACTORIAL EXPERIMENTS

Up to now we have considered one-way experimental designs. These are designs in which only one independent variable is manipulated. We shall now look briefly at a design called a factorial experiment in which two or more independent variables (called *factors*) are varied or manipulated at the same time. The treatments involved in a particular factor are called the *levels* of that factor. These levels may be qualitative or quantitative.

In a factorial experiment each level of a factor is combined with every level of all other factors. For example, if an experiment contains two factors, one with 5 levels and the other with 3 levels, there would be a total of $5 \times 3 = 15$ *treatment combinations* in the factional design; or if an experiment had four factors, two at 2 levels, one at 3 levels, and the other at 4 levels, there would be $2 \times 2 \times 3 \times 4 = 48$ treatment combinations. The number of factors in the experiment and the number of levels of each factor define the *dimensions* of the experiment. Thus, the first experiment in the sentence above is a 5×3 (read "five by three") factorial experiment, and the second one is a $2 \times 2 \times 3 \times 4$ factorial design. Many different designs may be constructed for factorial experiments. This discussion will be restricted to the simplest experiment, a two-dimensional design characterized as an $R \times C$ design (for rows and columns), with n independent observations per cell. That design is presented schematically in Table 9.24.

General Layout

The upper half of Table 9.24 shows the usual layout for an experiment. The table of totals at the bottom of Table 9.24 is new. This table contains the sum of the scores for each treatment combination within the table as well as the marginal totals which represent the overall effects of factors R and C. Three questions can be asked of this table:

1. Are there significant effects due to the C variable, regardless of the R variable?
2. Are there significant effects due to the R variable, regardless of the C variable?
3. Are there significant effects due to the unique combination of the R and C variables which cannot be accounted for from the information in 1 and 2 above?

TABLE 9.24 General layout for an $R \times C$ factorial design with n subjects per treatment combination cell

	C_1	C_2	$\cdots$	C_c
R_1	$X_{11\cdot1}$	$X_{21\cdot1}$		$X_{C_1\cdot1}$
	$X_{11\cdot2}$	$X_{21\cdot2}$		$X_{C_1\cdot2}$
	$\cdot$	$\cdot$		$\cdot$
	$X_{11\cdot n}$	$X_{21\cdot n}$		$X_{C_1\cdot n}$
	$\overline{\Sigma X_{C_1 R_1}}$	$\overline{\Sigma X_{C_2 R_1}}$		$\overline{\Sigma X_{C_c R_1}}$
R_2	$X_{12\cdot1}$	$X_{22\cdot1}$		$X_{C_2\cdot1}$
	$X_{12\cdot2}$	$X_{22\cdot2}$		$X_{C_2\cdot2}$
	$\cdot$	$\cdot$		$\cdot$
	$X_{12\cdot n}$	$X_{22\cdot n}$		$X_{C_2\cdot n}$
	$\overline{\Sigma X_{C_1 R_2}}$	$\overline{\Sigma X_{C_2 R_2}}$		$\overline{\Sigma X_{C_c R_2}}$
$\cdot$				
$\cdot$				
$\cdot$				
R_r	$X_{1R\cdot1}$	$X_{2R\cdot1}$		$X_{CR\cdot1}$
	$X_{1R\cdot2}$	$X_{2R\cdot2}$		$X_{CR\cdot2}$
	$\cdot$	$\cdot$		$\cdot$
	$X_{1R\cdot n}$	$X_{2R\cdot n}$		$X_{CR\cdot n}$
	$\overline{\Sigma X_{C_1 R_r}}$	$\overline{\Sigma X_{C_2 R_r}}$		$\overline{\Sigma X_{C_c R_r}}$

Table of totals

	C_1	C_2	$\cdots$	C_c	Total
R_1	$\Sigma X_{C_1 R_1}$	$\Sigma X_{C_2 R_1}$		$\Sigma X_{C_c R_1}$	ΣX_{R_1}
R_2	$\Sigma X_{C_1 R_2}$	$\Sigma X_{C_2 R_2}$		$\Sigma X_{C_c R_2}$	ΣX_{R_2}
$\cdot$					
$\cdot$					
$\cdot$					
R_r	$\Sigma X_{C_1 R_r}$	$\Sigma X_{C_2 R_r}$		$\Sigma X_{C_c R_r}$	ΣX_{R_r}
Total	ΣX_{C_1}	ΣX_{C_2}	$\cdots$	ΣX_{C_c}	ΣX

The first two questions are concerned with the *main effects* of the variables, that is, the evaluation of each variable considered separately. To answer these questions we work with the row and column marginal totals.

The third question concerns the *interaction* between the variables, that is, the effect due to the particular combination of the levels of R and the levels of C (the concept of synergy in biology is somewhat similar to the concept of interaction). This is a question that we have not asked before because it is necessary to vary at least two variables at the same time to determine if an interaction is present. To answer the interaction question we work with the treatment combination totals within the table of totals.

The analysis of variance summary table for an $R \times C$ factorial design is shown in Table 9.25. The first three sources of variance (C, R, and $R \times C$) are the statistical counterparts of the three questions asked above. The table of totals is used to compute the SS for factors R and C, and the $R \times C$ interaction, whereas the SS_{within} is obtained form the original data table seen in the upper half of Table 9.24. Table 9.26 shows how these various sources of variation are derived, and the general rules to obtain the SS for the $R \times C$ factorial design are given in Table 9.27.

Computational Example

To illustrate the use of a factorial design, consider a physiological psychologist who is interested in studying the effects of drugs on the memory consolidation process. He has one drug that he has reason to believe will facilitate consolidation and a second drug that should act to inhibit consolidation. He is also interested in determining whether the drugs have the same effect on fast and slow learners because the rate of learning may affect later retention. He has in his laboratory two genetic strains of rats, one selected for fast maze learning and the other selected for slow maze lerarning, and he decides to have animals from both strains learn a complex maze for food reward. He has,

TABLE 9.25 General summary table for an $R \times C$ factorial design

Source	df	SS	MS	F
Between columns (C)	$C-1$	SS_C	$\dfrac{SS_C}{C-1} = MS_C$	MS_C/MS_W
Between rows (R)	$R-1$	SS_R	$\dfrac{SS_R}{R-1} = MS_R$	MS_R/MS_W
$R \times C$ interaction	$(C-1)(R-1)$	$SS_{R \times C}$	$\dfrac{SS_{R \times C}}{(C-1)(R-1)} = MS_{R \times C}$	$MS_{R \times C}/MS_W$
Within	$CR(n-1)$	SS_{within}	$\dfrac{SS_{within}}{CR(n-1)} = MS_W$	

TABLE 9.26 Showing how the various sums of squares are obtained in an $R \times C$ factorial design

The addition of the SS from within each treatment combination $= SS_{\text{within}}$

	C_1	C_2	$\cdots$	C_c
R_1	$X_{11\cdot 1}$	$X_{21\cdot 1}$		$X_{C1\cdot 1}$
	$X_{11\cdot 2}$	$X_{21\cdot 2}$		$X_{C1\cdot 2}$
	$\cdot$	$\cdot$		$\cdot$
	$X_{11\cdot n}$	$X_{21\cdot n}$		$X_{C1\cdot n}$
	$\Sigma X_{C_1 R_1}$	$\Sigma X_{C_2 R_1}$		$\Sigma X_{C_c R_1}$
R_2	$X_{12\cdot 1}$	$X_{22\cdot 1}$		$X_{C2\cdot 1}$
	$X_{12\cdot 2}$	$X_{22\cdot 2}$		$X_{C2\cdot 2}$
	$\cdot$	$\cdot$		$\cdot$
	$X_{12\cdot n}$	$X_{22\cdot n}$		$X_{C2\cdot n}$
	$\Sigma X_{C_1 R_2}$	$\Sigma X_{C_2 R_2}$		$\Sigma X_{C_c R_2}$
$\cdot$				
R_r	$X_{1R\cdot 1}$	$X_{2R\cdot 1}$		$X_{CR\cdot 1}$
	$X_{1R\cdot 2}$	$X_{2R\cdot 2}$		$X_{CR\cdot 2}$
	$\cdot$	$\cdot$		$\cdot$
	$X_{1R\cdot n}$	$X_{2R\cdot n}$		$X_{CR\cdot n}$
	$\Sigma X_{C_1 R_r}$	$\Sigma X_{C_2 R_r}$		$\Sigma X_{C_c R_r}$

Table of Totals

The variation among the treatment combination totals $= SS_{\text{total treatment variation}}$

	C_1	C_2	$\cdots$	C_c	Total	
R_1	$\Sigma X_{C_1 R_1}$	$\Sigma X_{C_2 R_2}$		$\Sigma X_{C_c R_1}$	ΣX_{R_1}	The variation among the row totals $= SS_R$
R_2	$\Sigma X_{C_1 R_2}$	$\Sigma X_{C_2 R_2}$		$\Sigma X_{C_c R_2}$	ΣX_{R_2}	
$\cdot$						
R_r	$\Sigma X_{C_1 R_r}$	$\Sigma X_{C_2 R_r}$		$\Sigma X_{C_c R_r}$	ΣX_{R_r}	
Total	ΣX_{C_1}	ΣX_{C_2}	$\cdots$	ΣX_{C_c}	ΣX	

The variation among the column totals $= SS_C$

$$SS_{\text{total treatment variation}} - SS_R - SS_C = SS_{R \times C \text{ interaction}}$$

TABLE 9.27 General rules and formulas to obtain the sums of squares in an
$R \times C$ factorial design

The data table

See Table 9.24 for layout of data table.

Definitions

X = any raw score within the data table

$\Sigma X_{C_1 R_1}, \Sigma X_{C_2 R_2}, \cdots, \Sigma X_{C_c R_r}$ = totals of the scores for each $R \times C$ treatment combination

$\Sigma X_{C_1}, \Sigma X_{C_2}, \cdots, \Sigma X_{C_c}$ = totals of all the scores within treatments $C_1, C_2, \cdots, C_c$

$\Sigma X_{R_1}, \Sigma X_{R_2}, \cdots, \Sigma X_{R_r}$ = totals of all the scores within treatments $R_1, R_2, \cdots, R_r$

n = number of subjects within any $R \times C$ treatment combination cell. n is assumed to be the same for all cells in the factorial design

N = total number of subjects in the experiment = nRC

Obtain the following values

ΣX = sum of all the raw scores in the experiment

ΣX^2 = sum of the square of all the raw scores in the experiment

$(\Sigma X)^2 / N$ = correction term = CT

Rules and formulas

SS_{total} The total sum of squares of all the data is always obtained by subtracting the correction term from the sum of the squares of the raw scores:

$$SS_{total} = \Sigma X^2 - CT$$

$SS_{columns}$ To obtain the SS between columns, square each column total, add these values together, divide this amount by the number of observations making up a column total, and subtract the correction term:

$$SS_{col} = \frac{(\Sigma X_{C_1})^2 + (\Sigma X_{C_2})^2 + \cdots + (\Sigma X_{C_c})^2}{nR} - CT$$

SS_{rows} To obtain the rows SS, square each row total, add these values together, divide this amount by the number of observations making up a row total, and subtract the correction term:

$$SS_{rows} = \frac{(\Sigma X_{R_1})^2 + (\Sigma X_{R_2})^2 + \cdots + (\Sigma X_{R_r})^2}{nC} - CT$$

$SS_{R \times C}$ To obtain the SS for the interaction of the rows and columns, square each $R \times C$ treatment combination sum, add these values together, divide this amount by the number of observations making up a treatment combination total, and subtract the correction term. This value is the total variation among all treatment combinations. From this value subtract SS_{col} and SS_{rows} to get the remainder, which is the interaction:

TABLE 9.27 (*continued*) General rules and formulas to obtain the sums of squares in an $R \times C$ factorial design

$$SS_{R \times C} = \frac{(\Sigma X_{C_1 R_1})^2 + (\Sigma X_{C_2 R_2})^2 + \cdots + (\Sigma X_{C_c R_r})^2}{n} - CT$$
$$- SS_{\text{col}} - SS_{\text{rows}}$$

SS_{within}	The SS of subjects within treatment combinations is usually obtained by subtracting SS_{col}, SS_{rows}, and $SS_{R \times C}$ from SS_{total}:

$$SS_{\text{within}} = SS_{\text{total}} - SS_{\text{col}} - SS_{\text{rows}} - SS_{R \times C}$$

An alternate method: The sum of squares can be computed directly following the rules given at the end of Table 8.4.

therefore, a 3 × 2 factorial design with 3 levels of drug (facilitory, inhibitory, and saline) and 2 levels of genetics (fast learners and slow learners). He decides to test 3 animals in each cell of the design, thus using 18 animals in the experiment.

Each animal is tested individually in the maze until it reaches a criterion of 8 out of 10 errorless trials. The psychologist then randomly selects one of the three treatment combinations (saline, facilitory drug, inhibitory drug), injects the animal and puts it back in its home cage. This procedure continues until the researcher has obtained the 18 animals needed to fill in his experimental design. Twenty-four hours after original learning, each animal is retested until it achieves the same criterion of 8 out of 10 errorless trials. The criterion score is the number of trials needed to relearn the maze.

The experimental design and the relearning scores are given in Table 9.28. The calculations of the SS's, given in Table 9.29, are based on the general instructions given in Table 9.27. The final analysis of variance summary is given in Table 9.30.

The Treatment effect and the Treatment × Strains interaction were both significant beyond the .01 level, but the main effect for Strains was not significant. However, before drawing any conclusions about these data, it is necessary to examine the pattern of mean scores *because the interaction was significant.* A significant interaction immediately tells you that the patterning of means for one factor is *not* the same as the patterning for the other factor. That is, the profiles of the curves for the factors are *not* parallel. This is seen by inspecting the table of totals in Table 9.28 or the two curves in Figure 9.1, which represent the same data as are contained in Table 9.28. From Figure 9.1 we see that the drugs had essentially no effect on relearning for those rats who were slow original learners, whereas the fast learners were affected—the facilitory drug enhanced relearning and the inhibitory drug impaired relearning, relative to the saline control.

TABLE 9.28   An $R \times C$ factorial design showing number of trials to relearn a maze as a function of genetic strain and drug condition

Genetic strain	Saline injection	Facilitory drug	Inhibitory drug
Fast learners	8	5	14
	9	3	17
	7	1	18
	24	9	49
Slow learners	12	10	8
	9	11	10
	10	10	7
	31	31	25

Table of totals

	Saline	Facilitory	Inhibitory	Total
Fast	24	9	49	82
Slow	31	31	25	87
Total	55	40	74	169

TABLE 9.29   Computation of the sums of squares for the data in Table 9.28

$\Sigma X = 169$

$\Sigma X^2 = 1{,}897$

$N = 18$

$$CT = \frac{(\Sigma X)^2}{N} = 1{,}586.72$$

$$SS_{total} = \Sigma X^2 - CT = 310.28$$

$$SS_{treatments} = \frac{(55)^2 + (40)^2 + (74)^2}{6} - CT = 1{,}683.50 - CT = 96.78$$

$$SS_{strains} = \frac{(82)^2 + (87)^2}{9} - CT = 1{,}588.11 - CT = 1.39$$

$$SS_{trt \times strains} = \frac{(24)^2 + (9)^2 + \cdots + (25)^2}{3} - CT - SS_{trt} - SS_{levels} = 1{,}868.33 - CT$$
$$- SS_{trt} - SS_{strains} = 183.44$$

$$SS_{within} = SS_{total} - SS_{trt} - SS_{strains} - SS_{trt \times strains} = 28.67$$

TABLE 9.30 Summary table for the data in Table 9.28

Source of variation	df	SS	MS	F
Strains (R)	1	1.39	1.39	.58
Treatments (C)	2	96.78	48.39	20.25**
Strains × treatments (R × C)	2	183.44	91.72	38.38**
Within	12	28.67	2.39	
Total	17	310.28		

**p < .01.

Thus, even though there was a significant main effect for Drug Treatments, it would have been incorrect to draw any general conclusion about the drug factor because of the significant interaction. Similarly, the failure to find a significant main effect for Strain does not mean that this variable was without effect, again because of the interaction.

Some General Comments on Factorial Designs

One of the major uses of factorial experiments is that of assessing interactions between two or more factors. Often these interactions are the most important

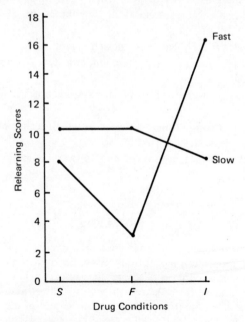

FIGURE 9.1 Showing the profile of scores for the data in Table 9.28.

part of the experiment. If interactions are significant, this limits the breadth of your generality, but, at the same time, you also find out the limits within which you are able to make generalizations. If interactions are not significant, and you find significant main effects, then the breadth of your generalization is much greater because you now know that your phenomenon is true across the levels of the other factors in your experiment.

This concludes our introduction to the analysis of variance and experimental design. It should now be apparent that the choice of an experimental design is not something done casually when preparing to do an experiment. Although the designs covered in these last two chapters will be sufficient for many of your research needs, they will not suffice for more elaborate experiments. For further discussions of the analysis of variance and experimental design the reader is referred to the texts by Hays (1963) and Winer (1971), who have written their books primarily for behaviorists, and to Cochran and Cox (1957) and Snedecor and Cochran (1967), who write primarily to the biologist.

SUMMARY OF DEFINITIONS

Confounding   When two or more variables are combined in such a manner that it is not possible to evaluate their separate effects.

Factor   An independent variable in an experimental design.

Factorial experiment   An experimental design in which two or more independent variables are systematically combined in all possible treatment combinations.

Hierarchical design   The same as a nested design.

Interaction   The unique effect of combining two or more variables at the same time.

Main effect   The overall effect of an experimental variable when evaluated separately.

Nested design   An experimental design in which certain effects are restricted or nested within the treatment conditions.

Nesting   When a particular effect is restricted only to one treatment condition.

PROBLEMS

1. A social psychologist was interested in the effects of instruction set or expectation upon one's perceptions. His experimental material consisted of a movie film which contained scenes of aggression and scenes of care and concern for others. He prepared three sets of instructions for viewing the film. One set was a neutral description of the plot of the movie; a second set used the same basic text as the first, but several changes were made to emphasize the aggressive portion of the film; the third set was changed to give

greater emphasis to the caring portion of the film. The film was shown during the class hour and so it was necessary to use the whole class as the statistical unit in a nested design. Twelve classes were included in the study, with four randomly assigned to each instruction set. When the movie was over, the students were asked to write a brief summary of the movie. The summaries of 10 students from each class were chosen at random and scored by two judges who determined the number of hostile or aggressive responses made by each student. The data summary is shown below.

	ΣX	ΣX^2
Neutral set		
Class 1	81	8,800
Class 2	65	6,372
Class 3	92	10,714
Class 4	86	9,607
Aggressive set		
Class 5	103	12,994
Class 6	95	13,408
Class 7	87	9,717
Class 8	112	14,766
Caring set		
Class 9	74	7,801
Class 10	81	8,709
Class 11	85	9,470
Class 12	58	5,610

(a) Do the analysis of variance of these data with classes nested within treatments and subjects nested within classes.

(b) Carry out a preliminary test on the model.

(c) Make the proper F test to evaluate the Treatment effect.

(d) Use Tukey's honestly significant difference procedure to compare the differences among the means.

(e) Assume that the experimenter had decided prior to collecting his data that he was going to compare his neutral (i.e., control) condition against each of the other two instruction sets. Use Dunnett's procedure to do this.

(f) Why do procedures (d) and (e) give different results?

2. To evaluate the effects of a drug, a researcher took two male mice from each of five litters, injected them with the drug, and recorded their activity in the open field. Controls consisted of two males from five litters which received an injection of saline. The following data were obtained:

Saline injection

Litter 1	83, 56
Litter 2	27, 64
Litter 3	92, 101
Litter 4	75, 46
Litter 5	35, 52

Drug injection

Litter 6	53, 27
Litter 7	45, 28
Litter 8	26, 76
Litter 9	24, 55
Litter 10	31, 33

Analyze these data including making a preliminary test on the model.

3. A researcher was interested in whether a particular task exhibited a learning function over repeated trials. He took five subjects and gave them each five trials on the task. His score was the number of correct units of work per trial. The data are as follows:

	Trial 1	Trial 2	Trial 3	Trial 4	Trial 5
Subject 1	83	100	109	111	117
Subject 2	72	92	98	109	106
Subject 3	67	88	77	102	92
Subject 4	64	76	69	81	84
Subject 5	61	55	60	60	73

(a) Analyze these data in a randomized blocks design.
(b) Take the Trials sum of squares and do a trend analysis, extracting the linear, quadratic, cubic, and quartic functions. Test each for significance.

4. A neuroendrocrinologist was interested in determining the breadth of the finding that castration of males in infancy would result in feminine behavior patterns in adulthood. He used mice from 10 strains and, at birth, two males were castrated while two other males received a sham operation. In adulthood all animals were tested in the open field, because this measure was known to discriminate between the sexes. The researcher obtained the following data:

Strain	Shams	Castrates
A	48	61
	57	75
B	34	48
	39	40
C	66	74
	74	80

Strain	Shams	Castrates
D	30	26
	26	45
E	72	65
	64	54
F	36	44
	58	89
G	51	55
	47	46
H	20	20
	29	38
I	51	64
	39	81
J	66	71
	55	64

Analyze this factorial design.

5. A research psychiatrist wishes to evaluate three therapeutic methods to see how these affect the behavior of individuals in four psychiatric categories. She selects 30 patients from each psychiatric category and randomly assigns 10 to each of the three therapeutic methods. At the end of the 4-week therapy program she administers a series of tests to the patients and also obtains data concerning their behavior on the wards. These scores are combined to yield one value to reflect "degree of adjustment." The higher score reflects better adjustment. The summary data are as follows:

Psychiatric category	Therapeutic methods		
	Method A	Method B	Method C
1	$\Sigma X = 502$	$\Sigma X = 745$	$\Sigma X = 707$
	$\Sigma X^2 = 35,000$	$\Sigma X^2 = 68,144$	$\Sigma X^2 = 64,306$
2	$\Sigma X = 450$	$\Sigma X = 390$	$\Sigma X = 816$
	$\Sigma X^2 = 30,125$	$\Sigma X^2 = 22,627$	$\Sigma X^2 = 77,884$
3	$\Sigma X = 391$	$\Sigma X = 511$	$\Sigma X = 535$
	$\Sigma X^2 = 24,157$	$\Sigma X^2 = 36,106$	$\Sigma X^2 = 34,934$
4	$\Sigma X = 223$	$\Sigma X = 751$	$\Sigma X = 430$
	$\Sigma X^2 = 11,963$	$\Sigma X^2 = 69,404$	$\Sigma X^2 = 26,701$

($n = 10$ per cell)

Analyze this factorial design and interpret the findings.

10 Determining the Linear
Relationship Between
Variables: The
Correlation Coefficient

Up to this point we have been using the experimental model as the basis for our discussion concerning the application of statistics. In this model the researcher imposes some external manipulation on his subjects and then obtains measurements of his endpoint or dependent variable. These data are then evaluated by means of the *t* test or the analysis of variance. The purpose of the experimental procedure is to see whether the independent variable affected the means of the experimental groups. To determine this, the variability within groups is used as the reference point (i.e., error variance) against which to assess the variability among the treatment means. When using the experimental model, the within-group variability is treated as a source of error, and one of our prime objectives in designing experiments is to minimize this variability and thus improve the power of our test of significance.

Thus, in the *experimental model* we *do something* to two or more groups of subjects and are interested in the means of those groups. A second model that is useful to us as experimental researchers is the *correlational model,* in which we *measure what is present* in a single group of subjects. At least two, and often more than two, variables are measured on each individual in the group. With the correlational model we do not view the within-group variability only as a source of error variance. Instead, we are interested in determining what proportion of the within-group variance can be systematically ascribed to the relationships among the two or more variables we have measured.

EXAMPLES OF CORRELATIONAL PROBLEMS
FOR EXPERIMENTALISTS

Following are some examples of typical research questions that an experimentalist may have that would be answered by correlational procedures.

Examples

1. A researcher who uses the open field as a dependent variable in many of his studies decides to automate the apparatus by means of photocells. He decides to build several units so that his technician can test several animals

at one time rather than a single animal. Therefore he takes one of his open fields and has an engineer add photocells and the appropriate electronic hardware to it so that a count of 1 is registered each time the animal breaks a photocell beam. The objective of the automation is to have an electronic system that *measures the same thing* as the technician measures when he tests animals in the open field and records the number of squares they enter. To be certain that the electronic count obtained is equivalent to the technician's count, the researcher would have the technician draw a random sample of animals from his colony, test each animal in the usual manner, and, at the same time, obtain the electronic score via the automated apparatus. A correlation coefficient (which will be defined and discussed below) would be computed between the scores obtained by the technician and the counts recorded by the electronic apparatus. If the apparatus is properly designed, there should be a very high degree of relationship between these two sets of scores.

2. A new blood assay is described in the literature which is a simpler and less expensive procedure than the assay currently being used by a research biologist. He would be willing to shift over to the new assay if the two techniques gave comparable results. To determine this he collects blood from 50 animals, then divides each sample in half and places each in a vial with an arbitrary code number. He gives his chemical technician one set of 50 vials and has him do the standard assay used in that laboratory. After this assay is completed, he then gives the technician the other 50 vials and has him assay the blood using the new procedures reported in the literature. If the two techniques are measuring the same biological material, there should be a high degree of relationship between the two sets of scores.

3. An ethologist is interested in observing social interactions among a group of monkeys maintained in a compound. To do this she has devised a check list of behaviors and has worked out a time sampling procedure for systematically observing each animal in the compound. To determine the reliability of this scoring system she has two observers watch the animals at the same time and independently score them. It is necessary to have a high degree of relationship between the scores of the two observers in order for this measurement system to be employed as a useful research tool.

4. A psychobiologist is interested in determining whether a relationship exists between aggression in mice and amount of testosterone present in the blood. The animals are given 10 days of aggression testing, and a single aggression index is obtained for each mouse. After this the animals are killed and their blood assayed for testosterone. The correlation is then obtained between the aggression scores and the plasma testosterone values.

5. In carrying out research on growth a biologist has been X-raying her animals and then measuring the length of a certain bone. She wonders whether a simpler measure, such as body weight or body length, can be substituted for the bone measurement. Because she has both body weight and length measurements for a number of animals, she correlates these variables separately against her measure of bone length.

6. A researcher is interested in determining whether there is significant variability among the litters in his colony. If so, then an experimental

design that uses littermate controls will be more efficient than a design that ignores this variable. To determine whether there is significant litter variability, he takes two adult males from each of 25 litters and tests them in the open field, which is his primary behavioral measure. Immediately after open-field testing a sample of blood is drawn and assayed for corticosterone, another endpoint used extensively by the researcher. The researcher then determines the correlation between the two open-field scores for each littermate pair and between the two corticosterone scores for each littermate pair. He then determines the correlation between the open-field activity scores and the corticosterone values.

A MEASURE OF LINEAR RELATIONSHIP: THE CORRELATION COEFFICIENT

In all the examples given above, two or more measures are obtained per animal for a single group of subjects. We are interested in determining the relationship between these variables. The statistical questions in which we are interested are the following:

1. Is there a significant linear relationship between the variables we have measured?
2. How strong is the relationship between these variables?

The statistic we shall use to answer these questions is known as the *Pearson product-moment correlation coefficient,* commonly called the correlation coefficient. Before discussing this coefficient in detail, there are several features about it that should be noted. First, this coefficient will measure only the *linear* relationship between any two variables. If there is a significant nonlinear correlation (e.g., a quadratic curve), this coefficient will not be sensitive to that relationship. Next, the coefficient is a pure number without any units. This is necessary because very rarely will the two variables we measure be in the same units, and therefore it is important to find a statistic that is unit-free. Finally, the range of the correlation coefficient is from -1 (which means a perfect negative relationship), through 0 (which means no relationship), to $+1$ (which refers to a perfect positive relationship).

There are several ways to present the principles underlying the correlation coefficient. We shall use both algebraic and geometric representations. Our first approach is to use the concept of variances with which you are familiar and to introduce a new concept, called covariance. In developing these principles we shall assume that we have two variables, which we will call X and Y, and that each subject in the study is measured on both X and Y.

MEASURING THE COVARIANCE OF X AND Y

The formula for the covariance of X and Y parallels the formula for the variances of X and Y. Recall that the variance of the X distribution is

$$s_x^2 = \frac{\Sigma x^2}{n-1} = \frac{\Sigma xx}{n-1}$$

The variance of the Y distribution is defined as follows:

$$s_y^2 = \frac{\Sigma y^2}{n-1} = \frac{\Sigma yy}{n-1}$$

We now define covariance as

$$\text{cov. } XY = \frac{\Sigma xy}{n-1} \qquad (10.1)$$

Formula 10.1 is the definition of the covariance. The equivalent computational formula is

$$\text{cov. } XY = \frac{n\,\Sigma XY - (\Sigma X)(\Sigma Y)}{n(n-1)} \qquad (10.2)$$

We now define our Pearson product-moment correlation coefficient, designated by the letter r, as follows:

$$r = \frac{\text{cov. } XY}{s_x s_y} \qquad (10.3)$$

That is, the correlation coefficient is the ratio of covariance between X and Y to the product of the standard deviations of X and Y. It can be shown through appropriate algebra that the maximum value that cov. XY can attain is $s_x s_y$, so that the maximum value of r is 1. If there is no linear relationship between X and Y, then the covariance is 0 and r is also 0.

It is not necessary to obtain the covariance of X and Y directly in order to compute the correlation coefficient. The most convenient computational formula is

$$r = \frac{n\Sigma XY - (\Sigma X)(\Sigma Y)}{\sqrt{[n\Sigma X^2 - (\Sigma X)^2]\,[n\Sigma Y^2 - (\Sigma Y)^2]}} \qquad (10.4)$$

SCATTERPLOTS OF CORRELATION COEFFICIENTS

The correlation coefficient indicates the degree of linear relationship between any two variables. One way of portraying this is by a geometrical representation called a *scatterplot* of the two variables on an X, Y coordinate system. Figures 10.1 through 10.5 present scatterplots of relationships between X and Y ranging from +1 to −1. In all the figures, the variance of X and the variance of Y are

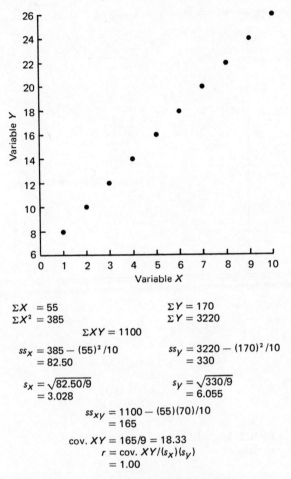

$\Sigma X = 55$ $\Sigma Y = 170$
$\Sigma X^2 = 385$ $\Sigma Y = 3220$

$\Sigma XY = 1100$

$ss_x = 385 - (55)^2/10$ $ss_y = 3220 - (170)^2/10$
$\quad = 82.50$ $\quad = 330$

$s_x = \sqrt{82.50/9}$ $s_y = \sqrt{330/9}$
$\quad = 3.028$ $\quad = 6.055$

$ss_{xy} = 1100 - (55)(70)/10$
$\quad = 165$

$\text{cov. } XY = 165/9 = 18.33$
$r = \text{cov. } XY/(s_x)(s_y)$
$\quad = 1.00$

FIGURE 10.1   Example of a perfect positive correlation.

kept the same; the only thing that changes is the covariance of X and Y. Thus, Figures 10.1 and 10.2 show perfect positive and negative relationships between X and Y. (Note that the covariance can be negative as well as positive, whereas variances must always be positive.) Figures 10.3 and 10.4 show reasonably high positive and negative correlations, respectively, and Figure 10.5 shows the scatterplot pattern when there is essentially no relationship between X and Y.

Inspection of Figures 10.1 through 10.5 should give you a feeling for the concept of covariance. This term represents the amount of covariation present in X and Y. When X and Y are highly related, then extreme values of X are associated with extreme values of Y (see Figures 10.3 and 10.4), thus resulting in a high numerical value for the covariance. When there is no relation between X

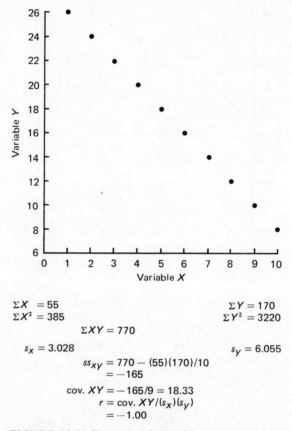

$\Sigma X = 55$

$\Sigma X^2 = 385$

$\Sigma Y = 170$

$\Sigma Y^2 = 3220$

$\Sigma XY = 770$

$s_x = 3.028$

$s_y = 6.055$

$$ss_{xy} = 770 - (55)(170)/10$$
$$= -165$$

$$\text{cov. } XY = -165/9 = 18.33$$
$$r = \text{cov. } XY/(s_x)(s_y)$$
$$= -1.00$$

FIGURE 10.2   Example of a perfect negative correlation.

and Y (Figure 10.5), a high value in one variable is not systematically associated with a high value in the other variable, and the numerical value of the covariance will tend toward zero.

ANOTHER DEFINITION OF THE CORRELATION COEFFICIENT

The scatterplot approach to portraying the correlation coefficient allows us to define the correlation coefficient by using a different geometrical concept. We can define r as *the slope of the line of best fit when the X and the Y distributions have been standardized.* In Figures 10.1 through 10.5, s_y (= 6.055) is twice as large as s_x (= 3.028), but the scale of units for variable Y in the scatterplot is twice as large as the scale for variable X. Thus in all five scatterplots the two standard deviations have been equated. We can let these standard deviations equal 1. Then the slope of the line of best fit is equal to the

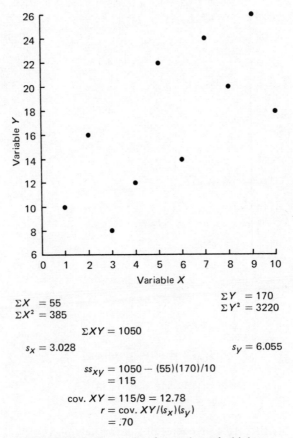

$$\Sigma X = 55$$
$$\Sigma X^2 = 385$$

$$\Sigma Y = 170$$
$$\Sigma Y^2 = 3220$$

$$\Sigma XY = 1050$$

$$s_X = 3.028$$

$$s_Y = 6.055$$

$$ss_{XY} = 1050 - (55)(170)/10$$
$$= 115$$

$$\text{cov. } XY = 115/9 = 12.78$$
$$r = \text{cov. } XY/(s_X)(s_Y)$$
$$= .70$$

FIGURE 10.3 Example of a moderately high positive correlation.

correlation coefficient. These five slopes are shown in Figure 10.6, where the X and Y axes are now scaled in terms of standard score units rather than raw scores units.

Figures 10.6(a) and 10.6(b), which show the best-fitting straight lines for the data in Figures 10.1 and 10.2, reveal that for each standard score increase of 1 in X, there is a corresponding change of 1 standard score unit in Y. Figures 10.6(c) and 10.6(d) show the best-fitting straight lines for the scatterplots of Figures 10.3 and 10.4. In Figure 10.6(c) an increase of 1 standard score unit in X brings about an increase of .7 standard score units in Y, and in Figure 10.6(d) an increase of 1 standard score unit in X corresponds to a decrease of .7 standard score units in Y. Figure 10.6(e), which plots the best-fitting straight line for the correlation of .03 from Figure 10.5, shows that a change of 1 standard score unit in X brings about a change of only .03 standard score units in Y.

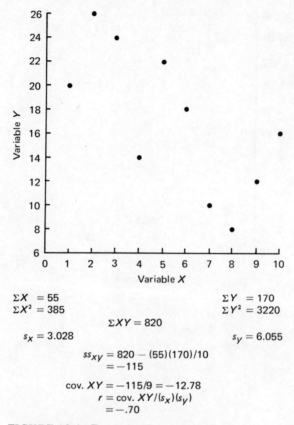

$\Sigma X = 55$

$\Sigma X^2 = 385$

$\Sigma Y = 170$

$\Sigma Y^2 = 3220$

$\Sigma XY = 820$

$s_x = 3.028$

$s_y = 6.055$

$$ss_{xy} = 820 - (55)(170)/10$$
$$= -115$$

$$\text{cov. } XY = -115/9 = -12.78$$
$$r = \text{cov. } XY/(s_x)(s_y)$$
$$= -.70$$

FIGURE 10.4   Example of a moderately high negative correlation.

THE LINE OF BEST FIT AND THE PRINCIPLE OF LEAST SQUARES

We have defined the correlation coefficient as the slope of the line of best fit when X and Y are in standard score units, but we have not specified what we mean by "best fit." Our criterion is the principle of least squares, which we discussed in Chapter 3. There we showed that taking deviations around the mean of a set of scores resulted in a minimum value for the sum of squares of the deviations (Formula 3.4). If we take deviations around any value other than the mean we shall get a numerically higher value for the sum. Thus the principle of least squares offers a unique solution in that there is only one number that can satisfy this criterion of achieving a minimal value. The application of this principle to the problem of correlation is to require that a straight line be found such that the sum of the squares of the deviations around that line is a minimal

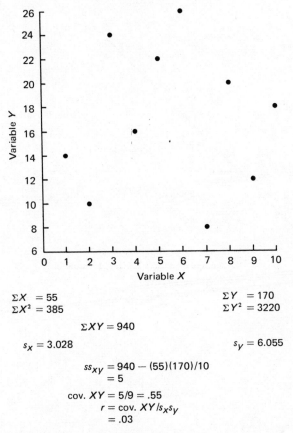

$\Sigma X = 55$ $\Sigma Y = 170$
$\Sigma X^2 = 385$ $\Sigma Y^2 = 3220$

$\Sigma XY = 940$

$s_x = 3.028$ $s_y = 6.055$

$ss_{xy} = 940 - (55)(170)/10$
$= 5$

cov. $XY = 5/9 = .55$
$r = $ cov. $XY/s_x s_y$
$= .03$

FIGURE 10.5 Example of a zero correlation.

value. This is illustrated in Figure 10.7, which plots the data for the correlation of .70 for Figure 10.6(c). Deviations from the line of best fit in Figure 10.7 are symbolized by the term $y \cdot x$, which signifies the distance of y from the point on the line of best fit that would be expected or predicted from our knowledge of x, which is y's paired score. If these values are squared and summed ($\Sigma y \cdot x^2$), the numerical value will be less than if any other straight line is used. It follows from this principle that the best-fitting straight line must always go through the intersection of the two means, $\bar{X}$ and $\bar{Y}$.

Because r has been defined as the slope of the best-fitting straight line when the X and Y distributions have been transformed to z-score (i.e., standard score) units, let us examine r from the perspective of the straight-line equation. The general equation is

$$Y' = a + bX$$

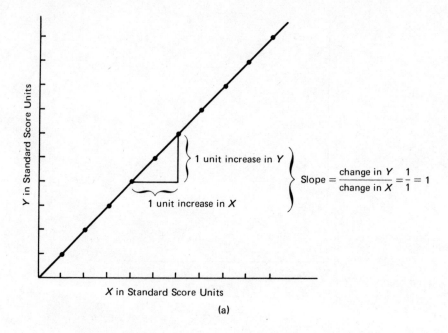

(a)

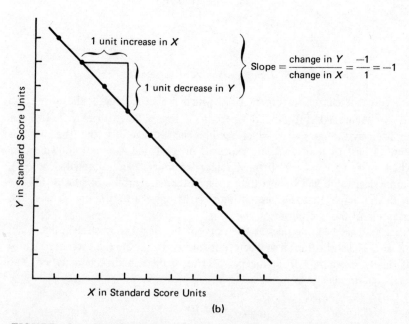

(b)

FIGURE 10.6   Best-fitting straight lines for Figures 10.1 through 10.5.

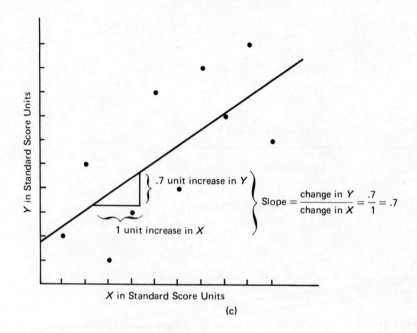

(c)

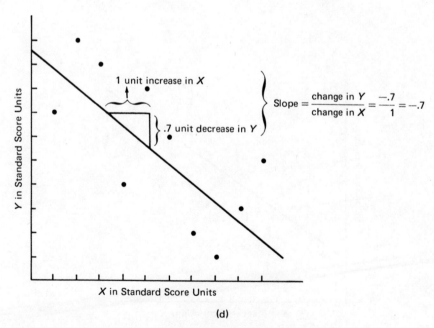

(d)

FIGURE 10.6 (*continued*) Best-fitting straight lines for Figures 10.1 through 10.5.

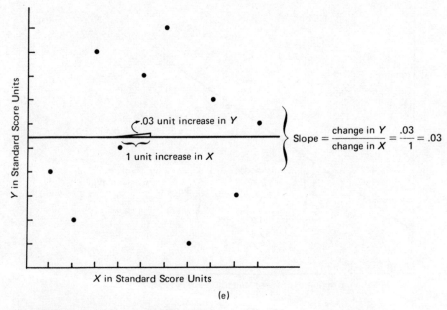

(e)

FIGURE 10.6 (*continued*) Best-fitting straight lines for Figures 10.1 through 10.5.

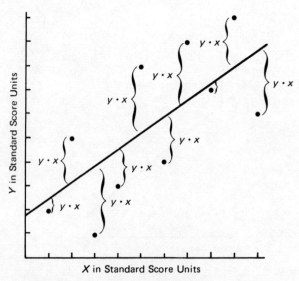

FIGURE 10.7 Showing deviations of each data point ($y \cdot x$) from the line of best fit.

where Y' = our expected or predicted score on the Y variable given our knowledge of X

a = intercept or the value of Y' when $X = 0$

b = slope

In z-score form the equation becomes

$$z'_y = a + rz_x$$

with r replacing b because both distributions have been standardized. In the equation above, if we let $z_x = 0$, then $z'_y = a$. But 0 is the mean of a z-score distribution, and we have already seen that our best-fitting line must go through the means of both distributions. Thus z'_y must equal 0 when $z_x = 0$. Therefore, $a = 0$, and the equation becomes

$$z'_y = rz_x \qquad (10.5)$$

In words, our best prediction about a subject's standard score on variable Y is obtained by multiplying his standard score on X by the correlation coefficient.

Because, in a correlation problem, the choice of which variable to call X and which Y is arbitrary, we have a symmetrical arrangement. Therefore, if we wish to predict X scores from a knowledge of Y, our equation becomes

$$z'_x = rz_y \qquad (10.6)$$

THE STANDARD ERROR OF ESTIMATE

The information in Figure 10.7 can be used to develop another concept about correlations. Suppose that there is no correlation between X and Y (i.e., $r_{xy} = 0$). Then given information about a value, X, the best prediction to make is the mean of the Y distribution, $\bar{Y}$. That is, the best-fitting line will have a slope of 0, which makes it parallel to the X axis. The height of that line (i.e., the a intercept) will be the mean of the Y distribution because, by the least-squares principle, the mean is the value to choose to minimize errors in prediction. This is shown in Figure 10.8, which uses the data of Figure 10.5. Deviations of each data point from the line of best fit are indicated by y, defined as $Y - \bar{Y}$. We know that if we square and sum those deviations, divide by $n - 1$, and take the square root of this number, we obtain the standard deviation of the Y distribution:

$$s_y = \sqrt{\frac{\Sigma y^2}{n - 1}}$$

This standard deviation is the maximum error possible because it represents the variability present when there is no relationship between X and Y. What

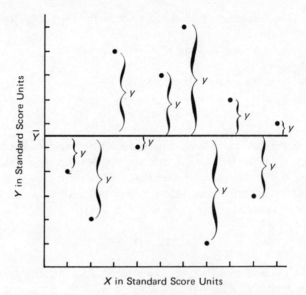

X in Standard Score Units

FIGURE 10.8   Showing deviations of each data point (y) from the mean of the Y distribution (Y) for a scatterplot with a correlation of zero.

happens to this standard deviation when the correlation is some value other than 0? This is shown in Figure 10.9, which represents the same data as in Figure 10.7 together with the line of best fit and the line based on no correlation ($\bar{Y}$). Each data point is shown in terms of two deviations: (1) from the line of the mean, and (2) from the line of best fit. Deviations from the line of the mean are called y as in Figure 10.8, whereas deviations from the line of best fit are called $y \cdot x$ as in Figure 10.7. We can take the $y \cdot x$ values and compute a standard deviation as follows:

$$s_{y \cdot x} = \sqrt{\frac{\Sigma y \cdot x^2}{n - 2}} \qquad (10.7)$$

This statistic is called the *standard error of estimate* and represents the variability or errors around the best-fitting straight line. Inspection of Figure 10.9 shows that $s_{y \cdot x}$ is numerically smaller than s_y, as it has to be because the criterion for selecting the best-fitting straight line was to minimize the sum of the squares of the deviations.

Because s_y^2 = total variance in the Y distribution, and $s_{y \cdot x}^2$ = the remaining variance taken around the line of best fit, then the statistic

$$\frac{s_y^2 - s_{y \cdot x}^2}{s_y^2}$$

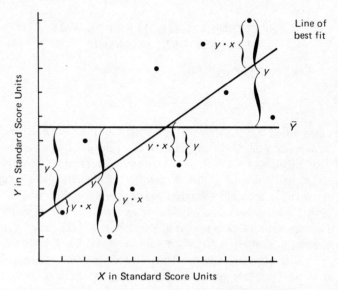

FIGURE 10.9 Comparing deviations taken from the best-fitting straight line $(y \cdot x)$ with deviations taken from the mean of the Y distribution (y).

equals the proportion of total variance accounted for by the line of best fit. That statistic is the square of the correlation coefficient, defined as follows:

$$r^2 = \frac{s_y^2 - s_{y \cdot x}^2}{s_y^2} \qquad (10.8)$$

By going through some algebra, we can now define $s_{y \cdot x}$ as follows:

$$s_{y \cdot x} = s_y \sqrt{1 - r^2} \qquad (10.9)$$

From Formula 10.9 we see that the factor $\sqrt{1 - r^2}$ is the proportion by which s_y is reduced, thus resulting in $s_{y \cdot x}$.

Because the choice of which variable is X and which Y is arbitrary, we again have a symmetrical arrangement and we can define the standard error for predicting X from a knowledge of Y as follows:

$$s_{x \cdot y} = s_x \sqrt{1 - r^2} \qquad (10.10)$$

Note that $s_{x \cdot y}$ is not the same as $s_{y \cdot x}$, in the same way that s_x is not the same as s_y.

THE CORRELATION COEFFICIENT AND
DEGREE OF RELATIONSHIP

If Formula 10.9 is squared, we get

$$s_{y \cdot x}^2 = s_y^2 - r^2 s_y^2 \tag{10.11}$$

This gives us another way to view r^2. The objective of any researcher, whether he uses the experimental model or the correlational model, is to account for as much of the variance of the endpoint or criterion measure as possible. When using experimental procedures, this is done by manipulating the independent variable or variables in such a fashion as to maximize the mean differences between groups. When using correlation procedures, this is done by finding significant linear relationships among variables. Let us say that, as experimenters, we are interested in a criterion variable X, and we know that X is related to Y by means of a significant correlation coefficient. Our question now becomes: What proportion of the variance of the X distribution can be accounted for by a knowledge of Y? We see from Formula 10.8 that the answer to that question is given by squaring the correlation coefficient. In Formula 10.11, which is algebraically equivalent to Formula 10.8, $r^2 s_y^2$ is the proportion of accountable variance in Y. When subtracted from s_y^2, the remainder is unaccountable variance, the square root of which is the standard error of estimate.

The square of r is often called the *coefficient of determination* and is defined below:

$$\begin{aligned} r^2 &= \text{coefficient of determination} \\ &= \text{proportion of variance in } Y \text{ which} \\ &\quad \text{can be accounted for by a knowledge} \\ &\quad \text{of } X, \text{ and vice versa} \end{aligned} \tag{10.12}$$

Interpreting r^2

Consider a situation where there is a correlation of .50 between variables X and Y. This means that we can account for $(.50)^2$ equals 25% of the variance of Y from a knowledge of X. Whether this is useful to us depends on the nature of our research problem. For example, if we wished to use X as a matching variable in an experimental design where we are interested in measuring Y as our endpoint, then a correlation of .5 is extremely valuable because this means that we can reduce the error variance by approximately 25% through the use of X. In this context it is clear that the precision and power of the experiment can be markedly improved through the use of X.

However, suppose that the correlation of .5 was obtained by the biologist when he correlated his technician's data for the two blood assay procedures. This r of .5 tells the biologist that only 25% of the total variance of either assay is related to the other assay. This is clearly unacceptable. The correlation between the values obtained by the two assays must be of the order of .90 to .95 (i.e., we must account for 80 to 90% of the known variance) before the biologist would be willing to shift from one procedure to the other. Similarly, when one is checking out an instrument to be certain that it is recording properly, the correlation between observations of behavior and the records made by the instrument should also be about .90 to .95.

On the other hand, suppose that a researcher found a correlation of .31 between a measure of learning in rats and the amount of a particular enzyme in the brain. Furthermore, suppose that the researcher's n was sufficiently large so that her correlation was significantly different from zero. Even though the correlation only accounts for approximately 10% of the variance of either the learning scores or the enzyme data $[(.31)^2 = 10.24\%]$, this is a potentially very important theoretical finding (if it can be replicated) and is worthy of further investigation.

DETERMINING THE SIGNIFICANCE OF THE CORRELATION COEFFICIENT

In testing for the significance of the correlation coefficient, we follow the same logical format as used before. We begin by assuming that there is a population of X, Y values in which X and Y are both normally distributed, and that we have drawn a random sample from that population. In the population the relationship between X and Y is represented by the Greek letter ρ (read *rho*). Our statistical hypothesis and its alternative are

$$H_0: \rho = 0$$

$$H_1: \rho \neq 0$$

The test of the hypothesis that ρ is equal to 0 is done via a t test as follows:

$$t = \frac{r\sqrt{n-2}}{\sqrt{1-r^2}} \tag{10.13}$$

This t test has $n - 2$ degrees of freedom, where n = number of paired observations. The reason the degrees of freedom are $n - 2$ is because two parameters are estimated to get the line of best fit.

It is not necessary to compute the value of t by Formula 10.13 for each correlation that you have. This is such a frequently used statistic that the critical values for significance have been tabled. Table XI in the Appendix lists the values of r required for significance at the .10, .05, .02, and .01 levels for varying degrees of freedom. Table XI is based on a two-tailed test on ρ. If you use a one-tailed test, then halve the probability values at the top of that table.

Inspection of Table XI reveals that the numerical value required for significance of the correlation coefficient decreases as the *df* increase. Thus with 52 pairs of cases, an *r* of .273 is significant at the .05 level; an *r* of .164 would be significant at the .05 level if you were making a one-tailed test with 102 cases in your sample. Thus, when we find out that our correlation coefficient is different from 0 at a specified significance level, this tells us that in the population there is a nonzero relationship between X and Y, but it tells us nothing about the degree of relationship. This is where we use r^2, the coefficient of determination.

A COMPUTATIONAL EXAMPLE

Table 10.1 gives the general rules involved in computing a correlation coefficient, and Table 10.2 is an example involving the correlation of an observer's recording of activity of rats in an open field versus the electronic counts obtained by a photocell system. The means of the X and Y distribution are not the same because the X distribution is based on the number of squares entered whereas the Y distribution is a measure of length traversed in the field. The correlation coefficient of .95 indicates that approximately 90% of the variance measured by the observer can be accounted for by the electronic sensing apparatus. This is a highly satisfactory correlation and establishes that the researcher can use the photocell system and be quite confident that he is measuring the same thing as the technician was measuring when he recorded number of squares entered by the animal.

USES OF CORRELATION IN EXPERIMENTAL RESEARCH

Correlation and the *t* Test for Matched Pairs

In Chapter 7 we discussed an experimental design involving matched pairs with the example of littermates raised in an impoverished or enriched environment. We pointed out that this method is useful only if the variable on which the animals are matched is related to the criterion measure being evaluated. The expression "is related to" in the sentence above actually means that there is a significant linear correlation between the paired subjects on the criterion measure. It was not necessary, however, to discuss the concept of correlation at that point because the formula for the matched *t* test (7.16) took the correlation into account via the use of difference scores as was shown in Table 7.2. In order to understand the matching procedure more fully, we shall now examine the problem from a correlational perspective.

The most general formula for the standard error of the difference between means is

$$s_{\bar{x}_c - \bar{x}_e} = \sqrt{s_{\bar{x}_c}^2 + s_{\bar{x}_e}^2 - 2r_{ce}s_{\bar{x}_c}s_{\bar{x}_e}} \qquad (10.14)$$

TABLE 10.1 General rules and procedures to compute a correlation coefficient

The data table

Subject	Scores on variable X	Scores on variable Y
1	X_1	Y_1
2	X_2	Y_2
.		
.		
.		
n	X_n	Y_n

Definitions

X, Y = paired scores for a subject. It is necessary to have both the X and the Y score for each subject

n = number of *paired* observations = number of subjects

Obtain the following values

ΣX = sum of the raw scores of the X distribution

ΣX^2 = sum of the squares of the raw scores of the X distribution

ΣY = sum of the raw scores of the Y distribution

ΣY^2 = sum of the squares of the raw scores of the Y distribution

ΣXY = sum of the cross products of X and Y

Rules and formulas

r: The Pearson product-moment correlation coefficient is obtained by the following formula:

$$r = \frac{n\Sigma XY - (\Sigma X)(\Sigma Y)}{\sqrt{[n\Sigma X^2 - (\Sigma X)^2][n\Sigma Y^2 - (\Sigma Y)^2]}}$$

r^2: The coefficient of determination, which tells the proportion of variance of the X variable accounted for by a knowledge of Y (and vice versa), is found by squaring the correlation coefficient.

where r_{ce} = the correlation coefficient based on the paired scores of the experimental and control subjects.

If the subjects in the experiment have been randomly assigned to the two treatments so that these are independent groups, then $r_{ce} = 0$, and the last part of the term, $2r_{ce}s_{\bar{x}_c}s_{\bar{x}_e}$, drops out of the formula. The remaining portion is identical with Formula 7.5 for the standard error of a difference between independent groups.

TABLE 10.2 Computational example to obtain the correlation coefficient

Subject number	$X =$ open-field activity measured by observer	$Y =$ open-field activity measured by electronic apparatus
1	67	190
2	52	140
3	77	213
4	82	232
5	50	147
6	96	270
7	68	188
8	92	260
9	57	163
10	100	261
11	77	218
12	80	228
13	110	259
14	52	145
15	104	240
16	68	200
17	70	192
18	87	240
19	61	163
20	74	215

$$\Sigma X = 1,524 \qquad \Sigma Y = 4,164$$
$$\Sigma X^2 = 122,138 \qquad \Sigma Y^2 = 899,508$$
$$\Sigma XY = 330,589$$

$$r = \frac{n\Sigma XY - (\Sigma X)(\Sigma Y)}{[n\Sigma X^2 - (\Sigma X)^2][n\Sigma Y^2 - (\Sigma Y)^2]}$$

$$= \frac{(20)(330,589) - (1,524)(4,164)}{\sqrt{[(20)(122,138) - (1,524)^2][(20)(899,508) - (4,164)^2]}}$$

$$= \frac{6,611,780 - 6,345,936}{\sqrt{(120,184)(651,264)}}$$

$$= \frac{265,844}{\sqrt{78,271,512,576}}$$

$$= .95$$

$$r^2 = (.95)^2 = .90$$

However, if the subjects have been matched, thus resulting in a positive correlation, this will act to reduce the standard error, thereby increasing the power of the t test. Formula 10.14 can be shown, by appropriate algebra, to be identical with Formula 7.15, which used difference scores (i.e., $X_c - X_e$). This can be demonstrated most easily by using Formula 10.14 to reanalyze the data in Table 7.2. Recall that littermate pairs were reared in an enriched or improverished environment during early life and then each animal's total cortex was weighed to the nearest milligram. The basic data from Table 7.2 is reproduced in Table 10.3 together with all necessary computations. It is first necessary to obtain the correlation between the littermate pairs. This turns out to be .9889. That value, along with the standard errors, are then used in Formula 10.14 to give a standard error of the difference of 7.23. When this is divided into the mean difference of 29.7, the resulting $t = 4.11$, which is within rounding error of the t value of 4.14 obtained via the difference formula in Table 7.2.

Correlation and Matching in a Factorial Design

It is not necessary to have subjects matched on a one-to-one basis, nor even to have them paired, to use the concept of correlation in designing experiments. All that is necessary is to know that there is a matching variable (X) that is significantly related to the endpoint (Y). Given that information, there are various ways to arrange the conditions of the experiment. One such way is to use littermate pairs, as was shown in the example above. This can easily be extended to three or four littermates, each receiving a different treatment condition in a randomized blocks design (see Tables 9.21, 9.22, and 9.23 for an example).

A matching variable can also be built into a factorial design. A common situation is one in which it is not possible to match subjects exactly, and so a grosser way of grouping is needed. A useful solution is to define "levels" of the matching variable and assign equal numbers of subjects within each level. These subjects are then randomly distributed among the treatment conditions and the experiment is analyzed as an $R \times C$ factorial.

We shall consider an example of this use of matching in the following context. Suppose that we are nutritionists interested in determining whether either of two specially prepared diets increases adult body weight of our experimental rats. In looking through our colony we find 30 males ranging in weight from 200 grams to 350 grams. We know that the lighter animals are younger and will show an increase in body weight over time as they grow. Thus there will be a correlation between initial body weight and the increment of weight gain at the end of the experiment. We realize that this will introduce considerable variability into our data which will act to inflate our error term. In order to avoid this problem we decide to use initial body weight as a classificational variable in a factorial design. Therefore, we weigh the animals and place them in rank order from lightest to heaviest in weight. We then take the six lightest animals and randomly assign two each to the Control Diet, Experimental

TABLE 10.3 Reanalysis of the data in Table 7.2 by use of Formula 10.14 rather than Formula 7.15

Pair number	Control	Experimental
1	620	681
2	835	862
3	427	447
4	770	765
5	536	574
6	617	617
7	698	740
8	811	866
9	473	485
10	755	802

$$\Sigma C = 6{,}542 \qquad \Sigma E = 6{,}839$$

$$\Sigma C^2 = 4{,}463{,}518 \qquad \Sigma E^2 = 4{,}877{,}989$$

$$\Sigma CE = 4{,}664{,}023$$

$$n = 10$$

$$r = \frac{n \Sigma CE - (\Sigma C)(\Sigma E)}{\sqrt{[n \Sigma C^2 - (\Sigma C)^2][n \Sigma E^2 - (\Sigma E)^2}}$$

$$= \frac{(10)(4{,}664{,}023) - (6{,}542)(6{,}839)}{\sqrt{[(10)(4{,}463{,}518) - (6{,}542)^2][(10)(4{,}877{,}989) - (6{,}839)^2]}}$$

$$= .9889$$

$$\bar{X}_c = 654.2 \qquad\qquad \bar{X}_e = 683.9$$

$$s_{\bar{x}_c}^2 = \frac{(10)(4{,}463{,}518) - (6{,}542)^2}{(100)(9)} \qquad s_{\bar{x}_e}^2 = \frac{(10)(4{,}877{,}989) - (6{,}839)^2}{(100)(9)}$$

$$= 2{,}041.5733 \qquad\qquad = 2{,}231.0767$$

$$s_{\bar{x}_c} = 45.18 \qquad\qquad s_{\bar{x}_c} = 47.23$$

$$t = \frac{\bar{X}_e - \bar{X}_c}{\sqrt{s_{\bar{x}_c}^2 + s_{\bar{x}_e}^2 - 2r s_{\bar{x}_c} s_{\bar{x}_e}}}$$

$$= \frac{683.9 - 654.2}{\sqrt{2{,}041.5733 + 2{,}231.0767 - (2)(.9889)(45.18)(47.23)}}$$

$$= \frac{29.7}{7.23} = 4.11$$

Diet 1, or Experimental Diet 2. We takes the next six heavier rats and repeat the identical procedure; and so forth. Each animal is then individually caged and placed on one of the three diets for two weeks. At the end of that time, all animals are reweighed, and their increase in weight is obtained.

Table 10.4 shows the weight increase in grams for each of the 30 animals. The computations for the analysis of variance are given at the bottom of that table. Inspection of Table 10.4 shows that we were wise to group our animals on initial weight level, because those animals that weighed the least at the beginning of the experiment gained the most during the two-week feeding interval. This is shown quantitatively in Table 10.5, where the vast majority of the variability is attributable to differences in weight levels. Because we wished to determine whether the special diets increased body weight, a one-tailed test is appropriate, and Dunnett's procedure is used to compare the control group against each of the experimental groups following the steps outlined in Table 8.11. We find that Experimental Diet 2 significantly increased body weight, as is shown at the bottom of Table 10.5.

CORRELATION AND REGRESSION

There are times when all we want to know is the degree of relationship between two or more variables. If so, then the correlation coefficient is sufficient for our purposes. However, we may also want to know the equation relating our two variables. This occurs quite frequently when X and Y are in different units and we wish to translate our data from one measurement system to another. For example, if the biologist had found a sufficiently high correlation between the two blood assay procedures, he then planned to switch to the newer method. If the two procedures did not have identical measurement scales, the biologist would need to know what a particular value on the new scale corresponded to on the scale he has used in the past. In a similar fashion, the researcher who had found a correlation of .95 between the score obtained from the electronic apparatus monitoring the open field and the score recorded by an observer will now use the photocell system to measure activity, but he may wish to transform the photocell counts to observer scores because he is more used to interpreting data in terms of number of squares entered by the animal.

The equations relating the two variables, X and Y, are called *regression equations*. There are two such equations for every pair of variables. If we are interested in predicting Y from a knowledge of X, the equation is

$$Y' = a_{y \cdot x} + b_{y \cdot x} X \qquad (10.15)$$

where Y' = predicted Y score
  $a_{y \cdot x}$ = the intercept or the value of Y' when $X = 0$
  $b_{y \cdot x}$ = the regression coefficient or slope of the best-fitting line

TABLE 10.4 Example showing weight gain (in grams) of rats fed a control diet or one of two experimental diets where initial weight level is used as a matching variable in an $R \times C$ factorial design

Level of weight	Control Diet	Experimental Diet 1	Experimental Diet 2	Total
1	11.4	10.5	12.0	66.5
	9.5	9.9	13.2	
2	7.3	6.8	7.1	44.1
	7.0	7.5	8.4	
3	10.6	10.9	10.0	60.1
	8.2	11.3	9.1	
4	5.7	4.9	5.8	30.4
	3.9	4.6	5.5	
5	6.1	4.7	8.2	31.4
	2.9	4.2	5.3	
Total	72.6	75.3	84.6	232.5

Table of totals

Level of weight	Control Diet	Experimental Diet 1	Experimental Diet 2	Total
1	20.9	20.4	25.2	66.5
2	14.3	14.3	15.5	44.1
3	18.8	22.2	19.1	60.1
4	9.6	9.5	11.3	30.4
5	9.0	8.9	13.5	31.4
Total	72.6	75.3	84.6	232.5

$\Sigma X = 232.5$

$\Sigma X^2 = 2,018.81$

$N = 30$

$$CT = \frac{(\Sigma X)^2}{N} = 1,801.875$$

$$SS_{total} = \Sigma X^2 - CT = 2,018.81 - 1,801.875 = 216.935$$

TABLE 10.4 (*continued*) Example showing weight gain (in grams) of rats fed a control diet or one of two experimental diets where initial weight level is used as a matching variable in an $R \times C$ factorial design

$$SS_{\text{treatments}} = \frac{(72.6)^2 + (75.3)^2 + (84.6)^2}{10} - CT = 1,809.801 - CT = 7.926$$

$$SS_{\text{levels}} = \frac{(66.5)^2 + \cdots + (31.4)^2}{6} - CT = 1,981.532 - CT = 179.657$$

$$SS_{\text{trt} \times \text{levels}} = \frac{(20.9)^2 + \cdots + (13.5)^2}{2} - CT - SS_{\text{trt}} - SS_{\text{levels}}$$

$$= 198.570 - CT - SS_{\text{trt}} - SS_{\text{levels}}$$

$$= 10.987$$

$$SS_{\text{within}} = SS_{\text{total}} - SS_{\text{trt}} - SS_{\text{levels}} - SS_{\text{trt} \times \text{levels}} = 18.365$$

The regression equation to predict X from a knowledge of Y is as follows:

$$X' = a_{x \cdot y} + b_{x \cdot y} Y \tag{10.16}$$

where X' = predicted X score

$a_{x \cdot y}$ = the intercept or value of X' when $Y = 0$

$b_{x \cdot y}$ = the regression coefficient or slope of the best-fitting straight line

These regression equations can be derived directly using the principle of least squares or they can be obtained by an algebraic transformation of Equation 10.5 ($z_y' = rz_x$) from z score units to raw score units. In fact Equation 10.5 can be used to predict X from Y and vice versa, but it is usually more convenient to have an equation that deals directly with raw scores rather than z scores.

In the z score equation the correlation coefficient, r, is also the regression coefficient. This is true only because the X and Y variables have been transformed to standard score units which equated their respective standard deviations. In raw score units it is rare that X and Y will have the same standard deviations, and so the two b coefficients will have different numerical values. However, there is an intimate relationship between the regression coefficients and the correlation coefficient as shown below:

$$r_{xy} = \sqrt{(b_{x \cdot y})(b_{y \cdot x})} \tag{10.17}$$

In words, the square root of the product of the two b coefficients will equal the correlation coefficient.

There is no need to make a significance test of the b coefficients because that test is equivalent to a test of the correlation coefficient. That is, the b coefficients always have the same level of significance (or nonsignificance) as the correlation coefficient. Table 10.6 gives the general rules and formulas to compute the a and b coefficients for each equation. To illustrate their use we

TABLE 10.5 Analysis of variance summary table for the data in Table 10.4, and Dunnett's test comparing the control group against the two experimental groups

Source	df	SS	MS
Levels	4	179.657	44.914
Treatments	2	7.926	3.963
Levels × trt	8	10.987	1.373
Within	15	18.365	1.224
Total	29	216.935	

Dunnett's Procedure

1. $\sqrt{2nMS_{within}} = \sqrt{(2)(10)(1.224)} = \sqrt{24.48} = 4.95$

2. $t_{.05}(3, 15) = 2.07$

3. Critical difference $= (4.95)(2.07) = 10.25$

4. Experimental Diet 1 $-$ Control $= 75.3 - 72.6 = 2.7$

   Experimental Diet 2 $-$ Control $= 84.6 - 72.6 = 12.0$

Experimental Diet 2 exceeds the critical difference at the .05 level using a one-tailed test.

TABLE 10.6 General rules and procedures to compute a linear regression equation

The *Data table* and *Definitions* are the same as in Table 10.1

Rules and formulas

To compute the regression equation to predict Y from knowledge of X:

$Y' = a_{y \cdot x} + b_{y \cdot x} X$ — The regression equation

$b_{y \cdot x}$: The regression coefficient is obtained by the following formula:

$$b_{y \cdot x} = \frac{n \Sigma XY - (\Sigma X)(\Sigma Y)}{n \Sigma X^2 - (\Sigma X)^2}$$

$a_{y \cdot x}$: The intercept is obtained by the following formula:

$$a_{y \cdot x} = \bar{Y} - b_{y \cdot x} \bar{X}$$

To compute the regression equation to predict X from knowledge of Y:

$X' = a_{x \cdot y} + b_{x \cdot y} Y$ — The regression equation

$b_{x \cdot y}$: The regression coefficient is obtained by the following formula:

$$b_{x \cdot y} = \frac{n \Sigma XY - (\Sigma X)(\Sigma Y)}{n \Sigma Y^2 - (\Sigma Y)^2}$$

$a_{x \cdot y}$: The intercept is obtained by the following formula:

$$a_{x \cdot y} = \bar{X} - b_{x \cdot y} \bar{Y}$$

shall return to the data in Table 10.2 involving the two ways of measuring open-field activity. Table 10.7 shows the computations for obtaining the regression equation to predict electronic apparatus scores from a knowledge of the observer's scores, and the equation predicting the observer's scores from the electronic information. The two regression coefficients are 2.2120 and .4082, and the square root of their product is .95, the value of the correlation

TABLE 10.7 Computational example to obtain the two regression equations for the data in Table 10.2

Predicting electronic apparatus score (Y') from knowledge of observer's score (X):

$$Y' = a_{y \cdot x} + b_{y \cdot x}X$$

$$b_{y \cdot x} = \frac{n \Sigma XY - (\Sigma X)(\Sigma Y)}{n \Sigma X^2 - (\Sigma X)^2}$$

$$= \frac{20(330,589) - (1,524)(4,164)}{20(122,138) - (1,524)^2}$$

$$= \frac{265,844}{120,184}$$

$$= 2.2120$$

$$a_{y \cdot x} = \bar{Y} - b_{y \cdot x}\bar{X}$$

$$= 208.20 - (2.2120)(76.20)$$

$$= 39.6456$$

$$Y' = 39.6456 + (2.2120)X$$

Predicting observer's score (X') from knowledge of electronic apparatus score (Y):

$$X' = a_{x \cdot y} + b_{x \cdot y}Y$$

$$b_{x \cdot y} = \frac{n \Sigma XY - (\Sigma X)(\Sigma Y)}{n \Sigma Y^2 - (\Sigma Y)^2}$$

$$= \frac{20(330,589) - (1,524)(4,164)}{(20)(899,508) - (4,164)^2}$$

$$= \frac{265,844}{651,264}$$

$$= .4082$$

$$a_{x \cdot y} = \bar{X} - b_{x \cdot y}\bar{Y}$$

$$= 76.20 - (.4082)(208.20)$$

$$= -8.7872$$

$$X' = -8.7872 + (.4082)Y$$

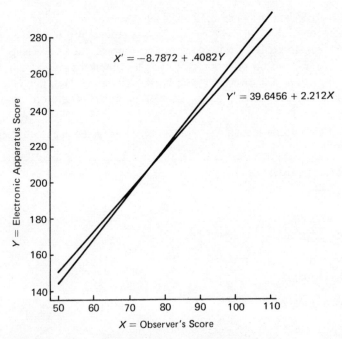

FIGURE 10.10   Showing the two regression lines from
Table 10.7 based on the data in Table 10.2.

coefficients in Table 10.2. Figure 10.10 plots the two equations. Note that the
two lines intersect at the point $(\bar{X}, \bar{Y})$.

COMPARING CORRELATION STUDIES AND
REGRESSION EXPERIMENTS

In Chapter 8 we discussed regression experiments and showed how to fit a
polynomial equation to a set of quantitative data. There are important
differences as well as similarities between regression experiments and correlation
studies using regression equations. First, the models of the two approaches are
quite different. In regression experiments there is a clearly specified independent
variable that the researcher manipulates in the context of a fixed-effects analysis
of variance model. In the correlation model there is no systematic manipulation
of an independent variable. Instead the researcher has a random sample of
subjects on which he has at least two measures, X and Y, both of which are
random variables. Next, in a regression experiment we are interested only in
predicting in one direction. It makes no sense to write an equation predicting the
level of the independent variable from a knowledge of the criterion measure.
However, in a correlation study it is often reasonable to compute both regression
equations, depending on the context of the experiment. Finally, in a regression

experiment the purpose is always to fit a function that may be linear or of a higher order, whereas the major purpose of a correlation study is to compute r. Whether one calculates a regression equation is usually dependent on the value of r obtained.

The two approaches are similar in terms of formulas and computational procedures. For example, the formulas given in Table 10.6 for the best-fitting straight line yield the identical results as the first-degree polynomial equation in Chapter 8. However, the orthogonal polynomial approach cannot be used in a correlational study because of the demand for equal n and equal spacing of the independent variable.

FINAL COMMENT

This concludes our introduction to correlation. Our discussion has been restricted to linear relationships. Thus, if there is a significant curvilinear relationship but with no major linear component, the correlation coefficient can be close to zero. An examination of the scatterplot of X and Y is a good way to detect gross deviations from linearity. There are quantitative methods to test for curvilinearity, but they are beyond the scope of this book.

We have also restricted the discussion to two variables. There are techniques called *multiple correlation* and *multiple regression* to analyze data consisting of more than two variables when we are working in the context of a correlational study. If we have measured several endpoints in an analysis of variance experimental design, there is a technique called *multivariate analysis of variance* which will simultaneously assess the effects of all endpoints, taking into account the correlations among them. For discussions of these topics, as well as more extensive discussion of linear correlation, the reader is referred to the books by Hays (1963) and Winer (1971).

SUMMARY OF DEFINITIONS

Regression equations The best-fitting straight-line equations obtained from a correlational study to predict Y from a knowledge of X, or to predict X from a knowledge of Y.

Scatterplot A graphic way of depicting the distribution of pairs of numbers by plotting them on X-Y coordinates.

PROBLEMS

1. An instructor gave his class a multiple-choice test for their first examination. The second examination was an essay test. The two sets of scores were as follows

Student	Test 1	Test 2
1	97	73
2	54	59
3	33	24
4	51	89
5	43	78
6	51	66
7	54	47
8	99	88
9	89	57
10	42	32
11	81	76
12	88	61
13	53	70
14	77	67
15	43	75
16	51	85

Determine the correlation between the two tests, and test to see if the correlation differs significantly from zero.

2. An educational researcher, interested in the relationship between the verbal ability of parents and their children, gave a vocabulary definition test to both parents and to one of their children. He added the scores of both parents and correlated the total score against the score of their child. The data are as follows:

Family	Parents' score	Child's score
1	28	11
2	12	12
3	26	11
4	34	13
5	17	10
6	24	10
7	24	12
8	43	27
9	24	15
10	15	18
11	57	23
12	20	14
13	40	23
14	22	6
15	34	8
16	22	12
17	36	8
18	26	15
19	36	9
20	17	11

(a) Determine the correlation and its level of significance.

(b) The researcher was interested in predicting the scores of the children from a knowledge of their parents' score. What is the regression equation?

(c) What is the coefficient of determination?

(d) What is the standard error of estimate in predicting the children's scores from a knowledge of their parents' scores?

3. A reseacher interested in sex differences in behavior took one male and one female rat from each of 15 litters and tested them in the open field. The following data were obtained:

$$\Sigma M = 665 \qquad \Sigma F = 1029$$

$$\Sigma M^2 = 44812 \quad \Sigma F^2 = 88467$$

$$\Sigma MF = 52903$$

$$n = 15$$

(a) What is the correlation between the littermates?

(b) Determine the value of t, taking the correlation into account. How many degrees of freedom are associated with the t test?

(c) Assume that the researcher ignored the correlational element. What would be his value of t and how many df would he use to assess this value?

4. A primatologist developed a scoring system to score the number of social interactions within a troop of baboons. To determine the reliability of the instrument he had two judges observe the troop and record the number of interactions that occurred within successive 3-minute intervals. The judges did this for 75 minutes and obtained the following data:

Time epoc	Judge A	Judge B
1	10	8
2	13	16
3	15	14
4	0	0
5	2	1
6	24	23
7	35	29
8	6	7
9	3	7
10	12	16
11	10	17
12	8	6
13	9	9
14	8	8
15	6	2
16	5	0

Time epoc	Judge A	Judge B
17	21	24
18	22	16
19	33	28
20	35	42
21	22	30
22	19	12
23	18	18
24	13	9
25	11	10

Determine the reliability of the two judges' scores.

5. The primatologist in Problem 4 was also interested in determining whether the two judges had the same average response, or whether one judge was systematically higher (or lower) than the other judge. To determine this he computed a correlated t test to compare their mean scores. What conclusion did he draw from this analysis?

CHAPTER 11

Chi-Square Tests of Nominal Data

Our discussions of statistical procedures so far have been based on the assumption that we are able to obtain a quantitative score of our dependent variable by using a measurement procedure involving an interval scale. There are times, however, when our variable of interest does not have a quantity that can be measured but is simply reported as categorical information. For example, animals are either male or female, they survive the effects of a pathological agent or they die from it, and they have either pigmented eyes or red eyes. The categorization does not have to be restricted to dichotomies, although this is very common. For example, eye color can be designated as black, brown, blue, gray, and "other"; aggressive behavior can be categorized as attacking and killing a conspecific, attacking but not killing, or not attacking; and animals can be classified as showing right paw preference, left paw preference, or no preference. These are examples of the nominal scale of measurement. This scale classifies our subject matter into two or more mutually exclusive and totally inclusive categories. This is a qualitative scale in that there is no implication of one category being better than or poorer than another category, although, as we pointed out in Chapter 2, experimenters, may choose to interpret their data within the context of better or worse.

Nominal data is described by reporting the frequency or relative frequency (proportion) of occurrence within each category of classification. Our concern in this chapter will be the kinds of inferences that can be drawn from these data. Before doing so, it is first necessary to introduce a new statistic called χ^2 (Greek chi, squared).

THE χ^2 STATISTIC

Our basic data when working with nominal information is the frequency of occurrence of an event as obtained by counting the number of subjects that are in each of our categories of classification. We shall designate the frequency of occurrence of an observed event by the symbol f_o. The sum of all the observed frequencies will be equal to the total number of subjects in our study. Therefore $\Sigma f_o = n$.

For each observed frequency we shall also obtain (by procedures to be discussed below) an expected or theoretical frequency, which we shall designate by the symbol f_e. The summation of all the expected frequencies will also add up to n, so therefore $\Sigma f_e = n$.

In general, the null hypothesis we shall be testing is that the distribution of observed frequencies does not differ significantly from the distribution of expected frequencies. To make a statistical test of this hypothesis, we work with the difference between a pair of observed and expected frequencies, that is, $f_o - f_e$. However, this difference is not a sufficient statistic, because the summation of these scores must equal 0 (because both Σf_o and $\Sigma f_e = n$). This difficulty can be bypassed by squaring $(f_o - f_e)^2$ because this will always be a positive number; the value of $(f_o - f_e)^2$ will be in rough agreement with the magnitude of discrepancy between the observed and expected distributions. However, in part $(f_o - f_e)^2$ is dependent on the numerical value of our expected frequencies. That is, $(f_o - f_e)^2$ looms much larger if the expected frequency is small than if it is large. We can adjust for this by dividing by the expected frequency, thus giving us the statistic $(f_o - f_e)^2/f_e$. If we now obtain this value for each of our categories and then sum these up, we have the χ^2 statistic, which can now be defined as follows:

$$\chi^2 = \sum \frac{(f_o - f_e)^2}{f_e} \tag{11.1}$$

Table XII in the Appendix lists a number of critical values of the χ^2 distribution. This table is organized in a manner similar to the t table in that you enter the table with the number of degrees of freedom and the value of χ^2, and you can then determine the approximate probability value of χ^2. The procedure for obtaining degrees of freedom will be discussed when we get into particular applications of the χ^2 statistic. The thing to note now in Table XII is that the numerical value of χ^2 for significance at a designated level (e.g., the .05 level) increases in amount as the number of degrees of freedom increase.

We shall now turn to the first application of χ^2.

COMPARING AN OBSERVED FREQUENCY DISTRIBUTION WITH A THEORETICAL FREQUENCY DISTRIBUTION: THE GOODNESS-OF-FIT TEST

An important use of the χ^2 test is to find out whether an observed distribution of cases could have arisen by random sampling from a population with a specified theoretical distribution. The test comparing the observed frequency distribution with the hypothetical frequency distribution is called a test for *goodness of fit*.

A Biological Example

As an illustration, consider the common problem in biology of investigating genetic dominance. Suppose that we wish to test the hypothesis that a particular biological character is independently determined by two simple Mendelian genes. The phenotype of the first gene will be scored as "present" if the genotype is either *AA* or *Aa*. Similarly, the phenotype for the second gene will be scored as "present" when the genotype is either *BB* or *Bb*. If the genotype is either *aa* or *bb*, then the phenotype will be scored as "absent." Given purebred parental stock, it can be shown by genetic theory that in the second filial generation the expected distribution of phenotypes will be in the ratio of 9:3:3:1. These four ratios correspond to the four phenotypic categories of *AB*, *Ab*, *aB*, *ab*. Suppose that we test our experimental hypothesis by taking two purebred strains of mice, crossing them to get the F_1 generation, and then crossing the animals within this generation to produce an F_2 generation. Suppose that we take 256 animals from the F_2 generation and categorize them with respect to one of the four phenotypes. Based on genetic theory, we would expect the 256 animals to be apportioned in the ratios of 9:3:3:1, or with frequencies of 144, 48, 48, and 16. Now suppose that our actual observed frequencies are 137, 41, 51, and 27. These data are summarized in Table 11.1, along with the calculations to determine χ^2.

Degrees of freedom The value of χ^2 obtained in Table 11.1 is 9.1111. Before we can evaluate that value, we must determine the degrees of freedom associated with it. We have imposed one restriction on our expected frequencies, namely, that the total of the expected frequencies must equal the *n* of our observed frequencies, or 256. Therefore, we can freely assign any numbers we wish to three of the four categories in Table 11.1, but the fourth number is automatically fixed in order to meet our restriction. Therefore, in general terms,

TABLE 11.1 Comparing an observed frequency distribution with the frequency distribution expected based on a simple dominant two-gene model

Category (phenotype)	f_o (observed frequency)	f_e (expected frequency)	$\dfrac{(f_o - f_e)^2}{f_e}$
AB	137	144	0.3403
Ab	41	48	1.0208
aB	51	48	0.1875
ab	27	16	7.5625
	256	256	$9.1111 = \chi^2$

the number of degrees of freedom in a test for goodness of fit is the number of categories minus 1, or C − 1.

In this problem there are four categories and therefore there are three degrees of freedom to evaluate the χ^2 of 9.1111. When we look this up in Table XII in the Appendix, we find that this value is between the .05 and .02 levels of significance. Therefore, if we had set the .05 level to reject the null hypothesis, we would conclude that our observed frequency distribution could not have arisen by random sampling from a population characterized by two independent dominant genes.

The effects of sample size To use the χ^2 statistic effectively, it is generally necessary to have large numbers of cases. This can be illustrated by supposing that we had done our experiment using 128 mice rather than 256. Suppose that we obtained the same proportion of animals in each phenotypic category. These data are presented in Table 11.2, and it can be seen that the value of χ^2 does not reach significance.

A caution in interpreting a nonsignificant χ^2 Note here that not rejecting the null hypothesis has a different consequence than not rejecting the null hypothesis when comparing two or more groups in the analysis of variance. In the latter situation we typically conclude that we have not isolated an experimental variable of interest, and we generally would not even consider publishing such results. In the instance of the χ^2 test for goodness of fit, however, not rejecting the null hypothesis leads us to conclude that the observed distribution is consistent with our theoretical model, and that kind of conclusion may be considered worthy of publication. The moral here is plain: When testing an observed distribution against a theoretical model, the lack of significance must be looked at with great caution unless the number of cases involved is very large. Otherwise, you run the risk of making a serious misinterpretation of your data.

TABLE 11.2 Testing the same hypothesis as in Table 11.1, but with only half the number of cases

Category (phenotype)	f_o (observed frequency)	f_e (expected frequency)	$\dfrac{(f_o - f_e)^2}{f_e}$
AB	68	72	0.2222
Ab	20	24	0.6667
aB	26	24	0.1667
ab	14	8	4.5000
	128	128	$5.5556 = \chi^2$

A Behavioral Example

The usual form of the null hypothesis when doing a t test is that the experimental treatment has had no effect. This is equivalent to stating that the experimental and control means in the population are identical. The parallel statement for frequency data would be that the proportion (or percentage) of subjects choosing different categories would be identical in the population. Thus, if three choices were possible, the theoretical expectation is that one-third would choose the first category, one-third the second category, and one-third the last category. As an example, consider a study in which rats are placed into an apparatus where they can go into one of three compartments: a dark box, a lighted box containing food, or a lighted box within which is another rat of the same sex housed within a wire mesh cage. We place an experimental animal into the apparatus and record the amount of time the animal spends in each of the three chambers during one hour. Each animal's preference is defined as the chamber within which he spent most time during the 60 minutes. The data are summarized in Table 11.3. The observed frequencies deviate markedly from the theoretical expectation of one-third, and the χ^2 of 31.61 is highly significant with two degrees of freedom. We would conclude that rats do not show an equal preference distribution for the three chambers; instead, there is a high preference for either the dark or the social chamber and a concomitant low preference for the food chamber.

A continuous distribution and discrete numbers An important thing to note about the example in Table 11.3 is that the expected frequencies do not have to be whole numbers, but can take on any value. This is because the χ^2 distribution is continuous rather than discrete. Because our observed frequencies must be discrete numbers, it is apparent that the χ^2 distribution does not have a perfect one-to-one relationship to our observed distribution. That is, the χ^2 distribution is an approximation of the distribution of discrete frequencies and, therefore, it is necessary to make several very important assumptions in order to use the χ^2 table. One of those assumptions is that we are able to use a theoretical

TABLE 11.3 Comparison of observed frequency distribution with the frequency distribution expected based on the null hypothesis of no difference

Category (chambers)	f_o (observed frequency)	f_e (expected frequency)	$\dfrac{(f_o - f_e)^2}{f_e}$
Dark chamber	49	33.3	7.40
Food chamber	7	33.3	20.77
Social chamber	44	33.3	3.44
	100	100	$31.61 = \chi^2$

continuous distribution to give us information about observed discrete events. The error that is made with this assumption is very small and not worth considering when there are at least three categories of classification for our data. However, when there are only two categories, then it is necessary to make an adjustment. We shall now consider this special case.

Two Categories of Classification

To illustrate the use of χ^2 with two categories of classification, let us return to our genetics experiment. Suppose that we now wish to make a test of the hypothesis that a single Mendelian dominant gene controls a specific character in our mouse population. We know that if we cross the two parental strains to get our F_1 and then breed within the F_1 to generate an F_2 distribution, the expectation is that 75% of the animals in the F_2 generation will exhibit the character and 25% will not if this is a single dominant gene. We carry out our breeding experiment, obtain 101 animals, and classify each one with respect to the presence or absence of the character. The results are shown in Table 11.4.

The first four columns of Table 11.4 are the same as we have seen in the previous tables, and the summation at the bottom of the fourth column gives the χ^2 value of 4.043. With one degree of freedom this is significant beyond the .05 level. However, this calculation is incorrect because, in a situation where there is one degree of freedom, the use of the continuous χ^2 distribution to approximate the data obtained from discrete observations results in a numerical value of χ^2 that is too large. Fortunately, it is quite easy to correct for this by subtracting .5 from the absolute difference between the observed and expected frequencies. This is known as *Yates' correction for continuity* and must *always* be used in any situation where there is one degree of freedom for a χ^2 test. The formula is as follows:

$$\chi^2 = \frac{(|f_o - f_e| - .5)^2}{f_e} \tag{11.2}$$

The corrected χ^2 value is shown in the last column of Table 11.4 and is found to be 3.595. With one degree of freedom this value is not significant, whereas the uncorrected value appears to be significant beyond the .05 level.

TABLE 11.4 Showing Yates' correction for continuity in a χ^2 with one degree of freedom

Category	f_o	f_e	$\dfrac{(f_o - f_e)^2}{f_e}$	$\dfrac{(|f_o - f_e| - .5)^2}{f_e}$
Present	67	75.75	1.011	0.899
Absent	34	25.25	3.032	2.696
	101	101	4.043	3.595

The previous example was based on theoretical expectations of 75% and 25%. If there is no theory to guide you, the usual form of the null hypothesis is to assume that both situations are equally likely to occur, which is the same as saying that the theoretical expectation for each event is 50%. Suppose, for example, that we wish to investigate whether rats were "right-handed" or "left-handed." Our procedure is to let animals run down the stem of a T maze and note whether they turn right or left at the choice point. Eighty-two animals are tested, and we find that 50 turn right and 32 turn left. The expected frequencies based on the null hypothesis is that 50%, or 41 animals, should turn right and the same number turn left. The analysis of the data, including Yates' correction for continuity, is summarized in Table 11.5. The χ^2 value of 3.524 falls between the .05 and .10 levels with one degree of freedom. Assuming that we had set the .05 level for rejecting the null hypothesis, we would conclude that there is no evidence that rats prefer to turn in one direction as compared with the other direction.

COMPARING TWO OBSERVED FREQUENCY DISTRIBUTIONS: TESTING FOR ASSOCIATION

We have discussed one common use of the χ^2 statistic, namely to see whether an observed distribution of frequencies is in correspondence with a theoretical distribution based on a particular hypothesis concerning the proportion or percentage of cases to be expected in each category. A second common use of the χ^2 statistic is to compare two empirically obtained distributions to see whether they are independent of or associated with each other. For example, we could take mice of a particular strain and classify them in one of several categories. We could then repeat this study with a different strain of mice. We would use the χ^2 test to determine whether the two strains are independent of each other or whether there is an association between strain and category of classification. Rather than use mice, we can think of the nominal classification of males and females and can then ask whether there is an association between an animal's sex and the categories of classification.

TABLE 11.5 Evaluating right or left turning preference against the hypothesis of equal preference

Category	f_o	f_e	$\dfrac{(|f_o - f_e| - 0.5)^2}{f_e}$
Right	50	41	1.762
Left	32	41	1.762
	82	82	$3.524 = \chi^2$

There are certain important distinctions between this use of the χ^2 test and the one previously discussed. In this situation the objective is not to test our data against a hypothesized distribution that was theoretically derived. Instead, our question is whether our two observed distributions differ from each other. If so, as indicated by a sufficiently high value of χ^2, we may conclude that there is an association between the variables we are investigating. This test is similar in principle to the analysis of variance where we are making a specific test of the means of two or more groups to see whether they may be considered to be from the same population. When we use χ^2, we are making a comparison of the overall distributions of each variable to see whether we can conclude that the distributions come from the same population.

We shall first discuss the use of this test for the general case where there are three or more categories for each level of classification and then we shall discuss this for the special case where each of the classifications has two levels.

The $R \times C$ Contingency Table

If we have two major ways of classifying our data, we can conveniently place this into a *contingency table* where there are R rows and C columns. To demonstrate the use of this table we shall consider the following research example. Suppose that we are endocrinologists interested in how hormonal factors influence maternal care in rabbits. We have injected various regimens of testosterone proprionate during different ages of gestation. There were four differing hormonal-age regimens, which we shall designate by the letters A, B, C, and D. Each animal was observed at the time of birth and was scored with respect to whether or not she built a maternal nest. In addition, they were observed thereafter and were classified with respect to whether or not the litter survived until weaning. This resulted in three categorical classifications as follows: nest built, litter survived; nest built, litter died; and nest not built, litter died. (The fourth logical category of nest not built, and litter survived never occurred.) The results of the experiment are shown in Table 11.6.

Inspection of the data suggests that there is an association between the variable of hormone-age regimen and the maternal behavior categories. In order to determine whether this is an association that is significantly greater than we could expect by chance alone, we need to find a method to obtain expected frequencies with which we can compare our observed frequencies. Because we do not have any theoretical basis for assigning expected frequencies to the cells, we need to find another method to obtain the expected values. To do this, we proceed as follows.

Obtaining the expected frequencies We first assume that the marginal totals are our best representation of the proportional distribution of frequencies in the population. For example, the three maternal behavior categories have, respectively, 51, 28, and 30 cases. Thus, we would expect in a population in which there are the four hormone-age regimens that $51/109 = .468$ cases would consist

TABLE 11.6 Observed frequencies in study relating maternal behavior to hormone treatment during pregnancy in rabbits

Hormone-age regimen	Maternal behavior categories			Total
	Nest built, litter survived	Nest built, litter died	Nest not built, litter died	
	(1)	(2)	(3)	
A	25	0	0	25
B	6	18	6	30
C	16	5	5	26
D	4	5	19	28
Total	51	28	30	109

of rabbits that built nests and successfully reared their litters. Similarly, we would expect that $28/109 = .257$ proportion of rabbits who received the four hormone-age regimens would be in the category of those that built nests but did not rear their litters successfully; and $30/109 = .275$ cases would be in the category of rabbits that neither built nests nor reared their young successfully. Now, if it is true that there is no association between the hormone-age regimens and the maternal behavior categories (i.e., if we assume the null hypothesis to be true), then the proportions of .468, .257, and .275 are what we would expect to find *within* each row of Table 11.6. Thus, to get the expected frequency for the cell in the upper left-hand corner of Table 11.6, we would multiply the expected proportion by the marginal total of 25 cases, or $(51/109) \times 25 = 11.697$. To get the second proportion in row A, we would take the expected proportion $(28/109)$ and multiply that by the marginal total of 25 cases, or $(28/109) \times 25 = 6.422$. The last expected frequency in that row is obtained in the same manner: $(30/109) \times 25 = 6.881$. The summation of these three expected frequencies equals the marginal total of 25, as it must.

The rule for obtaining the expected frequencies in an $R \times C$ table can be stated as follows: *The expected frequency in any cell is equal to the product of the two marginal totals associated with that cell divided by the total number of cases involved.* Table 11.7 shows the expected frequencies for the data of Table 11.6 obtained by following this rule.

Once the expected frequencies have been obtained, we proceed in the usual fashion of getting the difference between the observed and expected frequency, squaring that value, and dividing by f_e. The summation of all these values is our χ^2 statistic. These computations are shown in Table 11.8.

Degrees of freedom The one remaining thing we have to discuss is how to determine the number of degrees of freedom in an $R \times C$ table. As usual, we shall follow the rule that degrees of freedom specify the number of independent

TABLE 11.7 Expected frequencies of data in Table 11.6 based on the assumption that hormonal-age regimens and maternal behavior categories are independent of each other

| Hormonal-age regimen | Maternal behavior categories | | | Total |
	Nest built, litter survived (1)	Nest built, litter died (2)	Nest not built, litter died (3)	
A	11.697	6.422	6.881	25.000
B	14.037	7.706	8.257	30.000
C	12.165	6.679	7.156	26.000
D	13.101	7.193	7.706	28.000
Total	51.000	28.000	30.000	109.000

TABLE 11.8 Computation of χ^2 from the data in Tables 11.6 and 11.7

Hormonal-age regimen	Maternal behavior categories	f_o	f_e	$\dfrac{(f_o - f_e)^2}{f_e}$
A	1	25	11.697	15.130
	2	0	6.422	6.422
	3	0	6.881	6.881
B	1	6	14.037	4.602
	2	18	7.706	13.751
	3	6	8.257	0.616
C	1	16	12.165	1.209
	2	5	6.679	0.422
	3	5	7.156	0.650
D	1	4	13.101	6.322
	2	5	7.193	0.668
	3	19	7.706	16.553
		109	109.000	$73.227 = \chi^2$

ways that a set of data may vary. We have placed two restrictions on our data in obtaining our expected values: namely, that the summation of the row-wise marginal frequencies and the column-wise marginal frequencies must add up to n, or the total number of cases in the sample. Thus, it is possible for all but one of the row values to be any numerical number, and the same is true for the column values. Thus, *there are (R − 1) degrees of freedom for the rows and (C − 1) degrees of freedom for the columns, and so the degrees of freedom available in an R × C tables equals (R − 1) (C − 1).* In this particular instance the degrees of freedom are $(4 - 1)(3 - 1) = 6$. The χ^2 value in Table 11.8 is 73.227, and with 6 *df* this is significant well beyond the .01 level. Therefore, we would reject the null hypothesis that there is no association between the hormone-age regimens and the categories of maternal behavior and would conclude, instead, that the regimens had very marked effects on the distribution of animals among the different maternal behavior categories.

The 2 × 2 Contingency Table

When we have two classification variables and each has two categories (e.g., "present" or "absent"), we can arrange our data in a 2 × 2 contingency table, which is also called a *fourfold table*. Tha layout of such a table is shown in Table 11.9 where the letters *a, b, c,* and *d* refer to the observed frequencies within each cell and *n* is the total number of observations.

Because this table has 1 *df*, it is necessary to make Yates' correction in calculating χ^2. For a fourfold table there is a simple formula to compute χ^2 with Yates' correction. Based on the symbols in Table 11.9, the formula is

$$\chi^2 = \frac{n(|ad - bc| - n/2)^2}{(a + b)(c + d)(a + c)(b + d)} \tag{11.3}$$

This formula is identical with formula 11.2 but is computationally much simpler because each of the two classifications has only two categories.

As an example of the use of this formula, consider the following situation. A psychobiologist reared one group of rats with mice from the time of weaning until they were 57 days of age, at which time the experimental rats were removed and placed into a rat colony. Control rats were reared from the time of

TABLE 11.9 The 2 × 2 contingency table

	0	1	Total
1	*a*	*b*	*a + b*
0	*c*	*d*	*c + d*
Total	*a + c*	*b + d*	*n*

weaning in the rat colony. When they were 90 days old, each rat was exposed to several mice and the presence or absence of mouse killing was recorded. The data are summarized in Table 11.10. χ^2 computed by Formula 11.3 is

$$\chi^2 = \frac{60 \, (|360 - 0| - 30)^2}{(18)(22)(20)(40)} = 10.80$$

With 1 df, χ^2 is significant beyond the .01 level, and the psychobiologist concluded that there was a significant association between rats' early social experience and the likelihood that they would kill mice in adulthood.

UNPLANNED USES OF χ^2

In all the examples we have given so far we have deliberately chosen to use the χ^2 statistic for our research because of the nature of the variables that we wished to examine. There are times, however, when researchers are compelled to use this statistic because some event in their experiment goes awry.

Occurrences of Death

A situation that is bound to occur at least occasionally in any research laboratory is one in which some animals die during the course of the experiment. If this happens, a serious question is whether the deaths are associated with the experimental treatment or whether they are a random event. If the former, this would be interpreted as one of the consequences of the experimental treatment. However, this raises the serious issue as to the validity of testing the remaining animals on the original endpoint because the significant difference in death rate suggests that those that lived may be a biased sample. On the other hand, if the deaths appear to be randomly distributed, the researcher could continue his or her endpoint measurement and feel confident in the conclusions drawn from the findings.

As an example, suppose that we had administered Treatment A to 30 animals in our colony and Treatment B to 30 other animals. During the interval between

TABLE 11.10 The relationship between early social rearing and mouse killing in rats

	Reared with mice in early life	Not reared with mice in early life	Total
Killed mice	0	18	18
Did not kill mice	20	22	42
Total	20	40	60

termination of treatment and the initiation of measurement of the dependent variable we find that seven of the animals receiving Treatment A have died, whereas 12 of those receiving Treatment B have died. Is this difference statistically significant? The data are summarized in Table 11.11, and the χ^2 corrected for continuity is found to be 1.23. We may conclude that there is no evidence of an association between the treatments and the occurrence of death. Therefore, the remaining 23 animals in Treatment A and the 18 animals in Treatment B can be considered to be unbiased samples.

Nonresponding Animals

A situation that occurs in research laboratories where animals are given behavioral testing is that on some occasions the animals will not respond. The usual thing that is done in such a case is to give the animal a score of 0 if some sort of achievement performance is being measured, such as number of avoidance responses in a shuttlebox or number of squares crossed in an open field; or to give the animal the maximal error or time score as arbitrarily decided upon by the researcher, such as 180 seconds if the researcher allows each animal 3 minutes to initiate some sort of response. This is an acceptable procedure if two conditions are met: (1) The number of nonresponders is very small, and (2) there is no significant difference in the proportion of nonresponders in the various treatment groups.

Consider a situation where we have administered Treatment A to 25 rats and Treatment B to 25 other rats. We test the animals in the open field and find that 12 of the rats from Treatment A do not move at all whereas only 1 rat from Treatment B does not move. These data are summarized in Table 11.12.

The value of χ^2 with Yates' correction is 10.40, and with 1 df this is highly significant. We may conclude that there is an association between the experimental treatments and the probability that the animals will enter one or more squares in the open field.

What should we do with the information on the 13 animals from Treatment A and the 24 animals from Treatment B that were active in the open field? Even though the χ^2 test implies that these animals are from different populations, the sensible thing to do is to compare these two groups via a t test to determine whether the mean activity differs as a function of experimental treatment. We

TABLE 11.11 The relationship between two experimental treatments and the later occurrence of death

	Treatment A	Treatment B	Total
Not die	23	18	41
Die	7	12	19
Total	30	30	60

TABLE 11.12 The relationship between two experimental treatments and the occurrence or nonoccurrence of open-field activity

	Treatment A	Treatment B	Total
One or more squares entered	13	24	37
No squares entered	12	1	13
Total	25	25	50

could then summarize our statistic findings in a manner similar to the following: "The χ^2 test found that the rats receiving Treatment A had a much smaller probability of entering one or more squares than those animals receiving Treatment B. A subsequent t test, using only those animals from Treatment A and Treatment B that entered one or more squares, found a significant (or nonsignificant) difference between the group means."

ASSUMPTIONS UNDERLYING THE USE OF THE χ^2 TEST

The χ^2 statistic is one of the easiest to compute in the whole statistical literature. It is probably also the one that is misused most often because of violation of assumptions. These assumptions must be met in order for the χ^2 statistic to be a valid assessor of your experimental data. The assumptions are as follows:

1. The categories into which subjects are classified are mutually exclusive and totally inclusive. Therefore, there is a slot for every subject.

2. All observations must be independent of all other observations. Operationally this means that each subject is placed into one, and only one, category and contributes a count of 1 toward the final total. This is a commonly violated assumption. As an example of this, consider the experiment in which we categorized each rat with respect to which of the three chambers he preferred (Table 11.3). We might wish to record the two most preferred chambers and tally each animal twice. This violates the assumption of independence, because an animal is inherently correlated with itself. The χ^2 value obtained in these circumstances would be totally useless and probably very misleading.

3. When the χ^2 test is based on one degree of freedom, it is necessary to use Yates' correction for continuity. The smaller the number of cases in the sample, the larger is the effect of this correction. This assumption is also often violated.

4. It is necessary that the number of cases be sufficiently large to allow one to use the χ^2 test. This assumption applies to the *expected* number of cases within each category (not the observed number of cases). Statisticians differ

among themselves with respect to what is a satisfactory minimal number. A general rule of thumb which will include most recommendations by statisticians is that the minimal expected number should be 5 in any χ^2 test with one degree of freedom. When there is more than one degree of freedom, some expected frequencies may be less than 5, perhaps as low as 3, but expected frequencies as low as this should not occur more than once out of every five categories.

If you are designing an experiment in which the expected frequency in one or more of your cells will be approximately 5 or less, it is strongly recommended that you consult with a statistician beforehand to get his or her advice on the design and analysis of your data.

THE FISHER EXACT PROBABILITY TEST

An obvious question now is: What is there to do if the expected frequency is less than 5? Fortunately, there is an answer to this question using a procedure called the Fisher exact probability test. The Fisher test is not a χ^2 statistic but is included here because it is used whenever one needs to test for significance in a 2 X 2 contingency table and is not able to use the χ^2 statistic because expected values for one or more cells are less than 5. The rationale underlying the use of the Fisher exact probability test is based on the fact that we are able to compute the exact probability for any particular configuration of numbers in a fourfold table. There are only a certain limited number of configurations possible in such a table, and if one were to calculate the probability associated with each such configuration, the summation of those probabilities would equal 1.000. Thus, it is possible for a researcher to determine whether the particular configuration obtained, *or one more extreme*, could have arisen by chance alone.

In order to compute the probabilities of the configuration of numbers within the fourfold table, we have to keep the marginal totals constant following the same rationale as we developed in our discussion of the $R \times C$ table. The probability for any particular configuration is given by the formula

$$p = \frac{(a+b)!\,(c+d)!\,(a+c)!\,(b+d)!}{n!\,a!\,b!\,c!\,d!} \tag{11.4}$$

where the letters are defined in Table 11.9.

To illustrate the principles just discussed, consider a situation where eight cases are observed and one set of marginal frequencies is 4 and 4, whereas the other set is 6 and 2. There are only three possible configurations of numbers within a fourfold table that will yield those marginal frequencies. These are shown in Table 11.13 along with the probability of occurrence of each of those configurations as calculated by Formula 11.4. As can be seen, the summation of the probabilities equals 1.

A different configuration is shown in Table 11.14, where 10 cases are distributed with 5 and 5 frequencies along one axis and 6 and 4 frequencies

TABLE 11.13 Showing all possible configurations and their associated probabilities for the case of marginal totals of 4 and 4, and 6 and 2

Configuration			Probability
	0	1	
1	4	0	4	.2143
0	2	2	4	
	6	2	8	
	0	1	
1	3	1	4	.5714
0	3	1	4	
	6	2	8	
	0	1	
1	2	2	4	
0	4	0	4	.2143
	6	2	8	$\Sigma p = 1.0000$

along the other axis. In this instance there are five possible configurations that may occur, and again the summation of the probabilities equals 1.

Note in Tables 11.13 and 11.14 that the configurations are symmetrical. That is, the first and the last configurations have the same probability values associated with them, and the distribution of numbers within the four cells are symmetrical; this is also true for the second and fourth configurations in Table 11.14. This situation will always be true when either the row or column totals are identical. In Tables 11.13 and 11.14 we see that the two row frequencies are 4 and 4, or 5 and 5. However, if neither the row nor column frequencies are equal, then there will not be a symmetrical distribution of configurations. This is shown in Table 11.15, where the eight cases of Table 11.13 are now set with the restriction that the row frequencies equal 5 and 3, rather than 4 and 4. Again there are three configurations possible, but the first and third configuration yield very different probabilities than those shown in Table 11.13. The importance of this point will be seen when we discuss the use of one-tailed or two-tailed test of significance.

For a research example of the Fisher exact probability test we shall consider an experiment in which female mice received a testosterone injection in infancy while control mice received a sham injection. After reaching adulthood, mice were tested for aggression under a standardized set of conditions, and each

TABLE 11.14 Showing all possible configurations and their associated probabilities for the case of marginal totals of 5 and 5, and 6 and 4

Configuration				Probability
	0	1		
1	5	0	5	
0	1	4	5	
	6	4	10	.0238
	0	1		
1	4	1	5	
0	2	3	5	
	6	4	10	.2381
	0	1		
1	3	2	5	
0	3	2	5	
	6	4	10	.4762
	0	1		
1	2	3	5	
0	4	1	5	
	6	4	10	.2381
	0	1		
1	1	4	5	
0	5	0	5	
	6	4	10	.0238
				$\Sigma p = 1.0000$

mouse was scored with respect to whether or not she initiated an attack during the test session. The data are summarized in Table 11.16.

All six of the testosterone-treated mice attacked, whereas only two of the eight shams initiated an attack. To determine if this is significant, we compute the probability of obtaining this particular configuration, and the probability is found to equal .0093. Because there is no configuration that is more extreme than the one obtained, we may conclude that testosterone administered in infancy increases the probability of aggressive behavior among female mice in adulthood. The probability value of .0093 obtained in this experiment is for a

TABLE 11.15 Showing all possible configurations and their associated probabilities for the case of marginal totals of 5 and 3, and 6 and 2

	Configuration			Probability
	0	1		
1	5	0	5	
0	1	2	3	.1071
	6	2	8	
	0	1		
1	4	1	5	
0	2	1	3	.5357
	6	2	8	
	0	1		
1	3	2	5	
0	3	0	3	.3571
	6	2	8	$\Sigma p =$ .9999

one-tailed test. Given our present state of knowledge concerning the relationship of testosterone in infancy to later behavior, it is reasonable to use a one-tailed test. However, suppose that we wish to make a two-tailed test of the hypothesis? If either the row marginal frequencies or the column marginal frequencies had been equal, all that would be necessary would be to double the value of the one-tailed test (because of the symmetrical arrangement as shown in Tables 11.13 and 11.14). However, this is not true in the present situation, and so it is necessary to determine the probability of occurrence of the equivalently

TABLE 11.16 The relationship between testosterone treatment in infancy and aggressive behavior in adulthood for female mice

	Testosterone treatment in infancy	Sham treatment in infancy	
Attacks	6	2	8
No attacks	0	6	6
Total	6	8	14

extreme case in the opposite direction. By examining Table 11.16, we see that six out of six animals treated in infancy attacked. The equivalent opposite configuration would be if none of the six animals treated in infancy attacked. That configuration is shown in Table 11.17. The probability of that particular configuration occurring is .0003. These two probabilities are added to yield the probability value of .0096. This is the probability associated with the two-tailed hypothesis that testosterone treatment in infancy has no effect on adult aggressive behavior.

Our final example in this section is of a neuropsychology investigation in which we are interested in the relationship between the olfactory bulbs and maternal caretaking in primiparous and multiparous rats. In one of our experiments we took female rats that had never given birth or had had one successful litter, performed a bilateral olfactory bulbectomy, and then mated these females. We observed their maternal behavior after birth and recorded whether or not they cannibalized their young. The data are listed in Table 11.18.

Inspection of Table 11.18 shows that the expected frequencies for two of the cells are 4.5, and so the χ^2 test cannot be used. By applying the Fisher exact probability test, we find that the probability of this particular configuration occurring is .00268. However, we must also determine the probability of even more extreme configurations occurring. There is only one other configuration more extreme than the one shown in Table 11.18, and that is shown in Table 11.19. The probability of that particular configuration occurrring is .00006, and so the probability of obtaining frequencies as extreme or more extreme than that shown in Table 11.18 is the summation of these two probabilities or .00274. That is the probability value associated with the one-tailed test that multiparous mothers will have a lower incidence of cannibalism than primiparous mothers. However, the hypothesis actually tested was the two-tailed test that there is no association between parity and maternal behavior. Because our column frequencies are equal, the two-tailed probability value is obtained by doubling the one-tailed probability value, and is found to equal .0055. Therefore, we would reject the null hypothesis and conclude that there is a significant association between parity and maternal behavior as measured by cannibalism.

TABLE 11.17 The equivalent but opposite configuration to that shown in Table 11.16

	Testosterone treatment in infancy	Sham treatment in infancy	Total
Attack	0	8	8
No attack	6	0	6
	6	8	14

TABLE 11.18 The relationship betwen parity and cannibalism in rats without olfactory bulbs

	Primiparous	Multiparous	Total
Cannibalized young	8.	1	9
Did not cannibalize young	2	9	11
Total	10	10	20

Electronics and Statistics

The Fisher exact probability test is generally not emphasized in statistical textbooks because of the marked difficulty in computation. This was lessened somewhat when Siegel (1956) published a set of tables for n's up to 30 with the restriction that no marginal cell could have more than 15 cases. These tables helped considerably if one had data that met these restrictions, but it was still burdensome to use the tables and there were many research instances where the n's could not meet Siegel's requirements. Fortunately, advances in electronics and computer technology have made the Fisher exact probability test much more available to researchers. Many relatively inexpensive pocket calculators now come with a factorial key (usually designated as $x!$) as part of the circuitry. With such a key, one is able to determine a Fisher probability value in a matter of moments.

CORRELATED PROPORTIONS IN A 2 × 2 TABLE

In all the situations discussed so far, each subject has been observed once and placed into one of several mutually exclusive categories based on the observed performance. There are times, however, when subjects are observed on two

TABLE 11.19 The more extreme configuration from that in Table 11.18

	Primiparous	Multiparous	Total
Cannibalized young	9	0	9
Did not cannibalize young	1	10	11
Total	10	10	20

TABLE 11.20 The relationship between litter parity and the probability of killing offspring

		Second litter		
		Not kill	Kill one or more pups	Total
First litter	Kill one or more pups	10	12	22
	Not kill	17	1	18
	Total	27	13	40

occasions. A common example is where each subject is used as its own control: A baseline observation is obtained, then there is an experimental treatment, followed by a second measure that evaluates the effect of the treatment relative to the baseline level. This procedure is more commonly seen with a measured variable and a correlated t test than with the χ^2 statistic, but there are occasions when data handled by nominal classification can be examined in this fashion.

We shall illustrate the use of this statistic by the following example. In going through his records an animal breeder found that 55% of 40 female rabbits had killed one or more of their offspring when they had their first litter, whereas the incidence of killing dropped to 32.5% for these same animals when they had their second litter. Is this a significant reduction in the percentage of killers? His data are presented in Table 11.20.

Even though this looks like the usual 2 × 2 contingency table, it is not. The 40 animals in this study were observed both when they had their first litter and when they had their second litter, and thus the two percentages of 55% and 32.5% are correlated rather than independent as we have seen in all previous cases in this chapter. The principle involved in testing to determine whether the change of proportion is significant is as follows.

In Table 11.20 the critical information is contained in cells a and d (which has, respectively, 10 and 1 cases). The animals in cells b and c were consistent killers or nonkillers and therefore make no contribution to the change in proportion. Cell a informs us that 10 rabbits who killed one or more pups at the first litter did not kill any pups when they had their second litter, and cell d tells us that one rabbit who did not kill when she had her first litter did kill at her second litter. These numbers are what bring about the reduction in killing incidence from 55% to 32.5%. If this reduction is random, then our expectation is that the 11 cases involved in the cell $a + d$ (= 10 + 1 = 11) should be equal or, in this example, 5.5 cases. The question then becomes: Is the observed distribution of the scores in cells a and d significantly deviant from the expectation of equality in the two cells? We can test that question by the following formula:

$$\chi^2 = \frac{(|a - d| - 1)^2}{a + d} \tag{11.5}$$

Formula 11.5 has built into it Yates' correction for continuity and is a χ^2 test with one degree of freedom. In this particular example $\chi^2 = (|10 - 1| - 1)^2/(10 + 1) = 64/11 = 5.82$. This value of χ^2 is between the .02 and .01 values in the χ^2 table. If the breeder had set the .05 level for rejecting the null hypothesis, he would conclude that the incidence of killing is reduced from the first to the second litter. If the .01 level had been set, then he would conclude that there is no evidence of a significant reduction in the incidence of killing young.

MEASURES OF STRENGTH OF ASSOCIATION FOR TWO CLASSIFICATIONS

In Chapter 10 we described the Pearson product-moment correlation coefficient and pointed out that it was used to measure the degree of linear relationship between two variables. There is a parallel situation for nominal data, although the coefficients we shall present do not have the ease of interpretation nor the degree of generality that the Pearson correlation coefficient has. However, these coefficients do give the researcher a rough indication of strength of association.

The Phi Coefficient in a 2 × 2 Table

When each of two nominal scales are categorized as 0 and 1, the strength of association between the two scales is given by the phi (ϕ) coefficient, defined as follows:

$$\phi = \frac{(bc - ad)}{\sqrt{(a + b)(c + d)(a + c)(b + d)}} \tag{11.6}$$

One does not usually compute the ϕ coefficient itself, because it can be obtained directly from the value of χ^2. The relationship is

$$\chi^2 = n\phi^2 \quad \text{and} \quad \phi = \sqrt{\frac{\chi^2}{n}} \tag{11.7}$$

Therefore, the usual χ^2 test for independence of association of the two nominal scales is also a test of the null hypothesis that ϕ equals 0.

There is an interesting relationship between the Pearson correlation coefficient and the ϕ coefficient. If the data on each variable is "scored" with the numbers 0 and 1, and if you compute the Pearson product-moment correlation coefficient on those data, you will get the identical value as you would get if you computed the ϕ coefficient by means of Formula 11.6. That is, the ϕ coefficient is a Pearson correlation coefficient when the variables X and Y are scored as 0 and 1.

The major difficulty with the ϕ coefficient is that the theoretical upper limit of the value for ϕ can only be 1.00 when the row marginal totals are the same as the column marginal totals. Otherwise, the maximum value of ϕ is less than 1.00. This makes for difficulty in interpretation of the exact meaning of ϕ. For example, the ϕ coefficient may appear to be relatively low in numerical value, but if the maximum upper limit is much less than 1.00, then the strength of association is stronger than appears. There are approximation procedures to correct for this disparity, but they will not be discussed here because this is a rarely used technique in experimental research.

The Contingency Coefficient for the $R \times C$ Table

For the general case of an $R \times C$ table when either or both the row and column classification has more than two categories, the degree of association between the two nominal variables is obtained by the coefficient of contingency, designated by the symbol C. The formula is

$$C = \sqrt{\frac{\chi^2}{n + \chi^2}} \tag{11.8}$$

The usual test of χ^2 is also a test of the null hypothesis that the C coefficient $= 0$ in the population. This coefficient has the same limitation as does the ϕ coefficient in that the maximal value can only be 1.00 under limited circumstances.

SUMMARY OF DEFINITIONS

Contingency table A table containing R rows of nominal categories representing one classification and C columns of nominal categories representing a second classification. Within the table are listed the frequency of occurrence of the observed event.

Fourfold table A contingency table in which both the row and column classifications contain two categories, thus resulting in four cells for the table.

PROBLEMS

1. Given the following observed frequencies, test the hypothesis that the theoretical frequencies in the population are in the ratio of $1:2:1$ for Classes I, II, and III.

Class

	I	II	III	Total
f_o	216	496	285	997

2. A public opinion survey asked 174 people from one community whether they would vote for Candidate A or Candidate B in the forthcoming election. One hundred people responded that they would vote for A, and 74 said that they would vote for B. Test the hypothesis that Candidates A and B are both equally preferred in the population.

3. A researcher investigated the hypothesis that certain intelligence test items were biased against women. Based on her hypothesis, she selected several items that she expected to show bias. She obtained the following data for one of these items.

	Pass	Fail	Total
Males	42	21	63
Females	33	35	68
Total	75	56	131

Test her hypothesis.

4. For the data in Problem 3, determine the degree of association between sex and passing or failing the test item.

5. An animal behaviorist noted that one of his strains of mice exhibited one of four different behavior patterns when they were placed into a complex environment and allowed to explore freely. He wondered whether other mouse strains would show the same distribution. To investigate this he took 40 mice from each of three strains, placed each singly into the apparatus, and classified the animal with respect to which of the four behavior patterns characterized its exploratory behavior. His data are as follows. Test the hypothesis that there is no association between the distribution of behavior patterns and the strain of mouse studied.

Behavior pattern

	A	B	C	D	Total
Strain 1	12	7	18	3	40
Strain 2	9	6	4	21	40
Strain 3	1	23	10	6	40
Total	22	36	32	30	120

6. For Problem 5, determine the contingency coefficient.

7. Compute the Fisher exact probability for the following table.

	−	+	Total
+	0	4	4
−	4	1	5
Total	4	5	9

8. A social psychologist wished to determine whether seeing a particular movie would change attitudes toward capital punishment. She had each of the 113 students in her class state whether they were opposed to or in favor of capital punishment. A week later she showed a movie in class and then asked them the same question. Her data are as follows.

		After movie		
		Opposed to capital punishment	In favor of capital punishment	Total
Before Movie	In favor of capital punishment	17	22	39
	Opposed to capital punishment	41	33	74
	Total	58	55	113

Did the movie have an effect on changing attitudes toward capital punishment?

CHAPTER 12

Order Statistics for
Ranked Data

Order Statistics for Ranked Data

One of the scales of measurement described in Chapter 2 was the ordinal scale, in which the subject matter under investigation is ranked in order on some dimension or attibute of interest to the researcher. Generally, researchers use rank-order data for one of three reasons: (1) the measurement procedure yields only rank-order information; (2) even though the data are measured on a continuous dimension, researchers may not have confidence that their measurements approximate an equal interval scale, and therefore they prefer to use the lesser ordinal scale; or (3) researchers do not believe that the assumptions necessary to use the *t* test or analysis of variance procedures can be met with the scale of measurement that being used. We shall comment on each of these in turn.

The most valid reason for using order statistics is because the data are obtained as rank-order information. For example, if a researcher asks three judges to observe the behavior of six adolescent male monkeys interacting in a colony and to rank the monkeys from the most aggressive to the least aggressive, then the measured variable is intrinsically ordinal information and it is not only appropriate but necessary to use order statistics to evaluate this information. Note that because one uses an ordinal measurement procedure as a research tool, this does not mean that the variable being measured has only ordinal properties. It may well be that a different measurement procedure may yield a scale approximating that of equal intervals. However, if all that the researcher has in hand are data based on rank-order information, he or she has no choice except to use order statistics to evaluate these data. The methods discussed in this chapter are particularly appropriate for this objective.

A second situation commonly found in research is one in which the investigator has been able to devise a measurement procedure that yields varying numerical values, and that appears to approximate an equal interval scale. However, the researcher may have reason to believe that the measures do not accurately characterize the spacing between subjects, although they may clearly represent the proper rank ordering of the subjects. In a sense, this kind of measurement falls between the ordinal and interval scales, and leaves us with two major choices. We can decide to convert our data to rank-order information because we have considerable confidence that our scale is at that level of capability. This is a conservative procedure because we are throwing away some of the information from the original measurement scale, but the loss of that information does us little harm if we are able to find significant effects with our ordinal scale of measurement. However, it is difficult to interpret the lack of significant findings in this situation. The second choice that we have is to use the more powerful t test or analysis of variance procedures with our original scale of measurement even though we lack confidence in the scale itself. The difficulty here is that we may have problems in interpreting significant results because they may reflect some peculiarities of the scale values rather than the effects of our experimental variable. An independent replication of the experiment would be a way to partially resolve that dilemma.

An argument that has been advanced at times for the use of rank-order statistics is that the measurement scale is not able to meet the assumptions underlying the use of the t test or the analysis of variance. The lack of normality and the lack of homogeneity of variance are the two most common examples cited. We have already seen in our discussion of the central limit theorem that the normality assumption is involved with the distribution of means and not of the raw scores themselves, and that when the sample size is approximately 20-30, then the normality assumption is satisfied regardless of the form of the raw score distribution. We have also pointed out that the homogeneity of variance assumption is not an important limitation to the use of analysis of variance procedures. Thus, if one is able to obtain sufficient numbers of subjects in each cell of the experiment, this argument loses its validity. However, if it is necessary to use small sample sizes, then it may be necessary to treat the data as ordinal information even though the scale of measurement is somewhat better than that.

It is apparent, therefore, that researchers often will have a choice as to whether to use the original scale in which things are measured in a continuous manner, or to convert the data to discrete rank-order information. In considering this choice certain things should be kept in mind. The first thing to note is that converting the data to ordinal values throws away some information. That is, if all that we use is the rank data, we are using less of the information than we have available to us from the original measurement procedure. The second thing to note is that if the assumptions underlying our classical statistical procedures, such as the t test or analysis of variance, are met, then by using rank-order statistics we have a less powerful experiment. That is, with all assumptions met, a statistical procedure based on ordinal information has less probability of rejecting the null hypothesis when it is false than does an analysis of variance or t test procedure.

The final point concerning the relative merits of the two procedures is that the null hypothesis being tested when we use ordinal statistical procedures is not the same null hypothesis as is tested in the t test or analysis of variance procedures. In the latter case the specific hypothesis under examination is that the means in the population of the various treatment groups are identical. When using order statistics to compare groups, the null hypothesis is that the population distributions are identical. There are numerous ways whereby two population distributions may differ from each other, only one of which is that they differ in their means. Thus, when the null hypothesis is rejected via ordinal statistical procedures, this does not lead to a very definitive conclusion. That is, to state that we reject the null hypothesis that the means in the population are the same is more specific than to reject the null hypothesis that two population distributions are the same.

The purpose of this chapter is to present five kinds of order statistics for the five most commonly occurring research situations you will encounter: comparing two independent groups, comparing two matched groups, comparing three or more independent groups, comparing three or more matched groups, and determining the degree of association between two variables. These procedures parallel the material in previous chapters where we discussed the use of the t test for independent and matched groups, the analysis of variance for independent groups and for matched groups (in the randomized blocks design), and the use of the Pearson correlation coefficient to determine the degree of linear association between variables. For an extensive discussion of the use of order statistics, as well as other nonparametric procedures, the reader is referred to the text by Siegel (1956).

COMPARING TWO INDEPENDENT GROUPS:
THE MANN-WHITNEY TEST

When we have measured two independent groups of subjects on a continuous variable, we typically use the t test to evaluate the means. However, it may be that we obtained our data directly in terms of ranks, or we may prefer to convert from the original scores to ranks because we do not believe that the scores as we obtained them are an accurate representation of the underlying continuous variable in which we are interested. When the data are in ranks, the Mann-Whitney test allows us to test the null hypothesis that the two population distributions are identical. This test is a close approximation to the t test, and the results compare favorably with the results obtained in using the t test when the assumptions underlying both tests are met.

To use this test we pool the data from both groups of subjects and give the lowest-ranking subject the score of 1, the next-lowest-ranking subject the score of 2, and so on. In doing this we disregard the particular group from which each subject came. When we have finished assigning rank scores to all the subjects, we choose one of the samples, for example the sample containing n_1 subjects, and find the sum of the ranks for that group. We shall call this value T_1. We then find

$$U = n_1 n_2 + \frac{n_1(n_1 + 1)}{2} - T_1 \tag{12.1}$$

If we had selected the other group, containing n_2 subjects, the sum of the ranks within that group would equal T_2 and we would then find the statistic

$$U' = n_1 n_2 + \frac{n_2(n_2 + 1)}{2} - T_2 \qquad (12.2)$$

The test of the null hypothesis is done by evaluating U or U'. The procedure for evaluating U is dependent on the sample size. If the larger of the two samples has 20 or more cases in it, then we can use a normal curve approximation. If the sample size is less than 20, special tables are needed. These are given in Siegel's (1956) text. We shall consider here the situation where at least one of the groups has 20 subjects.

If the null hypothesis that the subjects have been randomly drawn from one population is true, then the expected value for the U statistic is

$$\mu_u = \frac{n_1 n_2}{2} \qquad (12.3)$$

The standard deviation of the distribution is given by the following formula:

$$\sigma_u = \sqrt{\frac{n_1 n_2 (n_1 + n_2 + 1)}{12}} \qquad (12.4)$$

For the situation where at least one sample size is 20, the sampling distribution of U is approximately normal, and so the null hypothesis of no difference in population distributions can be tested by

$$z = \frac{U - \mu_u}{\sigma_u} \qquad (12.5)$$

To illustrate the use of this procedure we shall consider the study of a primatologist who observed that some pregnant female monkeys were attacked by colony females and were sufficiently injured to require medical attention, whereas other pregnant monkeys were not bothered by the females in the colony. He speculated that there might be a relationship between whether or not a female was attacked and the sex of the fetus she was carrying. Therefore, he went through the colony records and obtained the files of all females that had given birth in the past 3 years. He copied the medical records of the animals during their pregnancy onto separate sheets of paper and gave the 48 sheets to a judge who had no knowledge of the purpose of the experiment. The judge was requested to rank the 48 animals in order from the one who had the least amount of injury during gestation to the one who had the greatest amount of injury. The primatologist then took the ranks of the judge, went back to the original file records, and determined the sex of the fetuses that the females were

TABLE 12.1 Rank order of
amount of injury suffered by
pregnant monkeys classified
by sex of fetus

Carrying male fetus	Carrying female fetus
1	8
2	10
3	11
4	16
5	17
6	19
7	22
9	23
12	29
13	30
14	31
15	32
18	33
20	37
21	40
24	41
25	42
26	43
27	44
28	45
34	46
35	47
36	48
38	$T_2 = 714$
39	
$T_1 = 462$	

carrying. Table 12.1 shows the rank order of the 25 monkeys carrying male
fetuses and the 23 monkeys carrying female fetuses.

If we designate the group carrying male fetuses as n_1, the total of the ranks in
that group, T_1, is 462, and the value of U is

$$U = (25)(23) + \frac{(25)(26)}{2} - 462 = 438$$

On the assumption that the null hypothesis is true, the expected value of U is

$$\mu_u = \frac{(25)(23)}{2} = 287.5$$

The standard deviation of the U distribution is given by

$$\sigma_u = \sqrt{\frac{(25)\,(23)\,(49)}{12}} = 48.46$$

Using Formula 12.5, we find the value of z to be

$$z = \frac{438 - 287.5}{48.46} = 3.11$$

Because z is normally distributed, 3.11 can be looked up in the normal curve table and is found to be well beyond the .01 level. The primatologist would conclude, therefore, that the two populations of monkeys are not identical.

If U' had been used instead of U, the same result would have been obtained. For the data in Table 12.1, $U' = 137$ and $z = (137 - 287.5)/48.46 = -3.11$. If a two-tailed test is used, the sign of z is ignored. However, if a one-tailed test is used, then the sign is taken into consideration when using the normal curve tables.

Remember that the procedure described above may be used only when the larger sample size is 20 or more. Siegel (1956) has an extensive discussion with appropriate tables for the conditions when neither sample size is as large as 20.

COMPARING TWO MATCHED GROUPS:
THE WILCOXEN TEST

The Wilcoxen test, like the matched t test, requires that the subjects in the two groups be matched on a pair-by-pair basis. In addition, the subjects have to be measured on a continuous variable rather than simply be rank ordered as in the previous example. The difference between the paired scores on the measured variable is obtained, and then the differences are assigned rank-order values disregarding the sign of the difference. Thus, the smallest absolute difference receives a rank of 1, the next larger absolute difference receives a rank of 2, and so forth. Any pair that has a difference of zero is not included in the analysis. After the ranking, the algebraic sign is assigned to the rank scores. If the two samples had been randomly drawn from the same population, then the summation of the positive ranks should be approximately equal to the summation of the negative ranks in the sample. To test the null hypothesis, we obtain the test statistic, T, which is defined as the sum of the ranks with the less frequent sign. When the number of pairs is 9 or larger, the T statistic is distributed approximately normally with the mean equal to

$$\mu_t = \frac{n(n + 1)}{4} \tag{12.6}$$

The standard deviation is

$$\sigma_t = \sqrt{\frac{n(n+1)(2n+1)}{24}} \tag{12.7}$$

Thus, t can be evaluated as a normal deviate by the formula

$$z = \frac{T - \mu_t}{\sigma_t} \tag{12.8}$$

For our example we shall return to the research experiment comparing the cortical brain weights of littermate pairs, one of whom was reared in an impoverished environment while the other was reared in an enriched environment. These data were first given in Table 7.2 and are presented again in Table 12.2. The second and third columns list the weights in milligrams, the fourth column gives the difference between each pair in brain weight, the fifth column lists the absolute ranks disregarding the pair with zero difference, and the last column reproduces the rank values, but assigns the sign to each rank. The less frequent sign is negative and the summation of the negative values is one, and therefore $T = 1$.

Using Formula 12.6, we find that the mean of the ranks in the population = $(9)(10)/4 = 22.5$. From Formula 12.7 we determine that the standard deviation is equal to $\sqrt{(9)(10)(19)/24} = 8.44$. We now use Formula 12.8 and find that

$$z = \frac{1 - 22.5}{8.44} = 2.55$$

When we enter the normal curve table with our value of 2.55, we find the probability level associated with this is slightly greater than .01. Therefore, if we had set the .05 level for rejecting the null hypothesis, we would then conclude that the two samples do not come from the same population distribution.

TABLE 12.2 Analysis of the data from Table 7.2 by use of the Wilcoxen test

Pair number	Control	Experimental	Difference	Rank	Signed rank
1	620	681	61	9	9
2	835	862	27	4	4
3	427	447	20	3	3
4	770	765	−5	1	−1
5	536	574	38	5	5
6	617	617	0	−	−
7	698	740	42	6	6
8	811	866	55	8	8
9	473	485	12	2	2
10	755	802	47	7	7

COMPARING THREE OR MORE INDEPENDENT GROUPS:
THE KRUSKAL–WALLIS SINGLE–CLASSIFICATION
ANALYSIS OF VARIANCE BY RANKS

In the same manner that the analysis of variance is a generalized version of the t test, the Kruskal-Wallis test is a generalization of the Mann-Whitney test comparing two groups. We assume that the underlying variable is continuously distributed. The original scores are then transformed to ranks and the analysis is done on the ranked data.

For example, suppose that a pharmacologist takes 18 mice from her colony, randomly divides them into three groups, and injects each animal with either drug A, B, or C. An hour later she places each animal singly into a standardized aggression-testing situation and records the number of attacks against a standard stimulus mouse in 15 minutes. Table 12.3 lists the number of attacks, and, in parentheses, the rank order of the 18 attack scores. At the bottom of the table, T_i indicates the rank total for each drug condition. The null hypothesis H_0, is that the three drug conditions are all from identical populations. This hypothesis can be tested by a statistic called H, which is defined as follows:

$$H = \frac{12}{n(n + 1)} \left(\Sigma \; \frac{T_i^2}{n_i} \right) - 3(n + 1) \tag{12.9}$$

where T_i = total for ranks of the ith treatment group
  n_i = number of subjects in the ith treatment group
  n = total number of subjects in the study $= \Sigma \, n_i$

If there are three or more treatment groups, and n_i for each group is greater than 5, then the H statistic can be evaluated by use of the χ^2 table with $k - 1$ degrees of freedom, where k = number of treatment groups. If the observed value of H is equal to or larger than the value of χ^2 with $k - 1$ df, then H_0 may be rejected at the previously specified level of significance.

TABLE 12.3   Number of attacks by mice after treatment with drugs A, B, or C (ranks given in parentheses)

	Drug A	Drug B	Drug C
	10 (9)	11 (10)	26 (18)
	6 (3.5)	9 (7.5)	13 (13)
	2 (1)	7 (5)	21 (17)
	6 (3.5)	13 (13)	13 (13)
	9 (7.5)	15 (15)	18 (16)
	4 (2)	8 (6)	12 (11)
T_i	26.5	56.5	88

For the data in Table 12.3, the value of H is

$$H = \frac{12}{(18)(18+1)} \left[\frac{(26.5)^2}{6} + \frac{(56.5)^2}{6} + \frac{(88)^2}{6} \right] - 3(18+1) = 11.06$$

Because there are six cases in each group, the χ^2 table can be used. With 2 df, the H of 11.06 is beyond the .01 level in the χ^2 table. Thus, whether the researcher set the .05 or .01 level for rejecting the null hypothesis, she would conclude that the three drugs do not come from the same population.

Correction for Ties

There are several ties in the data in Table 12.3: two values of 6, two values of 9, and three values of 13. Tied ranks are always assigned the mid-rank value. The effects of tied ranks is to reduce the numerical value of the H statistic. In this instance, because H was significant, it would not be necessary to make an adjustment for ties. However, if the H statistic was near borderline for significance, then a correction should be made. The correction formula is

$$C = 1 - \left[\frac{\Sigma(t_j^3 - t_j)}{n^3 - n} \right] \qquad (12.10)$$

where t_j = the number tied in any set j.

For the problem in Table 12.3,

$$C = 1 - \frac{(2^3 - 2) + (2^3 - 22) + (3^3 - 3)}{18^3 - 18} = .9938$$

The adjusted value of H is obtained by dividing by C. Thus, in this example $11.06/.9938 = 11.13$.

COMPARING THREE OR MORE MATCHED GROUPS: THE FRIEDMAN TEST

In the same way that the Kruskal-Wallis test is an extension of the Mann-Whitney test from the two-group situation to a larger number of independent groups, so is the Friedman test an extension of the Wilcoxin test from the situation of matched pairs to the general case where there are more than two matched groups. Thus, the Friedman test is analogous to the randomized blocks analysis of variance design discussed in Chapter 9, and can be used in essentially the same situations as you would use a randomized blocks design.

The layout for the Friedman test is the same as in a two-way analysis of variance. The columns represent the experimental treatments and the rows represent the randomized block or matching variable. There is one observation per cell. The data can be in the form of a measured variable for each subject and experimental treatment combination which is then converted to a rank-order score, or the original data can be a rank-order score. In either case the ranking must be within rows, so that the matched subjects (or randomized blocks) are ranked in order across the experimental conditions.

As an example, suppose that the pharmacologist who did the experiment described in Table 12.3 wishes to expand her study to include other strains of mice. In her first experiment, she used three drugs with mouse strain 1. She now wishes to add a fourth drug and to extend this study to 10 different mouse strains. Within each strain four animals are chosen and are randomly given drug A, B, C, or D. They are then tested for aggression. The number of attacks are listed in Table 12.4. These data are converted to ranks within each row (i.e., strain) by assigning the rank of 1 to the animal with the least number of attacks, and so on. The sum of the ranks for each drug condition is given at the bottom of Table 12.4.

The null hypothesis is that these four samples are drawn from identical populations. If so, then the sum of the ranks in the populations are identical. Therefore, we evaluate the sum of the sample ranks to see whether they are more deviant than can be expected by chance alone. For the situation when there are four or more columns and 10 or more rows, the distribution of rank totals can be evaluated by a χ^2 test with $J - 1$ degrees of freedom, where J is the number of columns. The χ^2 statistic is

$$\chi_r^2 = \frac{12}{KJ(J + 1)} (\Sigma T_i^2) - 3K(J + 1) \qquad (12.11)$$

TABLE 12.4 Number of attacks by mice of 10 different strains treated with drugs A, B, C, or D (ranks given in parentheses)

Strain	Drug A	Drug B	Drug C	Drug D
1	4 (1)	10 (2)	15 (4)	12 (3)
2	21 (1)	23 (2)	25 (3)	27 (4)
3	4 (1)	21 (2)	33 (4)	26 (3)
4	5 (1)	30 (3)	29 (2)	33 (4)
5	36 (4)	32 (3)	25 (2)	23 (1)
6	17 (1)	22 (3)	18 (2)	33 (4)
7	12 (1)	15 (2)	31 (4)	28 (3)
8	16 (3)	12 (2)	4 (1)	24 (4)
9	26 (4)	4 (1)	10 (2)	24 (3)
10	31 (4)	12 (2)	6 (1)	28 (3)
T_i	21	22	25	32

where K = number of rows
J = number of columns
T_i = sum of ranks in the ith column
For this example,

$$\chi_r^2 = \frac{12}{(10)(4)(5)} [(21)^2 + (22)^2 + (25)^2 + (32)^2] - (3)(10)(5) = 4.44$$

With $J - 1 = 3$ df, the obtained value of χ^2 falls between the .30 and .20 probability levels. The researcher would conclude, therefore, that there is no evidence that the four drugs come from different populations. This finding appears to contradict her prior experiment, where she did find differences among drugs A, B, and C when strain 1 was used. Indeed, the findings of that experiment are replicated in the first row of Table 12.4. Inspection of the data table suggests that there are some major strain $\times$ drug interactions. If so, this could readily account for the failure to find a significant effect, because it is not possible to evaluate the interaction term with rank-order data. If she wished to evaluate the interaction term, the researcher could set up a factorial analysis of variance design with strains as one factor and drugs as another factor, have at least two observations within each treatment combination cell, and thus be able to obtain a within-cell error term to be used to evaluate the interaction effect.

If there are less than four columns or less than 10 rows, the χ^2 test described above cannot be used. See Siegel (1956) for the procedures and tables to use in this situation.

A MEASURE OF RELATIONSHIP: THE SPEARMAN RANK CORRELATION COEFFICIENT

Suppose that we are interesting in measuring the degree of association between two variables. We draw a sample of n cases and measure each of our subjects on both variables. If our measures approximate an interval scale, we would compute the Pearson product-moment correlation coefficient. However, if our data are measured only in rank order, or if we convert our measured variables to ranks, we can then determine the degree of association by use of the Spearman rank correlation coefficient. As an example of the use of this statistic, consider a psychologist who is interested in the question of generalization of drive. He wants to find out whether those animals that are most successful in food competition are also successful when it comes to competition for water. He takes 14 rats from his colony, places them on a food-deprivation schedule, then pairs two animals in a chamber where there is one food pellet, and records which animal is successful in obtaining food. He does this for all possible pairs of animals, each time recording the winner. From these data he is able to obtain a ranking with the animal ranked 1 having won the most fights, down to animal

14. After this, the animals are kept undisturbed for a week with ample food and water in their cages. Thereafter, he removes the water for 12 hours and repeats his procedure, this time using a dental wad of cotton that has been soaked in water. He again obtains a rank order of success under the water competition situation, and these data are shown in Table 12.5.

If the Pearson correlation coefficient is computed for the two sets of ranks in Table 12.5, this will be identical with the Spearman correlation coefficient, which is designated by the symbol r_s. However, a simpler computational formula to obtain the Spearman correlation is

$$r_s = 1 - \frac{6 \Sigma D^2}{n(n^2 - 1)} \qquad (12.12)$$

where D = difference in ranks of paired scores.

Thus, all that is necessary to compute the correlation coefficient is to take the difference in ranks, which is shown in the last column of Table 12.5, square and sum the D values, and substitute them in Formula 12.12. When this is done for this problem, we find that

$$r_s = 1 - \frac{(6)(128)}{14(195)} = .719$$

The significance of r can be determined from Table XI in the Appendix when there are 14 or more subjects in the study. When n is less than 14, the table given in Siegel's (1956) text may be used. For this example there are 14 subjects and

TABLE 12.5   Rank orders of rats tested under food and water competition

Subject	Rank order in food competition situation	Rank order in water competition situation	D
A	9	7	2
B	7	8	1
C	10	12	2
D	13	10	3
E	14	14	0
F	2	1	1
G	6	4	2
H	4	9	5
I	5	6	1
J	12	5	7
K	8	13	5
L	1	3	2
M	11	11	0
N	3	2	1

thus 12 degrees of freedom, and we find that r_s is significant beyond the .01 level. Thus, the researcher would reject the null hypothesis that $r_s = 0$ in the population.

If there are ties in the data so that two or more individuals obtain the same rank score, this will tend to inflate the value of r_s. In such a situation, use the Pearson correlation coefficient formula to compute r_s, and this will make the appropriate correction for ties.

A FINAL COMMENT

It is important to emphasize again the strengths and weaknesses of rank statistics. If your data are originally collected in rank-order form, then you have no choice but to use the statistics described in this chapter. However, if your measurements have been taken on a scale approximating that of an equal interval and you convert your data to ranks, you must be aware that you are doing two things: (1) You are using less information than you have available in your original data, and (2) you are *not* making a test that the population means are identical; instead you are testing the null hypothesis that the two populations distributions (which include the means as one parameter) are identical. We have seen from the central limit theorem that when our sample size is approximately 20-30, we are able to meet the assumptions involved in using the t test and analysis of variance. Thus, if you are worried about the nature of your scale of measurement, you can choose to transform your data to ranks and use rank-order statistics, or you can increase the number of subjects within each treatment group until you have adequate numbers to satisfy the central limit theorem. If you do the latter, you have a more powerful experiment statistically, and you also use all the information available in your measurement scale. Because empirical research is always fraught with many difficulties, it may be wiser to increase the sample size and use the more powerful statistical procedures.

PROBLEMS

1. A researcher wished to test the hypothesis that the presence of testosterone in infancy will increase the adult aggressiveness of female mice. He injected 20 mice with testosterone on the second day of life, whereas controls received a saline injection. When adult, each animal was given an aggression test on three successive days and a composite aggression score was obtained. The researcher considered the score to represent the rank order of the mice, but nothing more. He therefore converted the scores to ranks, with 1 representing the least amount of aggression. His data are as follows:

Saline in infancy	Testosterone in infancy
1	5
2	8
3	9
4	10
6	14
7	15
11	16
12	17
13	22
18	23
19	24
20	29
21	30
25	33
26	35
27	36
28	37
31	38
32	39
34	40

Test the one-tailed hypothesis that the group receiving testosterone is more aggressive.

2. A school psychologist wished to test the hypothesis that girls are more susceptible to suggestion than are boys. In the school were nine pairs of brother-sister fraternal twins which he used as his subjects. Each person was given a suggestability test, and the following scores were obtained.

Pair	Brother	Sister
1	10	17
2	46	31
3	79	50
4	19	31
5	20	11
6	31	33
7	25	15
8	28	20
9	74	60

Use the Wilcoxen test to evaluate the one-tailed hypothesis.

3. Return to Problem 1 in Chapter 8, convert the data to ranks, and compute the Kruskal-Wallis analysis of variance.

4. A consumer psychologist writes five different advertisements for a product. She then has a group of 11 judges rank the advertisements from poorest to best. Her data are as follows:

Advertisement

Judge	A	B	C	D	E
1	1	3	2	4	5
2	2	1	3	5	4
3	2	5	1	4	3
4	1	2	3	4	5
5	4	2	1	3	5
6	3	1	2	5	4
7	1	2	5	4	3
8	2	4	1	5	3
9	2	5	1	3	4
10	3	1	2	5	4
11	1	4	2	3	5

Evaluate these data by means of the Friedman test.

5. Seventeen fifth-year graduate students in clinical psychology were ranked by the faculty in terms of their expected contribution to the research literature during the next 10 years. The combined Verbal and Quantitative Graduate Record Examination scores, taken prior to admission to graduate school, were converted to ranks for these students. Their data were as follows.

Student	GRE rank	Faculty rank
1	13	17
2	2	15
3	17	12
4	12	3
5	16	14
6	5	9
7	8	7
8	3	4
9	10	8
10	11	16
11	1	5
12	15	6
13	14	10
14	7	11
15	6	2
16	9	13
17	4	1

Compute the Spearman correlation between these two sets of ranks.

Appendix

298   Table I   Proportion of the area under the normal distribution curve and ordinates corresponding to given standard scores

300   Figure I   Relationship of the symbols in Table I to the normal curve

301   Table II   Table of t

301   Table IIa   Additional values of t at the 5 and the 1 per cent levels of significance

302   Table III   Power table to determine n needed to detect d by t test for different values of a

304   Table IV   The 5 and 1 per cent points for the distribution of F

308   Table IVa   The 25 per cent point for the distribution of F

310   Table V   Distribution of the q statistic

312   Table VI   Distribution of t statistic in comparing treatment means with a control

314   Table VII   Coefficients of orthogonal polynomials

315   Table VIII n to detect f by F test at $a = .01$

316   Table IX n to detect f by F test at $a = .05$

317   Table X n to detect f by F test at $a = .10$

318   Table XI   Values of r at the 5 and the 1 per cent levels of significance

319   Table XII   Table of χ^2

320   Table XIII   Table of random numbers

TABLE I Proportion of the area under the normal distribution curve and ordinates corresponding to given standard scores

z Standard score (x/σ)	A Area from mean to x/σ	B Area in larger portion	C Area in smaller portion	y Ordinate at x/σ
0.00	.0000	.5000	.5000	.3989
0.05	.0199	.5199	.4801	.3984
0.10	.0398	.5398	.4602	.3970
0.15	.0596	.5596	.4404	.3945
0.20	.0793	.5793	.4207	.3910
0.25	.0987	.5987	.4013	.3867
0.30	.1179	.6179	.3821	.3814
0.35	.1368	.6368	.3632	.3752
0.40	.1554	.6554	.3446	.3683
0.45	.1736	.6736	.3264	.3605
0.50	.1915	.6915	.3085	.3521
0.55	.2088	.7088	.2912	.3429
0.60	.2257	.7257	.2743	.3332
0.65	.2422	.7422	.2578	.3230
0.70	.2580	.7580	.2420	.3123
0.75	.2734	.7734	.2266	.3011
0.80	.2881	.7881	.2119	.2897
0.85	.3023	.8023	.1977	.2780
0.90	.3159	.8159	.1841	.2661
0.95	.3289	.8289	.1711	.2541
1.00	.3413	.8413	.1587	.2420
1.05	.3531	.8531	.1469	.2299
1.10	.3643	.8643	.1357	.2179
1.15	.3749	.8749	.1251	.2059
1.20	.3849	.8849	.1151	.1942
1.25	.3944	.8944	.1056	.1826
1.30	.4032	.9032	.0968	.1714
1.35	.4115	.9115	.0885	.1604
1.40	.4192	.9192	.0808	.1497
1.45	.4265	.9265	.0735	.1394
1.50	.4332	.9332	.0668	.1295
1.55	.4394	.9394	.0606	.1200
1.60	.4452	.9452	.0548	.1109
1.65	.4505	.9505	.0495	.1023
1.70	.4554	.9554	.0446	.0940

Note. From J. P. Guilford, *Fundamental Statistics in Psychology and Education,* New York: McGraw-Hill, 1956. Copyright © 1956 by the McGraw-Hill Book Company, Inc. Reprinted by permission.

z Standard score (x/σ)	A Area from mean to x/σ	B Area in larger portion	C Area in smaller portion	y Ordinate at x/σ
1.75	.4599	.9599	.0401	.0863
1.80	.4641	.9641	.0359	.0790
1.85	.4678	.9678	.0322	.0721
1.90	.4713	.9713	.0287	.0656
1.95	.4744	.9744	.0256	.0596
2.00	.4772	.9772	.0228	.0540
2.05	.4798	.9798	.0202	.0488
2.10	.4821	.9821	.0179	.0440
2.15	.4842	.9842	.0158	.0396
2.20	.4861	.9861	.0139	.0355
2.25	.4878	.9878	.0122	.0317
2.30	.4893	.9893	.0107	.0283
2.35	.4906	.9906	.0094	.0252
2.40	.4918	.9918	.0082	.0224
2.45	.4929	.9929	.0071	.0198
2.50	.4938	.9938	.0062	.0175
2.55	.4946	.9946	.0054	.0154
2.60	.4953	.9953	.0047	.0136
2.65	.4960	.9960	.0040	.0119
2.70	.4965	.9965	.0035	.0104
2.80	.4974	.9974	.0026	.0079
2.90	.4981	.9981	.0019	.0060
3.00	.49865	.99865	.00135	.0044
3.10	.49903	.99903	.00097	.0033
3.20	.49931	.99931	.00069	.0024
3.40	.49966	.99966	.00034	.0012
3.60	.49984	.99984	.00016	.00061
3.80	.499928	.999928	.000072	.00029
4.00	.4999683	.9999683	.0000317	.00013
4.50	.4999966	.9999966	.0000034	.000015
5.00	.49999971	.99999971	.00000029	.0000015
6.00	.499999999	.999999999	.000000001	.000000006

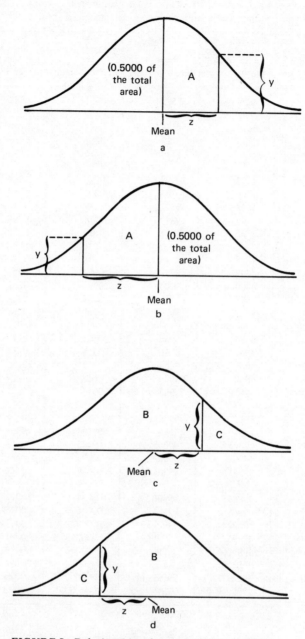

FIGURE I Relationship of the symbols in Table I to the normal curve. From J. P. Guilford, *Fundamental Statistics in Psychology and Education,* New York: McGraw-Hill, 1956. Copyright © 1956 by the McGraw-Hill Book Company, Inc. Reprinted by permission.

TABLE II Table of t

df P = .9	.8	.7	.6	.5	.4	.3	.2	.1	.05	.02	.01
1 .158	.325	.510	.727	1.000	1.376	1.963	3.078	6.314	12.706	31.821	63.657
2 .142	.289	.445	.617	.816	1.061	1.386	1.886	2.920	4.303	6.965	9.925
3 .137	.277	.424	.584	.765	.978	1.250	1.638	2.353	3.182	4.541	5.841
4 .134	.271	.414	.569	.741	.941	1.190	1.533	2.132	2.776	3.747	4.604
5 .132	.267	.408	.559	.727	.920	1.156	1.476	2.015	2.571	3.365	4.032
6 .131	.265	.404	.553	.718	.906	1.134	1.440	1.943	2.447	3.143	3.707
7 .130	.263	.402	.549	.711	.896	1.119	1.415	1.895	2.365	2.998	3.499
8 .130	.262	.399	.546	.706	.889	1.108	1.397	1.860	2.306	2.896	3.355
9 .129	.261	.398	.543	.703	.883	1.100	1.383	1.833	2.262	2.821	3.250
10 .129	.260	.397	.542	.700	.879	1.093	1.372	1.812	2.228	2.764	3.169
11 .129	.260	.396	.540	.697	.876	1.088	1.363	1.796	2.201	2.718	3.106
12 .128	.259	.395	.539	.695	.873	1.083	1.356	1.782	2.179	2.681	3.055
13 .128	.259	.394	.538	.694	.870	1.079	1.350	1.771	2.160	2.650	3.012
14 .128	.258	.393	.537	.692	.868	1.076	1.345	1.761	2.145	2.624	2.977
15 .128	.258	.393	.536	.691	.866	1.074	1.341	1.753	2.131	2.602	2.947
16 .128	.258	.392	.535	.690	.865	1.071	1.337	1.746	2.120	2.583	2.921
17 .128	.257	.392	.534	.689	.863	1.069	1.333	1.740	2.110	2.567	2.898
18 .127	.257	.392	.534	.688	.862	1.067	1.330	1.734	2.101	2.552	2.878
19 .127	.257	.391	.533	.688	.861	1.066	1.328	1.729	2.093	2.539	2.861
20 .127	.257	.391	.533	.687	.860	1.064	1.325	1.725	2.086	2.528	2.845
21 .127	.257	.391	.532	.686	.859	1.063	1.323	1.721	2.080	2.518	2.831
22 .127	.256	.390	.532	.686	.858	1.061	1.321	1.717	2.074	2.508	2.819
23 .127	.256	.390	.532	.685	.858	1.060	1.319	1.714	2.069	2.500	2.807
24 .127	.256	.390	.531	.685	.857	1.059	1.318	1.711	2.064	2.492	2.797
25 .127	.256	.390	.531	.684	.856	1.058	1.316	1.708	2.060	2.485	2.787
26 .127	.256	.390	.531	.684	.856	1.058	1.315	1.706	2.056	2.479	2.779
27 .127	.256	.289	.531	.684	.855	1.057	1.314	1.703	2.052	2.473	2.771
28 .127	.256	.389	.530	.683	.855	1.056	1.313	1.701	2.048	2.467	2.763
29 .127	.256	.389	.530	.683	.854	1.055	1.311	1.699	2.045	2.462	2.756
30 .127	.256	.389	.530	.683	.854	1.055	1.310	1.697	2.042	2.457	2.750
∞ .12566	.25335	.38532	.52440	.67449	.84162	1.03643	1.28155	1.64485	1.95996	2.32634	2.57582

Note. From R. A. Fisher, *Statistical Methods for Research Workers,* 11th Edition–Revised, New York: Hafner Publishing Company, 1950. Reprinted by permission.

TABLE IIa Additional values of t at the 5 and the 1 per cent levels of significance

df	5%	1%	df	5%	1%
35	2.030	2.724	70	1.994	2.648
40	2.021	2.704	80	1.989	2.638
45	2.014	2.690	90	1.986	2.630
50	2.008	2.678	100	1.982	2.625
55	2.004	2.669	120	1.980	2.617
60	2.000	2.660	∞	1.9600	2.5758

Note. Reprinted by permission from *Statistical Methods* by George W. Snedecor and William G. Cochran, sixth edition (c) 1967 by Iowa State University Press, Ames, Iowa.

TABLE III   Power table to determine n needed to detect d by t test for different values of a (a_2 = two-tailed test; a_1 = one-tailed test)

$a_2 = .01$ or $a_1 = .005$

Power	.10	.20	.30	.40	.50	.60	.70	.80	1.00	1.20	1.40
					d						
.25	725	183	82	47	31	22	17	13	9	7	6
.50	1329	333	149	85	55	39	29	22	15	11	9
.60	1603	402	180	102	66	46	34	27	18	13	10
2/3	1810	454	203	115	74	52	39	30	20	14	11
.70	1924	482	215	122	79	55	41	32	21	15	12
.75	2108	528	236	134	86	60	45	35	23	17	13
.80	2338	586	259	148	95	67	49	38	25	18	14
.85	2611	654	292	165	106	74	55	43	28	20	15
.90	2978	746	332	188	120	84	62	48	31	22	17
.95	3564	892	398	224	144	101	74	57	37	26	20
.99	4808	1203	536	302	194	136	100	77	50	35	26

$a_2 = .02$ or $a_1 = .01$

Power	.10	.20	.30	.40	.50	.60	.70	.80	1.00	1.20	1.40
					d						
.25	547	138	62	36	24	17	13	10	7	5	4
.50	1083	272	122	69	45	31	24	18	12	9	7
.60	1332	334	149	85	55	38	29	22	15	11	8
2/3	1552	382	170	97	62	44	33	25	17	12	9
.70	1627	408	182	103	66	47	35	27	18	13	10
.75	1803	452	202	114	74	52	38	30	20	14	11
.80	2009	503	224	127	82	57	42	33	22	15	12
.85	2263	567	253	143	92	64	48	37	24	17	13
.90	2605	652	290	164	105	74	55	42	27	20	15
.95	3155	790	352	198	128	89	66	51	33	23	18
.99	4330	1084	482	272	175	122	90	69	45	31	23

$a_2 = .05$ or $a_1 = .025$

Power	.10	.20	.30	.40	.50	.60	.70	.80	1.00	1.20	1.40
					d						
.25	332	84	38	22	14	10	8	6	5	4	3
.50	769	193	86	49	32	22	17	13	9	7	5
.60	981	246	110	62	40	28	21	16	11	8	6
2/3	1144	287	128	73	47	33	24	19	12	9	7
.70	1235	310	138	78	50	35	26	20	13	10	7
.75	1389	348	155	88	57	40	29	23	15	11	8
.80	1571	393	175	99	64	45	33	26	17	12	9
.85	1797	450	201	113	73	51	38	29	19	14	10
.90	2102	526	234	132	85	59	44	34	22	16	12
.95	2600	651	290	163	105	73	54	42	27	19	14
.99	3675	920	409	231	148	103	76	58	38	27	20

TABLE III (*continued*) Power table to determine n needed to detect d by t test for different values of a (a_2 = two-tailed test; a_1 = one-tailed test)

$a_2 = .10$ or $a_1 = .05$

Power	.10	.20	.30	.40	.50	.60	.70	.80	1.00	1.20	1.40
					d						
.25	189	48	21	12	8	6	5	4	3	2	2
.50	542	136	61	35	22	16	12	9	6	5	4
.60	721	181	81	46	30	21	15	12	8	6	5
2/3	862	216	96	55	35	25	18	14	9	7	5
.70	942	236	105	60	38	27	20	15	10	7	6
.75	1076	270	120	68	44	31	23	18	11	8	6
.80	1237	310	138	78	50	35	26	20	13	9	7
.85	1438	360	160	91	58	41	30	23	15	11	8
.90	1713	429	191	108	69	48	36	27	18	13	10
.95	2165	542	241	136	87	61	45	35	22	16	12
.99	3155	789	351	198	127	88	65	50	32	23	17

$a_2 = .20$ or $a_1 = .10$

Power	.10	.20	.30	.40	.50	.60	.70	.80	1.00	1.20	1.40
					d						
.25	74	19	9	5	3	3	2	2	2	2	2
.50	329	82	37	21	14	10	7	5	4	3	2
.60	471	118	53	30	19	14	10	8	5	4	3
2/3	586	147	65	37	24	17	12	10	6	4	3
.70	653	163	73	41	27	19	14	11	7	5	4
.75	766	192	85	48	31	22	16	13	8	6	4
.80	902	226	100	57	36	26	19	14	10	7	5
.85	1075	269	120	67	43	30	22	17	11	8	6
.90	1314	329	146	82	53	37	27	21	14	10	7
.95	1713	428	191	107	69	48	35	27	18	12	9
.99	2604	651	290	163	104	73	53	41	26	18	14

TABLE IV The 5 (Roman type) and 1 (boldface type) per cent points for the distribution of F

Numerator degrees of freedom (for greater mean square)

Denominator df	1	2	3	4	5	6	7	8	9	10	11	12	14	16	20	24	30	40	50	75	100	200	500	∞
1	161 **4,052**	200 **4,999**	216 **5,403**	225 **5,625**	230 **5,764**	234 **5,859**	237 **5,928**	239 **5,981**	241 **6,022**	242 **6,056**	243 **6,082**	244 **6,106**	245 **6,142**	246 **6,169**	248 **6,208**	249 **6,234**	250 **6,261**	251 **6,286**	252 **6,302**	253 **6,323**	253 **6,334**	254 **6,352**	254 **6,361**	254 **6,366**
2	18.51 **98.49**	19.00 **99.00**	19.16 **99.17**	19.25 **99.25**	19.30 **99.30**	19.33 **99.33**	19.36 **99.36**	19.37 **99.37**	19.38 **99.39**	19.39 **99.40**	19.40 **99.41**	19.41 **99.42**	19.42 **99.43**	19.43 **99.44**	19.44 **99.45**	19.45 **99.46**	19.46 **99.47**	19.47 **99.48**	19.47 **99.48**	19.48 **99.49**	19.49 **99.49**	19.49 **99.49**	19.50 **99.50**	19.50 **99.50**
3	10.13 **34.12**	9.55 **30.82**	9.28 **29.46**	9.12 **28.71**	9.01 **28.24**	8.94 **27.91**	8.88 **27.67**	8.84 **27.49**	8.81 **27.34**	8.78 **27.23**	8.76 **27.13**	8.74 **27.05**	8.71 **26.92**	8.69 **26.83**	8.66 **26.69**	8.64 **26.60**	8.62 **26.50**	8.60 **26.41**	8.58 **26.35**	8.57 **26.27**	8.56 **26.23**	8.54 **26.18**	8.54 **26.14**	8.53 **26.12**
4	7.71 **21.20**	6.94 **18.00**	6.59 **16.69**	6.39 **15.98**	6.26 **15.52**	6.16 **15.21**	6.09 **14.98**	6.04 **14.80**	6.00 **14.66**	5.96 **14.54**	5.93 **14.45**	5.91 **14.37**	5.87 **14.24**	5.84 **14.15**	5.80 **14.02**	5.77 **13.93**	5.74 **13.83**	5.71 **13.74**	5.70 **13.69**	5.68 **13.61**	5.66 **13.57**	5.65 **13.52**	5.64 **13.48**	5.63 **13.46**
5	6.61 **16.26**	5.79 **13.27**	5.41 **12.06**	5.19 **11.39**	5.05 **10.97**	4.95 **10.67**	4.88 **10.45**	4.82 **10.29**	4.78 **10.15**	4.74 **10.05**	4.70 **9.96**	4.68 **9.89**	4.64 **9.77**	4.60 **9.68**	4.56 **9.55**	4.53 **9.47**	4.50 **9.38**	4.46 **9.29**	4.44 **9.24**	4.42 **9.17**	4.40 **9.13**	4.38 **9.07**	4.37 **9.04**	4.36 **9.02**
6	5.99 **13.74**	5.14 **10.92**	4.76 **9.78**	4.53 **9.15**	4.39 **8.75**	4.28 **8.47**	4.21 **8.26**	4.15 **8.10**	4.10 **7.98**	4.06 **7.87**	4.03 **7.79**	4.00 **7.72**	3.96 **7.60**	3.92 **7.52**	3.87 **7.39**	3.84 **7.31**	3.81 **7.23**	3.77 **7.14**	3.75 **7.09**	3.72 **7.02**	3.71 **6.99**	3.69 **6.94**	3.68 **6.90**	3.67 **6.88**
7	5.59 **12.25**	4.74 **9.55**	4.35 **8.45**	4.12 **7.85**	3.97 **7.46**	3.87 **7.19**	3.79 **7.00**	3.73 **6.84**	3.68 **6.71**	3.63 **6.62**	3.60 **6.54**	3.57 **6.47**	3.52 **6.35**	3.49 **6.27**	3.44 **6.15**	3.41 **6.07**	3.38 **5.98**	3.34 **5.90**	3.32 **5.85**	3.29 **5.78**	3.28 **5.75**	3.25 **5.70**	3.24 **5.67**	3.23 **5.65**
8	5.32 **11.26**	4.46 **8.65**	4.07 **7.59**	3.84 **7.01**	3.69 **6.63**	3.58 **6.37**	3.50 **6.19**	3.44 **6.03**	3.39 **5.91**	3.34 **5.82**	3.31 **5.74**	3.28 **5.67**	3.23 **5.56**	3.20 **5.48**	3.15 **5.36**	3.12 **5.28**	3.08 **5.20**	3.05 **5.11**	3.03 **5.06**	3.00 **5.00**	2.98 **4.96**	2.96 **4.91**	2.94 **4.88**	2.93 **4.86**
9	5.12 **10.56**	4.26 **8.02**	3.86 **6.99**	3.63 **6.42**	3.48 **6.06**	3.37 **5.80**	3.29 **5.62**	3.23 **5.47**	3.18 **5.35**	3.13 **5.26**	3.10 **5.18**	3.07 **5.11**	3.02 **5.00**	2.98 **4.92**	2.93 **4.80**	2.90 **4.73**	2.86 **4.64**	2.82 **4.56**	2.80 **4.51**	2.77 **4.45**	2.76 **4.41**	2.73 **4.36**	2.72 **4.33**	2.71 **4.31**
10	4.96 **10.04**	4.10 **7.56**	3.71 **6.55**	3.48 **5.99**	3.33 **5.64**	3.22 **5.39**	3.14 **5.21**	3.07 **5.06**	3.02 **4.95**	2.97 **4.85**	2.94 **4.78**	2.91 **4.71**	2.86 **4.60**	2.82 **4.52**	2.77 **4.41**	2.74 **4.33**	2.70 **4.25**	2.67 **4.17**	2.64 **4.12**	2.61 **4.05**	2.59 **4.01**	2.56 **3.96**	2.55 **3.93**	2.54 **3.91**
11	4.84 **9.65**	3.98 **7.20**	3.59 **6.22**	3.36 **5.67**	3.20 **5.32**	3.09 **5.07**	3.01 **4.88**	2.95 **4.74**	2.90 **4.63**	2.86 **4.54**	2.82 **4.46**	2.79 **4.40**	2.74 **4.29**	2.70 **4.21**	2.65 **4.10**	2.61 **4.02**	2.57 **3.94**	2.53 **3.86**	2.50 **3.80**	2.47 **3.74**	2.45 **3.70**	2.42 **3.66**	2.41 **3.62**	2.40 **3.60**
12	4.75 **9.33**	3.88 **6.93**	3.49 **5.95**	3.26 **5.41**	3.11 **5.06**	3.00 **4.82**	2.92 **4.65**	2.85 **4.50**	2.80 **4.39**	2.76 **4.30**	2.72 **4.22**	2.69 **4.16**	2.64 **4.05**	2.60 **3.98**	2.54 **3.86**	2.50 **3.78**	2.46 **3.70**	2.42 **3.61**	2.40 **3.56**	2.36 **3.49**	2.35 **3.46**	2.32 **3.41**	2.31 **3.38**	2.30 **3.36**
13	4.67 **9.07**	3.80 **6.70**	3.41 **5.74**	3.18 **5.20**	3.02 **4.86**	2.92 **4.62**	2.84 **4.44**	2.77 **4.30**	2.72 **4.19**	2.67 **4.10**	2.63 **4.02**	2.60 **3.96**	2.55 **3.85**	2.51 **3.78**	2.46 **3.67**	2.42 **3.59**	2.38 **3.51**	2.34 **3.42**	2.32 **3.37**	2.28 **3.30**	2.26 **3.27**	2.24 **3.21**	2.22 **3.18**	2.21 **3.16**

Note. Reprinted by permission from *Statistical Methods* by George W. Snedecor and William G. Cochran, sixth edition (c) 1967 by Iowa State University Press, Ames, Iowa.

TABLE IV (*continued*)   The 5 (Roman type) and 1 (boldface type) per cent points for the distribution of F

Numerator degrees of freedom (for greater mean square)

Denominator df	1	2	3	4	5	6	7	8	9	10	11	12	14	16	20	24	30	40	50	75	100	200	500	∞
14	4.60 **8.86**	3.74 **6.51**	3.34 **5.56**	3.11 **5.03**	2.96 **4.69**	2.85 **4.46**	2.77 **4.28**	2.70 **4.14**	2.65 **4.03**	2.60 **3.94**	2.56 **3.86**	2.53 **3.80**	2.48 **3.70**	2.44 **3.62**	2.39 **3.51**	2.35 **3.43**	2.31 **3.34**	2.27 **3.26**	2.24 **3.21**	2.21 **3.14**	2.19 **3.11**	2.16 **3.06**	2.14 **3.02**	2.13 **3.00**
15	4.54 **8.68**	3.68 **6.36**	3.29 **5.42**	3.06 **4.89**	2.90 **4.56**	2.79 **4.32**	2.70 **4.14**	2.64 **4.00**	2.59 **3.89**	2.55 **3.80**	2.51 **3.73**	2.48 **3.67**	2.43 **3.56**	2.39 **3.48**	2.33 **3.36**	2.29 **3.29**	2.25 **3.20**	2.21 **3.12**	2.18 **3.07**	2.15 **3.00**	2.12 **2.97**	2.10 **2.92**	2.08 **2.89**	2.07 **2.87**
16	4.49 **8.53**	3.63 **6.23**	3.24 **5.29**	3.01 **4.77**	2.85 **4.44**	2.74 **4.20**	2.66 **4.03**	2.59 **3.89**	2.54 **3.78**	2.49 **3.69**	2.45 **3.61**	2.42 **3.55**	2.37 **3.45**	2.33 **3.37**	2.28 **3.25**	2.24 **3.18**	2.20 **3.10**	2.16 **3.01**	2.13 **2.96**	2.09 **2.98**	2.07 **2.86**	2.04 **2.80**	2.02 **2.77**	2.01 **2.75**
17	4.45 **8.40**	3.59 **6.11**	3.20 **5.18**	2.96 **4.67**	2.81 **4.34**	2.70 **4.10**	2.62 **3.93**	2.55 **3.79**	2.50 **3.68**	2.45 **3.59**	2.41 **3.52**	2.38 **3.45**	2.33 **3.35**	2.29 **3.27**	2.23 **3.16**	2.19 **3.08**	2.15 **3.00**	2.11 **2.92**	2.08 **2.86**	2.04 **2.79**	2.02 **2.76**	1.99 **2.70**	1.97 **2.67**	1.96 **2.65**
18	4.41 **8.28**	3.55 **6.01**	3.16 **5.09**	2.93 **4.58**	2.77 **4.25**	2.66 **4.01**	2.58 **3.85**	2.51 **3.71**	2.46 **3.60**	2.41 **3.51**	2.37 **3.44**	2.34 **3.37**	2.29 **3.27**	2.25 **3.19**	2.19 **3.07**	2.15 **3.00**	2.11 **2.91**	2.07 **2.83**	2.04 **2.78**	2.00 **2.71**	1.98 **2.68**	1.95 **2.62**	1.93 **2.59**	1.92 **2.57**
19	4.38 **8.18**	3.52 **5.93**	3.13 **5.01**	2.90 **4.50**	2.74 **4.17**	2.63 **3.94**	2.55 **3.77**	2.48 **3.63**	2.43 **3.52**	2.38 **3.43**	2.34 **3.36**	2.31 **3.30**	2.26 **3.19**	2.21 **3.12**	2.15 **3.00**	2.11 **2.92**	2.07 **2.84**	2.02 **2.76**	2.00 **2.70**	1.96 **2.63**	1.94 **2.60**	1.91 **2.54**	1.90 **2.51**	1.88 **2.49**
20	4.35 **8.10**	3.49 **5.85**	3.10 **4.94**	2.87 **4.43**	2.71 **4.10**	2.60 **3.87**	2.52 **3.71**	2.45 **3.56**	2.40 **3.45**	2.35 **3.37**	2.31 **3.30**	2.28 **3.23**	2.23 **3.13**	2.18 **3.05**	2.12 **2.94**	2.08 **2.86**	2.04 **2.77**	1.99 **2.69**	1.96 **2.63**	1.92 **2.56**	1.90 **2.53**	1.87 **2.47**	1.85 **2.44**	1.84 **2.42**
21	4.32 **8.02**	3.47 **5.78**	3.07 **4.87**	2.84 **4.37**	2.68 **4.04**	2.57 **3.81**	2.49 **3.65**	2.42 **3.51**	2.37 **3.40**	2.32 **3.31**	2.28 **3.24**	2.25 **3.17**	2.20 **3.07**	2.15 **2.99**	2.09 **2.88**	2.05 **2.80**	2.00 **2.72**	1.96 **2.63**	1.93 **2.58**	1.89 **2.51**	1.87 **2.47**	1.84 **2.42**	1.82 **2.38**	1.81 **2.36**
22	4.30 **7.94**	3.44 **5.72**	3.05 **4.82**	2.82 **4.31**	2.66 **3.99**	2.55 **3.76**	2.47 **3.59**	2.40 **3.45**	2.35 **3.35**	2.30 **3.26**	2.26 **3.18**	2.23 **3.12**	2.18 **3.02**	2.13 **2.94**	2.07 **2.83**	2.03 **2.75**	1.98 **2.67**	1.93 **2.58**	1.91 **2.53**	1.87 **2.46**	1.84 **2.42**	1.81 **2.37**	1.80 **2.33**	1.78 **2.31**
23	4.28 **7.88**	3.42 **5.66**	3.03 **4.76**	2.80 **4.26**	2.64 **3.94**	2.53 **3.71**	2.45 **3.54**	2.38 **3.41**	2.32 **3.30**	2.28 **3.21**	2.24 **3.14**	2.20 **3.07**	2.14 **2.97**	2.10 **2.89**	2.04 **2.78**	2.00 **2.70**	1.96 **2.62**	1.91 **2.53**	1.88 **2.48**	1.84 **2.41**	1.82 **2.37**	1.79 **2.32**	1.77 **2.28**	1.76 **2.26**
24	4.26 **7.82**	3.40 **5.61**	3.01 **4.72**	2.78 **4.22**	2.62 **3.90**	2.51 **3.67**	2.43 **3.50**	2.36 **3.36**	2.30 **3.25**	2.26 **3.17**	2.22 **3.09**	2.18 **3.03**	2.13 **2.93**	2.09 **2.85**	2.02 **2.74**	1.98 **2.66**	1.94 **2.58**	1.89 **2.49**	1.86 **2.44**	1.82 **2.36**	1.80 **2.33**	1.76 **2.27**	1.74 **2.23**	1.73 **2.21**
25	4.24 **7.77**	3.38 **5.57**	2.99 **4.68**	2.76 **4.18**	2.60 **3.86**	2.49 **3.63**	2.41 **3.46**	2.34 **3.32**	2.28 **3.21**	2.24 **3.13**	2.20 **3.05**	2.16 **2.99**	2.11 **2.89**	2.06 **2.81**	2.00 **2.70**	1.96 **2.62**	1.92 **2.54**	1.87 **2.45**	1.84 **2.40**	1.80 **2.32**	1.77 **2.29**	1.74 **2.23**	1.72 **2.19**	1.71 **2.17**
26	4.22 **7.72**	3.37 **5.53**	2.98 **4.64**	2.74 **4.14**	2.59 **3.82**	2.47 **3.59**	2.39 **3.42**	2.32 **3.29**	2.27 **3.17**	2.22 **3.09**	2.18 **3.02**	2.15 **2.96**	2.10 **2.86**	2.05 **2.77**	1.99 **2.66**	1.95 **2.58**	1.90 **2.50**	1.85 **2.41**	1.82 **2.36**	1.78 **2.28**	1.76 **2.25**	1.72 **2.19**	1.70 **2.15**	1.69 **2.13**

TABLE IV (*continued*) The 5 (Roman type) and 1 (boldface type) per cent points for the distribution of F

Denominator df	1	2	3	4	5	6	7	8	9	10	11	12	14	16	20	24	30	40	50	75	100	200	500	∞
									Numerator degrees of freedom (for greater mean square)															
27	4.21 **7.68**	3.35 **5.49**	2.96 **4.60**	2.73 **4.11**	2.57 **3.79**	2.46 **3.56**	2.37 **3.39**	2.30 **3.26**	2.25 **3.14**	2.20 **3.06**	2.16 **2.98**	2.13 **2.93**	2.08 **2.83**	2.03 **2.74**	1.97 **2.63**	1.93 **2.55**	1.88 **2.47**	1.84 **2.38**	1.80 **2.33**	1.76 **2.25**	1.74 **2.21**	1.71 **2.16**	1.68 **2.12**	1.67 **2.10**
28	4.20 **7.64**	3.34 **5.45**	2.95 **4.57**	2.71 **4.07**	2.56 **3.76**	2.44 **3.53**	2.36 **3.36**	2.29 **3.23**	2.24 **3.11**	2.19 **3.03**	2.15 **2.95**	2.12 **2.90**	2.06 **2.80**	2.02 **2.71**	1.96 **2.60**	1.91 **2.52**	1.87 **2.44**	1.81 **2.35**	1.78 **2.30**	1.75 **2.22**	1.72 **2.18**	1.69 **2.13**	1.67 **2.09**	1.65 **2.06**
29	4.18 **7.60**	3.33 **5.42**	2.93 **4.54**	2.70 **4.04**	2.54 **3.73**	2.43 **3.50**	2.35 **3.33**	2.28 **3.20**	2.22 **3.08**	2.18 **3.00**	2.14 **2.92**	2.10 **2.87**	2.05 **2.77**	2.00 **2.68**	1.94 **2.57**	1.90 **2.49**	1.85 **2.41**	1.80 **2.32**	1.77 **2.27**	1.73 **2.19**	1.71 **2.15**	1.68 **2.10**	1.65 **2.06**	1.64 **2.03**
30	4.17 **7.56**	3.32 **5.39**	2.92 **4.51**	2.69 **4.02**	2.53 **3.70**	2.42 **3.47**	2.34 **3.30**	2.27 **3.17**	2.21 **3.06**	2.16 **2.98**	2.12 **2.90**	2.09 **2.84**	2.04 **2.74**	1.99 **2.66**	1.93 **2.55**	1.89 **2.47**	1.84 **2.38**	1.79 **2.29**	1.76 **2.24**	1.72 **2.16**	1.69 **2.13**	1.66 **2.07**	1.64 **2.03**	1.62 **2.01**
32	4.15 **7.50**	3.30 **5.34**	2.90 **4.46**	2.67 **3.97**	2.51 **3.66**	2.40 **3.42**	2.32 **3.25**	2.25 **3.12**	2.19 **3.01**	2.14 **2.94**	2.10 **2.86**	2.07 **2.80**	2.02 **2.70**	1.97 **2.62**	1.91 **2.51**	1.86 **2.42**	1.82 **2.34**	1.76 **2.25**	1.74 **2.20**	1.69 **2.12**	1.67 **2.08**	1.64 **2.02**	1.61 **1.98**	1.59 **1.96**
34	4.13 **7.44**	3.28 **5.29**	2.88 **4.42**	2.65 **3.93**	2.49 **3.61**	2.38 **3.38**	2.30 **3.21**	2.23 **3.08**	2.17 **2.97**	2.12 **2.89**	2.08 **2.82**	2.05 **2.76**	2.00 **2.66**	1.95 **2.58**	1.89 **2.47**	1.84 **2.38**	1.80 **2.30**	1.74 **2.21**	1.71 **2.15**	1.67 **2.08**	1.64 **2.04**	1.61 **1.98**	1.59 **1.94**	1.57 **1.91**
36	4.11 **7.39**	3.26 **5.25**	2.86 **4.38**	2.63 **3.89**	2.48 **3.58**	2.36 **3.35**	2.28 **3.18**	2.21 **3.04**	2.15 **2.94**	2.10 **2.86**	2.06 **2.78**	2.03 **2.72**	1.98 **2.62**	1.93 **2.54**	1.87 **2.43**	1.82 **2.35**	1.78 **2.26**	1.72 **2.17**	1.69 **2.12**	1.65 **2.04**	1.62 **2.00**	1.59 **1.94**	1.56 **1.90**	1.55 **1.87**
38	4.10 **7.35**	3.25 **5.21**	2.85 **4.34**	2.62 **3.86**	2.46 **3.54**	2.35 **3.32**	2.26 **3.15**	2.19 **3.02**	2.14 **2.91**	2.09 **2.82**	2.05 **2.75**	2.02 **2.69**	1.96 **2.59**	1.92 **2.51**	1.85 **2.40**	1.80 **2.32**	1.76 **2.22**	1.71 **2.14**	1.67 **2.08**	1.63 **2.00**	1.60 **1.97**	1.57 **1.90**	1.54 **1.86**	1.53 **1.84**
40	4.08 **7.31**	3.23 **5.18**	2.84 **4.31**	2.61 **3.83**	2.45 **3.51**	2.34 **3.29**	2.25 **3.12**	2.18 **2.99**	2.12 **2.88**	2.07 **2.80**	2.04 **2.73**	2.00 **2.66**	1.95 **2.56**	1.90 **2.49**	1.84 **2.37**	1.79 **2.29**	1.74 **2.20**	1.69 **2.11**	1.66 **2.05**	1.61 **1.97**	1.59 **1.94**	1.55 **1.88**	1.53 **1.84**	1.51 **1.81**
42	4.07 **7.27**	3.22 **5.15**	2.83 **4.29**	2.59 **3.80**	2.44 **3.49**	2.32 **3.26**	2.24 **3.10**	2.17 **2.96**	2.11 **2.86**	2.06 **2.77**	2.02 **2.70**	1.99 **2.64**	1.94 **2.54**	1.89 **2.46**	1.82 **2.35**	1.78 **2.26**	1.73 **2.17**	1.68 **2.08**	1.64 **2.02**	1.60 **1.94**	1.57 **1.91**	1.54 **1.85**	1.51 **1.80**	1.49 **1.78**
44	4.06 **7.24**	3.21 **5.12**	2.82 **4.26**	2.58 **3.78**	2.43 **3.46**	2.31 **3.24**	2.23 **3.07**	2.16 **2.94**	2.10 **2.84**	2.05 **2.75**	2.01 **2.68**	1.98 **2.62**	1.92 **2.52**	1.88 **2.44**	1.81 **2.32**	1.76 **2.24**	1.72 **2.15**	1.66 **2.06**	1.63 **2.00**	1.58 **1.92**	1.56 **1.88**	1.52 **1.82**	1.50 **1.78**	1.48 **1.75**
46	4.05 **7.21**	3.20 **5.10**	2.81 **4.24**	2.57 **3.76**	2.42 **3.44**	2.30 **3.22**	2.22 **3.05**	2.14 **2.92**	2.09 **2.82**	2.04 **2.73**	2.00 **2.66**	1.97 **2.60**	1.91 **2.50**	1.87 **2.42**	1.80 **2.30**	1.75 **2.22**	1.71 **2.13**	1.65 **2.04**	1.62 **1.98**	1.57 **1.90**	1.54 **1.86**	1.51 **1.80**	1.48 **1.76**	1.46 **1.72**
48	4.04 **7.19**	3.19 **5.08**	2.80 **4.22**	2.56 **3.74**	2.41 **3.42**	2.30 **3.20**	2.21 **3.04**	2.14 **2.90**	2.08 **2.80**	2.03 **2.71**	1.99 **2.64**	1.96 **2.58**	1.90 **2.48**	1.86 **2.40**	1.79 **2.28**	1.74 **2.20**	1.70 **2.11**	1.64 **2.02**	1.61 **1.96**	1.56 **1.88**	1.53 **1.84**	1.50 **1.78**	1.47 **1.73**	1.45 **1.70**

TABLE IV (*continued*) The 5 (Roman type) and 1 (boldface type) per cent points for the distribution of *F*

Numerator degrees of freedom (for greater mean square)

Denominator df	1	2	3	4	5	6	7	8	9	10	11	12	14	16	20	24	30	40	50	75	100	200	500	∞
50	4.03 **7.17**	3.18 **5.06**	2.79 **4.20**	2.56 **3.72**	2.40 **3.41**	2.29 **3.18**	2.20 **3.02**	2.13 **2.88**	2.07 **2.78**	2.02 **2.70**	1.98 **2.62**	1.95 **2.56**	1.90 **2.46**	1.85 **2.39**	1.78 **2.26**	1.74 **2.18**	1.69 **2.10**	1.63 **2.00**	1.60 **1.94**	1.55 **1.86**	1.52 **1.82**	1.48 **1.76**	1.46 **1.71**	1.44 **1.68**
55	4.02 **7.12**	3.17 **5.01**	2.78 **4.16**	2.54 **3.68**	2.38 **3.37**	2.27 **3.15**	2.18 **2.98**	2.11 **2.85**	2.05 **2.75**	2.00 **2.66**	1.97 **2.59**	1.93 **2.53**	1.88 **2.43**	1.83 **2.35**	1.76 **2.23**	1.72 **2.15**	1.67 **2.06**	1.61 **1.96**	1.58 **1.90**	1.52 **1.82**	1.50 **1.78**	1.46 **1.71**	1.43 **1.66**	1.41 **1.64**
60	4.00 **7.08**	3.15 **4.98**	2.76 **4.13**	2.52 **3.65**	2.37 **3.34**	2.25 **3.12**	2.17 **2.95**	2.10 **2.82**	2.04 **2.72**	1.99 **2.63**	1.95 **2.56**	1.92 **2.50**	1.86 **2.40**	1.81 **2.32**	1.75 **2.20**	1.70 **2.12**	1.65 **2.03**	1.59 **1.93**	1.56 **1.87**	1.50 **1.79**	1.48 **1.74**	1.44 **1.68**	1.41 **1.63**	1.39 **1.60**
65	3.99 **7.04**	3.14 **4.95**	2.75 **4.10**	2.51 **3.62**	2.36 **3.31**	2.24 **3.09**	2.15 **2.93**	2.08 **2.79**	2.02 **2.70**	1.98 **2.61**	1.94 **2.54**	1.90 **2.47**	1.85 **2.37**	1.80 **2.30**	1.73 **2.18**	1.68 **2.09**	1.63 **2.00**	1.57 **1.90**	1.54 **1.84**	1.49 **1.76**	1.46 **1.71**	1.42 **1.64**	1.39 **1.60**	1.37 **1.56**
70	3.98 **7.01**	3.13 **4.92**	2.74 **4.08**	2.50 **3.60**	2.35 **3.29**	2.23 **3.07**	2.14 **2.91**	2.07 **2.77**	2.01 **2.67**	1.97 **2.59**	1.93 **2.51**	1.89 **2.45**	1.84 **2.35**	1.79 **2.28**	1.72 **2.15**	1.67 **2.07**	1.62 **1.98**	1.56 **1.88**	1.53 **1.82**	1.47 **1.74**	1.45 **1.69**	1.40 **1.62**	1.37 **1.56**	1.35 **1.53**
80	3.96 **6.96**	3.11 **4.88**	2.72 **4.04**	2.48 **3.56**	2.33 **3.25**	2.21 **3.04**	2.12 **2.87**	2.05 **2.74**	1.99 **2.64**	1.95 **2.55**	1.91 **2.48**	1.88 **2.41**	1.82 **2.32**	1.77 **2.24**	1.70 **2.11**	1.65 **2.03**	1.60 **1.94**	1.54 **1.84**	1.51 **1.78**	1.45 **1.70**	1.42 **1.65**	1.38 **1.57**	1.35 **1.52**	1.32 **1.49**
100	3.94 **6.90**	3.09 **4.82**	2.70 **3.98**	2.46 **3.51**	2.30 **3.20**	2.19 **2.99**	2.10 **2.82**	2.03 **2.69**	1.97 **2.59**	1.92 **2.51**	1.88 **2.43**	1.85 **2.36**	1.79 **2.26**	1.75 **2.19**	1.68 **2.06**	1.63 **1.98**	1.57 **1.89**	1.51 **1.79**	1.48 **1.73**	1.42 **1.64**	1.39 **1.59**	1.34 **1.51**	1.30 **1.46**	1.28 **1.43**
125	3.92 **6.84**	3.07 **4.78**	2.68 **3.94**	2.44 **3.47**	2.29 **3.17**	2.17 **2.95**	2.08 **2.79**	2.01 **2.65**	1.95 **2.56**	1.90 **2.47**	1.86 **2.40**	1.83 **2.33**	1.77 **2.23**	1.72 **2.15**	1.65 **2.03**	1.60 **1.94**	1.55 **1.85**	1.49 **1.75**	1.45 **1.68**	1.39 **1.59**	1.36 **1.54**	1.31 **1.46**	1.27 **1.40**	1.25 **1.37**
150	3.91 **6.81**	3.06 **4.75**	2.67 **3.91**	2.43 **3.44**	2.27 **3.14**	2.16 **2.92**	2.07 **2.76**	2.00 **2.62**	1.94 **2.53**	1.89 **2.44**	1.85 **2.37**	1.82 **2.30**	1.76 **2.20**	1.71 **2.12**	1.64 **2.00**	1.59 **1.91**	1.54 **1.83**	1.47 **1.72**	1.44 **1.66**	1.37 **1.56**	1.34 **1.51**	1.29 **1.43**	1.25 **1.37**	1.22 **1.33**
200	3.89 **6.76**	3.04 **4.71**	2.65 **3.88**	2.41 **3.41**	2.26 **3.11**	2.14 **2.90**	2.05 **2.73**	1.98 **2.60**	1.92 **2.50**	1.87 **2.41**	1.83 **2.34**	1.80 **2.28**	1.74 **2.17**	1.69 **2.09**	1.62 **1.97**	1.57 **1.88**	1.52 **1.79**	1.45 **1.69**	1.42 **1.62**	1.35 **1.53**	1.32 **1.48**	1.26 **1.39**	1.22 **1.33**	1.19 **1.28**
400	3.86 **6.70**	3.02 **4.66**	2.62 **3.83**	2.39 **3.36**	2.23 **3.06**	2.12 **2.85**	2.03 **2.69**	1.96 **2.55**	1.90 **2.46**	1.85 **2.37**	1.81 **2.29**	1.78 **2.23**	1.72 **2.12**	1.67 **2.04**	1.60 **1.92**	1.54 **1.84**	1.49 **1.74**	1.42 **1.64**	1.38 **1.57**	1.32 **1.47**	1.28 **1.42**	1.22 **1.32**	1.16 **1.24**	1.13 **1.19**
1000	3.85 **6.66**	3.00 **4.62**	2.61 **3.80**	2.38 **3.34**	2.22 **3.04**	2.10 **2.82**	2.02 **2.66**	1.95 **2.53**	1.89 **2.43**	1.84 **2.34**	1.80 **2.26**	1.76 **2.20**	1.70 **2.09**	1.65 **2.01**	1.58 **1.89**	1.53 **1.81**	1.47 **1.71**	1.41 **1.61**	1.36 **1.54**	1.30 **1.44**	1.26 **1.38**	1.19 **1.28**	1.13 **1.19**	1.08 **1.11**
∞	3.84 **6.64**	2.99 **4.60**	2.60 **3.78**	2.37 **3.32**	2.21 **3.02**	2.09 **2.80**	2.01 **2.64**	1.94 **2.51**	1.88 **2.41**	1.83 **2.32**	1.79 **2.24**	1.75 **2.18**	1.69 **2.07**	1.64 **1.99**	1.57 **1.87**	1.52 **1.79**	1.46 **1.69**	1.40 **1.59**	1.35 **1.52**	1.28 **1.41**	1.24 **1.36**	1.17 **1.25**	1.11 **1.15**	1.00 **1.00**

TABLE IVa The 25 per cent point for the distribution of F

df for denominator	df for numerator											
	1	2	3	4	5	6	7	8	9	10	11	12
1	5.83	7.50	8.20	8.58	8.82	8.98	9.10	9.19	9.26	9.32	9.36	9.41
2	2.57	3.00	3.15	3.23	3.28	3.31	3.34	3.35	3.37	3.38	3.39	3.39
3	2.02	2.28	2.36	2.39	2.41	2.42	2.43	2.44	2.44	2.44	2.45	2.45
4	1.81	2.00	2.05	2.06	2.07	2.08	2.08	2.08	2.08	2.08	2.08	2.08
5	1.69	1.85	1.88	1.89	1.89	1.89	1.89	1.89	1.89	1.89	1.89	1.89
6	1.62	1.76	1.78	1.79	1.79	1.78	1.78	1.77	1.77	1.77	1.77	1.77
7	1.57	1.70	1.72	1.72	1.71	1.71	1.70	1.70	1.69	1.69	1.69	1.68
8	1.54	1.66	1.67	1.66	1.66	1.65	1.64	1.64	1.64	1.63	1.63	1.62
9	1.51	1.62	1.63	1.63	1.62	1.61	1.60	1.60	1.59	1.59	1.58	1.58
10	1.49	1.60	1.60	1.59	1.59	1.58	1.57	1.56	1.56	1.55	1.55	1.54
11	1.47	1.58	1.58	1.57	1.56	1.55	1.54	1.53	1.53	1.52	1.52	1.51
12	1.46	1.56	1.56	1.55	1.54	1.53	1.52	1.51	1.51	1.50	1.50	1.49
13	1.45	1.54	1.54	1.53	1.52	1.51	1.50	1.49	1.49	1.48	1.47	1.47
14	1.44	1.53	1.53	1.52	1.51	1.50	1.48	1.48	1.47	1.46	1.46	1.45
15	1.43	1.52	1.52	1.51	1.49	1.48	1.47	1.46	1.46	1.45	1.44	1.44
16	1.42	1.51	1.51	1.50	1.48	1.48	1.47	1.46	1.45	1.45	1.44	1.44
17	1.42	1.51	1.50	1.49	1.47	1.46	1.45	1.44	1.43	1.43	1.42	1.41
18	1.41	1.50	1.49	1.48	1.46	1.45	1.44	1.43	1.42	1.42	1.41	1.40
19	1.41	1.49	1.49	1.47	1.46	1.44	1.43	1.42	1.41	1.41	1.40	1.40
20	1.40	1.49	1.48	1.46	1.45	1.44	1.42	1.42	1.41	1.40	1.39	1.39
22	1.40	1.48	1.47	1.45	1.44	1.42	1.41	1.40	1.39	1.39	1.38	1.37
24	1.39	1.47	1.46	1.44	1.43	1.41	1.40	1.39	1.38	1.38	1.37	1.36
26	1.38	1.46	1.45	1.44	1.42	1.41	1.40	1.39	1.37	1.37	1.36	1.35
28	1.38	1.46	1.45	1.43	1.41	1.40	1.39	1.38	1.37	1.36	1.35	1.34
30	1.38	1.45	1.44	1.42	1.41	1.39	1.38	1.37	1.36	1.35	1.35	1.34
40	1.36	1.44	1.42	1.40	1.39	1.37	1.36	1.35	1.34	1.33	1.32	1.31
60	1.35	1.42	1.41	1.38	1.37	1.35	1.33	1.32	1.31	1.30	1.29	1.29
120	1.34	1.40	1.39	1.37	1.35	1.33	1.31	1.30	1.29	1.28	1.27	1.26
200	1.33	1.39	1.38	1.36	1.34	1.32	1.31	1.29	1.28	1.27	1.26	1.25
∞	1.32	1.39	1.37	1.35	1.33	1.31	1.29	1.28	1.27	1.25	1.24	1.24

Note. From E. S. Pearson and H. O. Hartley (eds.), *Biometrika Tables for Statisticians*, vol. 1, 2nd ed., London: Cambridge University Press, 1958. Copyright 1958 by the Cambridge University Press. Reprinted by permission.

| | | | | | df for numerator | | | | | | |
15	20	24	30	40	50	60	100	120	200	500	∞
9.49	9.58	9.63	9.67	9.71	9.74	9.76	9.78	9.80	9.82	9.84	9.85
3.41	3.43	3.43	3.44	3.45	3.45	3.46	3.47	3.47	3.48	3.48	3.48
2.46	2.46	2.46	2.47	2.47	2.47	2.47	2.47	2.47	2.47	2.47	2.47
2.08	2.08	2.08	2.08	2.08	2.08	2.08	2.08	2.08	2.08	2.08	2.08
1.89	1.88	1.88	1.88	1.88	1.88	1.87	1.87	1.87	1.87	1.87	1.87
1.76	1.76	1.75	1.75	1.75	1.75	1.74	1.74	1.74	1.74	1.74	1.74
1.68	1.67	1.67	1.66	1.66	1.66	1.65	1.65	1.65	1.65	1.65	1.65
1.62	1.61	1.60	1.60	1.59	1.59	1.59	1.58	1.58	1.58	1.58	1.58
1.57	1.56	1.56	1.55	1.55	1.54	1.54	1.53	1.53	1.53	1.53	1.53
1.53	1.52	1.52	1.51	1.51	1.50	1.50	1.49	1.49	1.49	1.48	1.48
1.50	1.49	1.49	1.48	1.47	1.47	1.47	1.46	1.46	1.46	1.45	1.45
1.48	1.47	1.46	1.45	1.45	1.44	1.44	1.43	1.43	1.43	1.42	1.42
1.46	1.45	1.44	1.43	1.42	1.42	1.42	1.41	1.41	1.40	1.40	1.40
1.44	1.43	1.42	1.41	1.41	1.40	1.40	1.39	1.39	1.39	1.38	1.38
1.43	1.41	1.41	1.40	1.39	1.39	1.38	1.38	1.37	1.37	1.36	1.36
1.41	1.40	1.39	1.38	1.37	1.37	1.36	1.36	1.35	1.35	1.34	1.34
1.40	1.39	1.38	1.37	1.36	1.35	1.35	1.34	1.34	1.34	1.33	1.33
1.39	1.38	1.37	1.36	1.35	1.34	1.34	1.33	1.33	1.32	1.32	1.32
1.38	1.37	1.36	1.35	1.34	1.33	1.33	1.32	1.32	1.31	1.31	1.30
1.37	1.36	1.35	1.34	1.33	1.33	1.32	1.31	1.31	1.30	1.30	1.29
1.36	1.34	1.33	1.32	1.31	1.31	1.30	1.30	1.30	1.29	1.29	1.28
1.35	1.33	1.32	1.31	1.30	1.29	1.29	1.28	1.28	1.27	1.27	1.26
1.34	1.32	1.31	1.30	1.29	1.28	1.28	1.26	1.26	1.26	1.25	1.25
1.33	1.31	1.30	1.29	1.28	1.27	1.27	1.26	1.25	1.25	1.24	1.24
1.32	1.30	1.29	1.28	1.27	1.26	1.26	1.25	1.24	1.24	1.23	1.23
1.30	1.28	1.26	1.25	1.24	1.23	1.22	1.21	1.21	1.20	1.19	1.19
1.27	1.25	1.24	1.22	1.21	1.20	1.19	1.17	1.17	1.16	1.15	1.15
1.24	1.22	1.21	1.19	1.18	1.17	1.16	1.14	1.13	1.12	1.11	1.10
1.23	1.21	1.20	1.18	1.16	1.14	1.12	1.11	1.10	1.09	1.08	1.06
1.22	1.19	1.18	1.16	1.14	1.13	1.12	1.09	1.08	1.07	1.04	1.00

TABLE V Distribution of the q statistic

df for MS_{within}	α	2	3	4	5	6	7	8	9	10	11	12	13	14	15
								k = number of treatment groups							
1	.05	18.0	27.0	32.8	37.1	40.4	43.1	45.4	47.4	49.1	50.6	52.0	53.2	54.3	55.4
	.01	90.0	135	164	186	202	216	227	237	246	253	260	266	272	277
2	.05	6.09	8.3	9.8	10.9	11.7	12.4	13.0	13.5	14.0	14.4	14.7	15.1	15.4	15.7
	.01	14.0	19.0	22.3	24.7	26.6	28.2	29.5	30.7	31.7	32.6	33.4	34.1	34.8	35.4
3	.05	4.50	5.91	6.82	7.50	8.04	8.48	8.85	9.18	9.46	9.72	9.95	10.2	10.4	10.5
	.01	8.26	10.6	12.2	13.3	14.2	15.0	15.6	16.2	16.7	17.1	17.5	17.9	18.2	18.5
4	.05	3.93	5.04	5.76	6.29	6.71	7.05	7.35	7.60	7.83	8.03	8.21	8.37	8.52	8.66
	.01	6.51	8.12	9.17	9.96	10.6	11.1	11.5	11.9	12.3	12.6	12.8	13.1	13.3	13.5
5	.05	3.64	4.60	5.22	5.67	6.03	6.33	6.58	6.80	6.99	7.17	7.32	7.47	7.60	7.72
	.01	5.70	6.97	7.80	8.42	8.91	9.32	9.67	9.97	10.2	10.5	10.7	10.9	11.1	11.2
6	.05	3.46	4.34	4.90	5.31	5.63	5.89	6.12	6.32	6.49	6.65	6.79	6.92	7.03	7.14
	.01	5.24	6.33	7.03	7.56	7.97	8.32	8.61	8.87	9.10	9.30	9.49	9.65	9.81	9.95
7	.05	3.34	4.16	4.69	5.06	5.36	5.61	5.82	6.00	6.16	6.30	6.43	6.55	6.66	6.76
	.01	4.95	5.92	6.54	7.01	7.37	7.68	7.94	8.17	8.37	8.55	8.71	8.86	9.00	9.12
8	.05	3.26	4.04	4.53	4.89	5.17	5.40	5.60	5.77	5.92	6.05	6.18	6.29	6.39	6.48
	.01	4.74	5.63	6.20	6.63	6.96	7.24	7.47	7.68	7.87	8.03	8.18	8.31	8.44	8.55
9	.05	3.20	3.95	4.42	4.76	5.02	5.24	5.43	5.60	5.74	5.87	5.98	6.09	6.19	6.28
	.01	4.60	5.43	5.96	6.35	6.66	6.91	7.13	7.32	7.49	7.65	7.78	7.91	8.03	8.13
10	.05	3.15	3.88	4.33	4.65	4.91	5.12	5.30	5.46	5.60	5.72	5.83	5.93	6.03	6.11
	.01	4.48	5.27	5.77	6.14	6.43	6.67	6.87	7.05	7.21	7.36	7.48	7.60	7.71	7.81
11	.05	3.11	3.82	4.26	4.57	4.82	5.03	5.20	5.35	5.49	5.61	5.71	5.81	5.90	5.99
	.01	4.39	5.14	5.62	5.97	6.25	6.48	6.67	6.84	6.99	7.13	7.26	7.36	7.46	7.56
12	.05	3.08	3.77	4.20	4.51	4.75	4.95	5.12	5.27	5.40	5.51	5.62	5.71	5.80	5.88
	.01	4.32	5.04	5.50	5.84	6.10	6.32	6.51	6.67	6.81	6.94	7.06	7.17	7.26	7.36

Note. From H. Leon Harter, Donald S. Clemm, and Eugene H. Guthrie, The probability integrals of the range and of the studentized range, *WADC Tech. Rep.* 58-484, 1959, 2. Reprinted by permission.

TABLE V (*continued*) Distribution of the q statistic

df for MS_{within}	α	k = number of treatment groups													
		2	3	4	5	6	7	8	9	10	11	12	13	14	15
13	.05	3.06	3.73	4.15	4.45	4.69	4.88	5.05	5.19	5.32	5.43	5.53	5.63	5.71	5.79
	.01	4.26	4.96	5.40	5.73	5.98	6.19	6.37	6.53	6.67	6.79	6.90	7.01	7.10	7.19
14	.05	3.03	3.70	4.11	4.41	4.64	4.83	4.99	5.13	5.25	5.36	5.46	5.55	5.64	5.72
	.01	4.21	4.89	5.32	5.63	5.88	6.08	6.26	6.41	6.54	6.66	6.77	6.87	6.96	7.05
16	.05	3.00	3.65	4.05	4.33	4.56	4.74	4.90	5.03	5.15	5.26	5.35	5.44	5.52	5.59
	.01	4.13	4.78	5.19	5.49	5.72	5.92	6.08	6.22	6.35	6.46	6.56	6.66	6.74	6.82
18	.05	2.97	3.61	4.00	4.28	4.49	4.67	4.82	4.96	5.07	5.17	5.27	5.35	5.43	5.50
	.01	4.07	4.70	5.09	5.38	5.60	5.79	5.94	6.08	6.20	6.31	6.41	6.50	6.58	6.65
20	.05	2.95	3.58	3.96	4.23	4.45	4.62	4.77	4.90	5.01	5.11	5.20	5.28	5.36	5.43
	.01	4.02	4.64	5.02	5.29	5.51	5.69	5.84	5.97	6.09	6.19	6.29	6.37	6.45	6.52
24	.05	2.92	3.53	3.90	4.17	4.37	4.54	4.68	4.81	4.92	5.01	5.10	5.18	5.25	5.32
	.01	3.96	4.54	4.91	5.17	5.37	5.54	5.69	5.81	5.92	6.02	6.11	6.19	6.26	6.33
30	.05	2.89	3.49	3.84	4.10	4.30	4.46	4.60	4.72	4.83	4.92	5.00	5.08	5.15	5.21
	.01	3.89	4.45	4.80	5.05	5.24	5.40	5.54	5.56	5.76	5.85	5.93	6.01	6.08	6.14
40	.05	2.86	3.44	3.79	4.04	4.23	4.39	4.52	4.63	4.74	4.82	4.91	4.98	5.05	5.11
	.01	3.82	4.37	4.70	4.93	5.11	5.27	5.39	5.50	5.60	5.69	5.77	5.84	5.90	5.96
60	.05	2.83	3.40	3.74	3.98	4.16	4.31	4.44	4.55	4.65	4.73	4.81	4.88	4.94	5.00
	.01	3.76	4.28	4.60	4.82	4.99	5.13	5.25	5.36	5.45	5.53	5.60	5.67	5.73	5.79
120	.05	2.80	3.36	3.69	3.92	4.10	4.24	4.36	4.48	4.56	4.64	4.72	4.78	4.84	4.90
	.01	3.70	4.20	4.50	4.71	4.87	5.01	5.12	5.21	5.30	5.38	5.44	5.51	5.56	5.61
∞	.05	2.77	3.31	3.63	3.86	4.03	4.17	4.29	4.39	4.47	4.55	4.62	4.68	4.74	4.80
	.01	3.64	4.12	4.40	4.60	4.76	4.88	4.99	5.08	5.16	5.23	5.29	5.35	5.40	5.45

TABLE VI Distribution of t statistic in comparing treatment means with a control

df for MS_{within}	α	k = number of means (including control)								
		2	3	4	5	6	7	8	9	10
5	.10	2.02	2.44	2.68	2.85	2.98	3.08	3.16	3.24	3.03
	.05	2.57	3.03	3.29	3.48	3.62	3.73	3.82	3.90	3.97
	.02	3.36	3.90	4.21	4.43	4.60	4.73	4.85	4.94	5.03
	.01	4.03	4.63	4.98	5.22	5.41	5.56	5.69	5.80	5.89
6	.10	1.94	2.34	2.56	2.71	2.83	2.92	3.00	3.07	3.12
	.05	2.45	2.86	3.10	3.26	3.39	3.49	3.57	3.64	3.71
	.02	3.14	3.61	3.88	4.07	4.21	4.33	4.43	4.51	4.59
	.01	3.71	4.21	4.51	4.71	4.87	5.00	5.10	5.20	5.28
7	.10	1.89	2.27	2.48	2.62	2.73	2.82	2.89	2.95	3.01
	.05	2.36	2.75	2.97	3.12	3.24	3.33	3.41	3.47	3.53
	.02	3.00	3.42	3.66	3.83	3.96	4.07	4.15	4.23	4.30
	.01	3.50	3.95	4.21	4.39	4.53	4.64	4.74	4.82	4.89
8	.10	1.86	2.22	2.42	2.55	2.66	2.74	2.81	2.87	2.92
	.05	2.31	2.67	2.88	3.02	3.13	3.22	3.29	3.35	3.41
	.02	2.90	3.29	3.51	3.67	3.79	3.88	3.96	4.03	4.09
	.01	3.36	3.77	4.00	4.17	4.29	4.40	4.48	4.56	4.62
9	.10	1.83	2.18	2.37	2.50	2.60	2.68	2.75	2.81	2.86
	.05	2.26	2.61	2.81	2.95	3.05	3.14	3.20	3.26	3.32
	.02	2.28	3.19	3.40	3.55	3.66	3.75	3.82	3.89	3.94
	.01	3.25	3.63	3.85	4.01	4.12	4.22	4.30	4.37	4.43
10	.10	1.81	2.15	2.34	2.47	2.56	2.64	2.70	2.76	2.81
	.05	2.23	2.57	2.76	2.89	2.99	3.07	3.14	3.19	3.24
	.02	2.76	3.11	3.31	3.45	3.56	3.64	3.71	3.78	3.83
	.01	3.17	3.53	3.74	3.88	3.99	4.08	4.16	4.22	4.28
11	.10	1.80	2.13	2.31	2.44	2.53	2.60	2.67	2.72	2.77
	.05	2.20	2.53	2.72	2.84	2.94	3.02	3.08	3.14	3.19
	.02	2.72	3.06	3.25	3.38	3.48	3.56	3.63	3.69	3.74
	.01	3.11	3.45	3.65	3.79	3.89	3.98	4.05	4.11	4.16
12	.10	1.78	2.11	2.29	2.41	2.50	2.58	2.64	2.69	2.74
	.05	2.18	2.50	2.68	2.81	2.90	2.98	3.04	3.09	3.14
	.02	2.68	3.01	3.19	3.32	3.42	3.50	3.56	3.62	3.67
	.01	3.05	3.39	3.58	3.71	3.81	3.89	3.96	4.02	4.07
13	.10	1.77	2.09	2.27	2.39	2.48	2.55	2.61	2.66	2.71
	.05	2.16	2.48	2.65	2.78	2.87	2.94	3.00	3.06	3.10
	.02	2.65	2.97	3.15	3.27	3.37	3.44	3.51	3.56	3.61
	.01	3.01	3.33	3.52	3.65	3.74	3.82	3.89	3.94	3.99

Note. From: (1) "A Multiple Comparison Procedure for Comparing Several Treatments with a Control" by C. W. Dunnett, *Journal of the American Statistical Association*, 1955, *50*, 1096–1121. Copyright 1955 by the American Statistical Association. Reprinted by permission. (2) "New Tables for Multiple Comparisons with a Control" by C. W. Dunnett, *Biometrics*, 1964, *20*, 482–491. Copyright 1964 by The Biometrics Society. Reprinted by permission.

TABLE VI (*continued*) Distribution of *t* statistic in comparing treatment means with a control

df for MS_{within}	α	*k* = number of means (including control)								
		2	3	4	5	6	7	8	9	10
14	.10	1.76	2.08	2.25	2.37	2.46	2.53	2.59	2.64	2.69
	.05	2.14	2.46	2.63	2.75	2.84	2.91	2.97	3.02	3.07
	.02	2.62	2.94	3.11	3.23	3.32	3.40	3.46	3.51	3.56
	.01	2.98	3.29	3.47	3.59	3.69	3.76	3.83	3.88	3.93
16	.10	1.75	2.06	2.23	2.34	2.43	2.50	2.56	2.61	2.65
	.05	2.12	2.42	2.59	2.71	2.80	2.87	2.92	2.97	3.02
	.02	2.58	2.88	3.05	3.17	3.26	3.33	3.39	3.44	3.48
	.01	2.92	3.22	3.39	3.51	3.60	3.67	3.73	3.78	3.83
18	.10	1.73	2.04	2.21	2.32	2.41	2.48	2.53	2.58	2.62
	.05	2.10	2.40	2.56	2.68	2.76	2.83	2.89	2.94	2.98
	.02	2.55	2.84	3.01	3.12	3.21	3.27	3.33	3.38	3.42
	.01	2.88	3.17	3.33	3.44	3.53	3.60	3.66	3.71	3.75
20	.10	1.72	2.03	2.19	2.30	2.39	2.46	2.51	2.56	2.60
	.05	2.09	2.38	2.54	2.65	2.73	2.80	2.86	2.90	2.95
	.02	2.53	2.81	2.97	3.08	3.17	3.23	3.29	3.34	3.38
	.01	2.85	3.13	3.29	3.40	3.48	3.55	3.60	3.65	3.69
24	.10	1.71	2.01	2.17	2.28	2.36	2.43	2.48	2.53	2.57
	.05	2.06	2.35	2.51	2.61	2.70	2.76	2.81	2.86	2.90
	.02	2.49	2.77	2.92	3.03	3.11	3.17	3.22	3.27	3.31
	.01	2.80	3.07	3.22	3.32	3.40	3.47	3.52	3.57	3.61
30	.10	1.70	1.99	2.15	2.25	2.33	2.40	2.45	2.50	2.54
	.05	2.04	2.32	2.47	2.58	2.66	2.72	2.77	2.82	2.86
	.02	2.46	2.72	2.87	2.97	3.05	3.11	3.16	3.21	3.24
	.01	2.75	3.01	3.15	3.25	3.33	3.39	3.44	3.49	3.52
40	.10	1.68	1.97	2.13	2.23	2.31	2.37	2.42	2.47	2.51
	.05	2.02	2.29	2.44	2.54	2.62	2.68	2.73	2.77	2.81
	.02	2.42	2.68	2.82	2.92	2.99	3.05	3.10	3.14	3.18
	.01	2.70	2.95	3.09	3.19	3.26	3.32	3.37	3.41	3.44
60	.10	1.67	1.95	2.10	2.21	2.28	2.35	2.39	2.44	2.48
	.05	2.00	2.27	2.41	2.51	2.58	2.64	2.69	2.73	2.77
	.02	2.39	2.64	2.78	2.87	2.94	3.00	3.04	3.08	3.12
	.01	2.66	2.90	3.03	3.12	3.19	3.25	3.29	3.33	3.37
120	.10	1.66	1.93	2.08	2.18	2.26	2.32	2.37	2.41	2.45
	.05	1.98	2.24	2.38	2.47	2.55	2.60	2.65	2.69	2.73
	.02	2.36	2.60	2.73	2.82	2.89	2.94	2.99	3.03	3.06
	.01	2.62	2.85	2.97	3.06	3.12	3.18	3.22	3.26	3.29
∞	.10	1.64	1.92	2.06	2.16	2.23	2.29	2.34	2.38	2.42
	.05	1.96	2.21	2.35	2.44	2.51	2.57	2.61	2.65	2.69
	.02	2.33	2.56	2.68	2.77	2.84	2.89	2.93	2.97	3.00
	.01	2.58	2.79	2.92	3.00	3.06	3.11	3.15	3.19	3.22

TABLE VII Coefficients of orthogonal polynomials

Number of groups	Polynomial	$X = 1$	2	3	4	5	6	7	8	9	10	Σc^2	λ
3	Linear	−1	0	1								2	1
	Quadratic	1	−2	1								6	3
	Linear	−3	−1	1	3							20	2
4	Quadratic	1	−1	−1	1							4	1
	Cubic	−1	3	−3	1							20	$10\frac{1}{3}$
	Linear	−2	−1	0	1	2						10	1
5	Quadratic	2	−1	−2	−1	2						14	1
	Cubic	−1	2	0	−2	1						10	$\frac{5}{6}$
	Quartic	1	−4	6	−4	1						70	$\frac{35}{12}$
	Linear	−5	−3	−1	1	3	5					70	2
6	Quadratic	5	−1	−4	−4	−1	5					84	$\frac{3}{2}$
	Cubic	−5	7	4	−4	−7	5					180	$\frac{5}{3}$
	Quartic	1	−3	2	2	−3	1					28	$\frac{7}{12}$
	Linear	−3	−2	−1	0	1	2	3				28	1
7	Quadratic	5	0	−3	−4	−3	0	5				84	1
	Cubic	−1	1	1	0	−1	−1	1				6	$\frac{1}{6}$
	Quartic	3	−7	1	6	1	−7	3				154	$\frac{7}{12}$
	Linear	−7	−5	−3	−1	1	3	5	7			168	2
	Quadratic	7	1	−3	−5	−5	−3	1	7			168	1
8	Cubic	−7	5	7	3	−3	−7	−5	7			264	$\frac{2}{3}$
	Quartic	7	−13	−3	9	9	−3	−13	7			616	$\frac{7}{12}$
	Quintic	−7	23	−17	−15	15	17	−23	7			2184	$\frac{7}{10}$
	Linear	−4	−3	−2	−1	0	1	2	3	4		60	1
	Quadratic	28	7	−8	−17	−20	−17	−8	7	28		2772	3
9	Cubic	−14	7	13	9	0	−9	−13	−7	14		990	$\frac{5}{6}$
	Quartic	14	−21	−11	9	18	9	−11	−21	14		2002	$\frac{7}{12}$
	Quintic	−4	11	−4	−9	0	9	4	−11	4		468	$\frac{3}{20}$
	Linear	−9	−7	−5	−3	−1	1	3	5	7	9	330	2
	Quadratic	6	2	−1	−3	−4	−4	−3	−1	2	6	132	$\frac{1}{2}$
10	Cubic	−42	14	35	31	12	−12	−31	−35	−14	42	8580	$\frac{5}{3}$
	Quartic	18	−22	−17	3	18	18	3	−17	−22	18	2860	$\frac{5}{12}$
	Quintic	−6	14	−1	−11	−6	6	11	1	−14	6	780	$\frac{1}{10}$

Note. From B. J. Winer, *Statistical Principles in Experimental Design*, New York: McGraw-Hill, 1971. Copyright © 1962, 1971 by McGraw-Hill, Inc. Reprinted by permission.

TABLE VIII n to detect f by F test at $a = .01$

Numerator df	Power	\(f \) .05	.10	.15	.20	.25	.30	.35	.40	.50	.60	.70	.80
1	.10	336	85	39	22	15	11	9	7	5	4	4	3
	.50	1329	333	149	85	55	39	29	22	15	11	9	7
	.70	1924	482	215	122	79	55	41	32	21	15	12	9
	.80	2338	586	259	148	95	67	49	38	25	18	14	11
	.90	2978	746	332	188	120	84	62	48	31	22	17	13
	.95	3564	892	398	224	144	101	74	57	37	26	20	16
	.99	4808	1203	536	302	194	136	100	77	50	35	26	21

Numerator df	Power	\(f \) .05	.10	.15	.20	.25	.30	.35	.40	.50	.60	.70	.80
2	.10	307	79	36	21	14	10	8	6	5	4	3	3
	.50	1093	275	123	70	45	32	24	19	13	9	7	6
	.70	1543	387	173	98	63	44	33	26	17	12	10	8
	.80	1851	464	207	117	76	53	39	30	20	14	11	9
	.90	2325	582	260	147	95	66	49	38	25	18	14	11
	.95	2756	690	308	174	112	78	58	45	29	21	16	12
	.99	3658	916	408	230	148	103	76	59	38	27	20	16

Numerator df	Power	\(f \) .05	.10	.15	.20	.25	.30	.35	.40	.50	.60	.70	.80
3	.10	278	71	32	19	13	9	7	6	4	3	3	2
	.50	933	234	105	59	38	27	20	16	11	8	6	5
	.70	1299	326	146	83	53	37	28	22	14	10	8	7
	.80	1548	388	175	98	63	44	33	25	17	12	9	8
	.90	1927	483	215	122	78	55	41	31	21	15	11	9
	.95	2270	568	253	143	92	64	48	37	24	17	13	10
	.99	2986	747	333	188	121	84	62	48	31	22	17	13

Numerator df	Power	\(f \) .05	.10	.15	.20	.25	.30	.35	.40	.50	.60	.70	.80
4	.10	253	64	29	17	12	8	7	5	4	3	3	2
	.50	820	206	92	52	34	24	18	14	10	7	6	5
	.70	1128	283	127	72	46	33	24	19	13	9	7	6
	.80	1341	336	150	85	55	38	29	22	15	11	8	7
	.90	1661	416	186	105	68	47	35	27	18	13	10	8
	.95	1948	488	218	123	79	55	41	32	21	15	11	9
	.99	2546	640	286	160	103	76	53	41	27	19	14	11

Note. From J. Cohen, *Statistical Power Analysis for the Behavioral Sciences,* New York: Academic, 1969. Copyright © 1969 by Academic Press, Inc. Reprinted by permission.

TABLE IX n to detect f by F test at $a = .05$

Numerator df	Power	.05	.10	.15	.20	.25	.30	.35	.40	.50	.60	.70	.80
							f						
1	.10	84	22	10	6	5	4	3	3	2	--	--	--
	.50	769	193	86	49	32	22	17	13	9	7	5	4
	.70	1235	310	138	78	50	35	26	20	13	10	7	6
	.80	1571	393	175	99	64	45	33	26	17	12	9	7
	.90	2102	526	234	132	85	59	44	34	22	16	12	9
	.95	2600	651	290	163	105	73	54	42	27	19	14	11
	.99	3675	920	409	231	148	103	76	58	38	27	20	15

Numerator df	Power	.05	.10	.15	.20	.25	.30	.35	.40	.50	.60	.70	.80
							f						
2	.10	84	22	10	6	5	4	3	3	2	--	--	--
	.50	662	166	74	42	27	19	15	11	8	6	5	4
	.70	1028	258	115	65	42	29	22	17	11	8	6	5
	.80	1286	322	144	81	52	36	27	21	14	10	8	6
	.90	1682	421	188	106	68	48	35	27	18	13	10	8
	.95	2060	515	230	130	83	58	43	33	22	15	12	9
	.99	2855	714	318	179	115	80	59	46	29	21	16	12

Numerator df	Power	.05	.10	.15	.20	.25	.30	.35	.40	.50	.60	.70	.80
							f						
3	.10	79	21	10	6	4	3	3	2	2	--	--	--
	.50	577	145	65	37	24	16	13	10	7	5	4	3
	.70	881	221	99	56	36	25	19	15	10	7	6	5
	.80	1096	274	123	69	45	31	23	18	12	9	7	5
	.90	1415	354	158	89	58	40	30	23	15	11	8	7
	.95	1718	430	192	108	70	49	36	28	18	13	10	8
	.99	2353	589	262	148	95	66	49	38	24	17	13	10

Numerator df	Power	.05	.10	.15	.20	.25	.30	.35	.40	.50	.60	.70	.80
							f						
4	.10	74	19	9	6	4	3	2	2	--	--	--	--
	.50	514	129	58	33	21	15	11	9	6	5	4	3
	.70	776	195	87	49	32	22	17	13	9	6	5	4
	.80	956	240	107	61	39	27	20	16	10	8	6	5
	.90	1231	309	138	78	50	35	26	20	13	10	7	6
	.95	1486	372	166	94	60	42	31	24	16	11	9	7
	.99	2021	506	225	127	82	57	42	33	21	15	11	9

Note. From J. Cohen, *Statistical Power Analysis for the Behavioral Sciences,* New York: Academic, 1969. Copyright © 1969 by Academic Press, Inc. Reprinted by permission.

TABLE X n to detect f by F test at $a = .10$

Numerator df	Power	f											
		.05	.10	.15	.20	.25	.30	.35	.40	.50	.60	.70	.80
1	.50	542	136	61	35	22	16	12	9	6	5	4	3
	.70	942	236	105	60	38	27	20	15	10	7	6	5
	.80	1237	310	138	78	50	35	26	20	13	9	7	6
	.90	1713	429	191	108	69	48	36	27	18	13	10	8
	.95	2165	542	241	136	87	61	45	35	22	16	12	9
	.99	3155	789	351	198	127	88	65	50	32	23	17	13

Numerator df	Power	f											
		.05	.10	.15	.20	.25	.30	.35	.40	.50	.60	.70	.80
2	.50	475	119	53	30	20	14	11	8	6	4	3	3
	.70	797	200	89	50	32	23	17	13	9	6	5	4
	.80	1029	258	115	65	41	29	22	17	11	8	6	5
	.90	1395	349	156	88	57	40	29	23	15	11	8	6
	.95	1738	435	194	109	70	49	36	28	18	13	10	8
	.99	2475	619	276	155	100	70	51	33	21	15	11	9

Numerator df	Power	f											
		.05	.10	.15	.20	.25	.30	.35	.40	.50	.60	.70	.80
3	.50	419	105	47	27	18	12	9	7	5	4	3	3
	.70	690	173	77	43	28	20	15	11	8	6	4	4
	.80	883	221	99	56	36	25	19	15	10	7	5	4
	.90	1180	296	132	74	48	34	25	19	13	9	7	5
	.95	1458	365	163	92	59	41	30	24	15	11	8	7
	.99	2051	513	229	129	83	58	43	33	21	15	11	9

Numerator df	Power	f											
		.05	.10	.15	.20	.25	.30	.35	.40	.50	.60	.70	.80
4	.50	376	95	43	24	16	11	9	7	5	4	3	3
	.70	612	154	68	38	25	18	13	10	7	5	4	3
	.80	773	193	87	49	32	22	17	13	9	6	5	4
	.90	1031	258	115	65	42	29	22	17	11	8	6	5
	.95	1267	317	141	80	51	36	27	21	13	10	7	6
	.99	1768	443	197	111	71	50	37	28	19	13	10	8

Note. From J. Cohen, *Statistical Power Analysis for the Behavioral Sciences,* New York: Academic, 1969. Copyright © 1969 by Academic Press, Inc. Reprinted by permission.

TABLE XI Values of *r* at the 5 and the 1 per cent levels of significance

Degrees of Freedom	5%	1%	Degrees of Freedom	5%	1%
1	.997	1.000	24	.388	.496
2	.950	.990	25	.381	.487
3	.878	.959	26	.374	.478
4	.811	.917	27	.367	.470
5	.754	.874	28	.361	.463
6	.707	.834	29	.355	.456
7	.666	.798	30	.349	.449
8	.632	.765	35	.325	.418
9	.602	.735	40	.304	.393
10	.576	.708	45	.288	.372
11	.553	.684	50	.273	.354
12	.532	.661	60	.250	.325
13	.514	.641	70	.232	.302
14	.497	.623	80	.217	.283
15	.482	.606	90	.205	.267
16	.468	.590	100	.195	.254
17	.456	.575	125	.174	.228
18	.444	.561	150	.159	.208
19	.433	.549	200	.138	.181
20	.423	.537	300	.113	.148
21	.413	.526	400	.098	.128
22	.404	.515	500	.088	.115
23	.396	.505	1000	.062	.081

Note. From R. A. Fisher, *Statistical Methods for Research Workers,* 11th Edition–Revised, New York: Hafner Publishing Company, 1950. Reprinted by permission.

The probabilities given are for a two-tailed test of significance ignoring the sign of *r*. To make a one-tailed test, halve the probability values. The degrees of freedom is $N - 2$, where $N =$ number of paired observations.

TABLE XII Table of χ^2

Degrees of Freedom df	$P = .99$	.98	.95	.90	.80	.70	.50	.30	.20	.10	.05	.02	.01
1	.000157	.000628	.00393	.0158	.0642	.148	.455	1.074	1.642	2.706	3.841	5.412	6.635
2	.0201	.0404	.103	.211	.446	.713	1.386	2.408	3.219	4.605	5.991	7.824	9.210
3	.115	.185	.352	.584	1.005	1.424	2.366	3.665	4.642	6.251	7.815	9.837	11.341
4	.297	.429	.711	1.064	1.649	2.195	3.357	4.878	5.989	7.779	9.488	11.668	13.277
5	.554	.752	1.145	1.610	2.343	3.000	4.351	6.064	7.289	9.236	11.070	13.388	15.086
6	.872	1.134	1.635	2.204	3.070	3.828	5.348	7.231	8.558	10.645	12.592	15.033	16.812
7	1.239	1.564	2.167	2.833	3.822	4.671	6.346	8.383	9.803	12.017	14.067	16.622	18.475
8	1.646	2.032	2.733	3.490	4.594	5.527	7.344	9.524	11.030	13.362	15.507	18.168	20.090
9	2.088	2.532	3.325	4.168	5.380	6.393	8.343	10.656	12.242	14.684	16.919	19.679	21.666
10	2.558	3.059	3.940	4.865	6.179	7.267	9.342	11.781	13.442	15.987	18.307	21.161	23.209
11	3.053	3.609	4.575	5.578	6.989	8.148	10.341	12.899	14.631	17.275	19.675	22.618	24.725
12	3.571	4.178	5.226	6.304	7.807	9.034	11.340	14.011	15.812	18.549	21.026	24.054	26.217
13	4.107	4.765	5.892	7.042	8.634	9.926	12.340	15.119	16.985	19.812	22.362	25.472	27.688
14	4.660	5.368	6.571	7.790	9.467	10.821	13.339	16.222	18.151	21.064	23.685	26.873	29.141
15	5.229	5.985	7.261	8.547	10.307	11.721	14.339	17.322	19.311	22.307	24.996	28.259	30.578
16	5.812	6.614	7.962	9.312	11.152	12.624	15.338	18.418	20.465	23.542	26.296	29.633	32.000
17	6.408	7.255	8.672	10.085	12.002	13.531	16.338	19.511	21.615	24.769	27.587	30.995	33.409
18	7.015	7.906	9.390	10.865	12.857	14.440	17.338	20.601	22.760	25.989	28.869	32.346	34.805
19	7.633	8.567	10.117	11.651	13.716	15.352	18.338	21.689	23.900	27.204	30.144	33.687	36.191
20	8.260	9.237	10.851	12.443	14.578	16.266	19.337	22.775	25.038	28.412	31.410	35.020	37.566
21	8.897	9.915	11.591	13.240	15.445	17.182	20.337	23.858	26.171	29.615	32.671	36.343	38.932
22	9.542	10.600	12.338	14.041	16.314	18.101	21.337	24.939	27.301	30.813	33.924	37.659	40.289
23	10.196	11.293	13.091	14.848	17.187	19.021	22.337	26.018	28.429	32.007	35.172	38.968	41.638
24	10.856	11.992	13.848	15.659	18.062	19.943	23.337	27.096	29.553	33.196	36.415	40.270	42.980
25	11.524	12.697	14.611	16.473	18.940	20.867	24.337	28.172	30.675	34.382	37.652	41.566	44.314
26	12.198	13.409	15.379	17.292	19.820	21.792	25.336	29.246	31.795	35.563	38.885	42.856	45.642
27	12.879	14.125	16.151	18.114	20.703	22.719	26.336	30.319	32.912	36.741	40.113	44.140	46.963
28	13.565	14.847	16.928	18.939	21.588	23.647	27.336	31.391	34.027	37.916	41.337	45.419	48.278
29	14.256	15.574	17.708	19.768	22.475	24.577	28.336	32.461	35.139	39.087	42.557	46.693	49.588
30	14.953	16.306	18.493	20.599	23.364	25.508	29.336	33.530	36.250	40.256	43.773	47.962	50.892

Note. From R. A. Fisher, *Statistical Methods for Research Workers*, 11th Edition–Revised, New York: Hafner Publishing Company, 1950. Reprinted by permission.

TABLE XIII   Table of random numbers

Row	00000 01234	00000 56789	11111 01234	11111 56789	22222 01234	22222 56789	33333 01234	33333 56789
				Column Number				
				1st Thousand				
00	23157	54859	01837	25993	76249	70886	95230	36744
01	05545	55043	10537	43508	90611	83744	10962	21343
02	14871	60350	32404	36223	50051	00322	11543	80834
03	38976	74951	94051	75853	78805	90194	32428	71695
04	97312	61718	99755	30870	94251	25841	54882	10513
05	11742	69381	44339	30872	32797	33118	22647	06850
06	43361	28859	11016	45623	93009	00499	43640	74036
07	93806	20478	38268	04491	55751	18932	58475	52571
08	49540	13181	08429	84187	69538	29661	77738	09527
09	36768	72633	37948	21569	41959	68670	45274	83880
10	07092	52392	24627	12067	06558	45344	67338	45320
11	43310	01081	44863	80307	52555	16148	89742	94647
12	61570	06360	06173	63775	63148	95123	35017	46993
13	31352	83799	10779	18941	31579	76448	62584	86919
14	57048	86526	27795	93692	90529	56546	35065	32254
15	09243	44200	68721	07137	30729	75756	09298	27650
16	97957	35018	40894	88329	52230	82521	22532	61587
17	93732	59570	43781	98885	56671	66826	95996	44569
18	72621	11225	00922	68264	35666	59434	71687	58167
19	61020	74418	45371	20794	95917	37866	99536	19378
20	97839	85474	33055	91718	45473	54144	22034	23000
21	89160	97192	22232	90637	35055	45489	88438	16361
22	25966	88220	62871	79265	02823	52862	84919	54883
23	81443	31719	05049	54806	74690	07567	65017	16543
24	11322	54931	42362	34386	08624	97687	46245	23245

Note. From "Randomness and Random Numbers" by M. G. Kendall and B. B. Smith, *Journal of the Royal Statistical Society*, 1938, *101*, 164–166. Copyright 1938 by the Royal Statistical Society. Reprinted by permission.

TABLE XIII (*continued*)   Table of random numbers

Row	COLUMN NUMBER							
	00000 01234	00000 56789	11111 01234	11111 56789	22222 01234	22222 56789	33333 01234	33333 56789
				2nd Thousand				
00	64755	83885	84122	25920	17696	15655	95045	95947
01	10302	52289	77436	34430	38112	49067	07348	23328
02	71017	98495	51308	50374	66591	02887	53765	69149
03	60012	55605	88410	34879	79655	90169	78800	03666
04	37330	94656	49161	42802	48274	54755	44553	65090
05	47869	87001	31591	12273	60626	12822	34691	61212
06	38040	42737	64167	89578	39323	49324	88434	38706
07	73508	30908	83054	80078	86669	30295	56460	45336
08	32623	46474	84061	04324	20628	37319	32356	43969
09	97591	99549	36630	35106	62069	92975	95320	57734
10	74012	31955	59790	96982	66224	24015	96749	07589
11	56754	26457	13351	05014	90966	33674	69096	33488
12	49800	49908	54831	21998	08528	26372	92923	65026
13	43584	89647	24878	56670	00221	50193	99591	62377
14	16653	79664	60325	71301	35742	83636	73058	87229
15	48502	69055	65322	58748	31446	80237	31252	96367
16	96765	54692	36316	86230	48296	38352	23816	64094
17	38923	61550	80357	81784	23444	12463	33992	28128
18	77958	81694	25225	05587	51073	01070	60218	61961
19	17928	28065	25586	08771	02641	85064	65796	48170
20	94036	85978	02318	04499	41054	10531	87431	21596
21	47460	60479	56230	48417	14372	85167	27558	00368
22	47856	56088	51992	82439	40644	17170	13463	18288
23	57616	34653	92298	62018	10375	76515	62986	90756
24	08300	92704	66752	66610	57188	79107	54222	22013

TABLE XIII (*continued*) Table of random numbers

Row	00000 01234	00000 56789	11111 01234	11111 56789	22222 01234	22222 56789	33333 01234	33333 56789
				3rd Thousand				
00	89221	02362	65787	74733	51272	30213	92441	39651
01	04005	99818	63918	29032	94012	42363	01261	10650
02	98546	38066	50856	75045	40645	22841	53254	44125
03	41719	84401	59226	01314	54581	40398	49988	65579
04	28733	72489	00785	25843	24613	49797	85567	84471
05	65213	83927	77762	03086	80742	24395	68476	83792
06	65553	12678	90906	90466	43670	26217	69900	31205
07	05668	69080	73029	85746	58332	78231	45986	92998
08	39302	99718	49757	79519	27387	76373	47262	91612
09	64592	32254	45879	29431	38320	05981	18067	87137
10	07513	48792	47314	83660	68907	05336	82579	91582
11	86593	68501	56638	99800	82839	35148	56541	07232
12	83735	22599	97977	81248	36838	99560	32410	67614
13	08595	21826	54655	08204	87990	17033	56258	05384
14	41273	27149	44293	69458	16828	63962	15864	35431
15	00473	75908	56238	12242	72631	76314	47252	06347
16	86131	53789	81383	07868	89132	96182	07009	86432
17	33849	78359	08402	03586	03176	88663	08018	22546
18	61870	41657	07468	08612	98083	97349	20775	45091
19	43898	65923	25078	86129	78491	97653	91500	80786
20	29939	39123	04548	45985	60952	06641	28726	46473
21	38505	85555	14388	55077	18657	94887	67831	70819
22	31824	38431	67125	25511	72044	11562	53279	82268
23	91430	03767	13561	15597	06750	92552	02391	38753
24	38635	68976	25498	97526	96458	03805	04116	63514

TABLE XIII (continued) Table of random numbers

	COLUMN NUMBER							
Row	00000 01234	00000 56789	11111 01234	11111 56789	22222 01234	22222 56789	33333 01234	33333 56789
				4th Thousand				
00	02490	54122	27944	39364	94239	72074	11679	54082
01	11967	36469	60627	83701	09253	30208	01385	37482
02	48256	83465	49699	24079	05403	35154	39613	03136
03	27246	73080	21481	23536	04881	89977	49484	93071
04	32532	77265	72430	70722	86529	18457	92657	10011
05	66757	98955	92375	93431	43204	55825	45443	69265
06	11266	34545	76505	97746	34668	26999	26742	97516
07	17872	39142	45561	80146	93137	48924	64257	59284
08	62561	30365	03408	14754	51798	08133	61010	97730
09	62796	30779	35497	70501	30105	08133	00097	91970
10	75510	21771	04339	33060	42757	62223	87565	48468
11	87439	01691	63517	26590	44437	07217	98706	39032
12	97742	02621	10748	78803	38337	65226	92149	59051
13	98811	06001	21571	02875	21828	83912	85188	61624
14	51264	01852	64607	92553	29004	26695	78583	62998
15	40239	93376	10419	68610	49120	02941	80035	99317
16	26936	59186	51667	27645	46329	44681	94190	66647
17	88502	11716	98299	40974	42394	62200	69094	81646
18	63499	38093	25593	61995	79867	80569	01023	38374
19	36379	81206	03317	78710	73828	31083	60509	44091
20	93801	22322	47479	57017	59334	30647	43061	26660
21	29856	87120	56311	50053	25365	81265	22414	02431
22	97720	87931	88265	13050	71017	15177	06957	92919
23	85237	09105	74601	46377	59938	15647	34177	92753
24	75746	75268	31727	95773	72364	87324	36879	06802

TABLE XIII (*continued*) Table of random numbers

Row	00000 01234	00000 56789	11111 01234	11111 56789	22222 01234	22222 56789	33333 01234	33333 56789
				5th Thousand				
00	29935	06971	63175	52579	10478	89379	61428	21363
01	15114	07126	51890	77787	75510	13103	42942	48111
02	03870	43225	10589	87629	22039	94124	38127	65022
03	79390	39188	40756	45269	65959	20640	14284	22960
04	30035	06915	79196	54428	64819	52314	48721	81594
05	29039	99861	28759	79802	68531	39198	38137	24373
06	78196	08108	24107	49777	09599	43569	84820	94956
07	15847	85493	91442	91351	80130	73752	21539	10986
08	36614	62248	49194	97209	92587	92053	41021	80064
09	40549	54884	91465	43862	35541	44466	88894	74180
10	40878	08997	14286	09982	90308	78007	51587	16658
11	10229	49282	41173	31468	59455	18756	08908	06660
12	15918	76787	30624	25928	44124	25088	31137	71614
13	13403	18796	49909	94404	64979	41462	18155	98335
14	66523	94596	74908	90271	10009	98648	17640	68909
15	91665	36469	68343	17870	25975	04662	21272	50620
16	67415	87515	08207	73729	73201	57593	96917	69699
17	76527	96996	23724	33448	63392	32394	60887	90617
18	19815	47789	74348	17147	10954	34355	81194	54407
19	25592	53587	76384	72575	84347	68918	05739	57222
20	55902	45539	63646	31609	95999	82887	40666	66692
21	02470	58376	79794	22482	42423	96162	47491	17264
22	18630	53263	13319	97619	35859	12350	14632	87659
23	89673	38230	16063	92007	59503	38402	76450	33333
24	62986	67364	06595	17427	84623	14565	82860	57300

Column Number

Answers to Problems

CHAPTER 3

1. (a) $\bar{X} = 14.4$
   (b) $\Sigma x = 0$
   (c) Sum of squares $= \Sigma x^2 = 177.2$
   (d) Variance $= \Sigma x^2 / n = s^2 = 44.3$
   (e) $s^2 = 44.3$

2. $\bar{X} = 48.78$
   $Md = 62$
   $SS = 8{,}569.5552$
   $s^2 = 1{,}071.1944$
   $s = 32.73$
   Negative skewness

3.
Subject	$\bar{X}$	Md
A	350.0	250
B	307.2	290
C	387.8	370
D	327.0	301
E	237.6	203
F	237.6	207

The median for each subject is the average to use because all distributions are positively skewed, and inspection of the raw data shows that one reaction time value for each subject is unusually large.

4.
	Section 1	Section 2	Section 3
$\bar{X}$	79.00	67.04	83.26
SS	3,888.0000	8,175.9149	9,497.4359
s^2	144.0000	177.7373	256.5641
s	12.00	13.33	16.02

$s_{within} = 13.94$

5. $\bar{X} = 57.96$
   $SS = 3,517.9184$
   $s^2 = 73.2900$
   $s = 8.56$

6. $\bar{X} = 6.01$
   $SS = 474.9917$
   $s^2 = 3.9583$
   $s = 1.99$

CHAPTER 5

1.

Raw score	Standard score	Proportion
24	−2.25	.9878
30	−1.50	.9332
34	−1.00	.8413
40	−0.25	.5987
42	0	.5000
46	0.50	.3085
48	0.75	.2266
56	1.75	.0401
66	3.00	.00135

2. (a) .1587
   (b) 6.68%
   (c) .62%
   (d) 58.59%
   (e) 545
   (f) 665

3. (a) 44 grams
   (b) 69 grams
   (c) Approximately 39%
   (d) .0035
   (e) 52–64 grams

4. $\sigma_{\bar{x}} = 8.5/\sqrt{25} = 1.70$
   95% limits: 54.67–61.33
   99% limits: 53.61–62.39
   90% limits: 55.20–60.81

5. $\sigma_{\bar{x}} = 100/\sqrt{334} = 5.47$
   Verbal GRE
      95% limits: 564.28–585.72
      99% limits: 560.89–589.11

Quantitative GRE
   95% limits: 536.28–557.72
   99% limits: 532.89–561.11

CHAPTER 6

1. (a) 8
   (b) 32
   (c) 126
   (d) 502
   (e) 784

2. (a) $s_{\bar{x}} = 7.1/\sqrt{10} = 2.245$
   (b) 95% limits: $38.6 \pm (2.262)(2.245) = 33.52 - 43.68$
       99% limits: $38.6 \pm (3.250)(2.245) = 31.30 - 45.90$
   (c) $n = 17$

3.

n	s	$s_{\bar{x}}$	$t_{.05}(df)$	Lower 95% limit	Upper 95% limit	$t_{.01}(df)$	Lower 99% limit	Upper 99% limit
10	13.2	4.17	2.262	86.27	105.13	3.250	82.15	109.25
15	13.2	3.41	2.145	88.39	103.01	2.977	85.55	105.85
20	13.2	2.95	2.093	89.53	101.87	2.861	87.26	104.14
25	13.2	2.64	2.064	90.25	101.15	2.797	88.32	103.08
50	13.2	1.87	2.009	91.94	99.46	2.680	90.69	100.71
100	13.2	1.32	1.984	93.08	98.32	2.626	92.23	99.17
200	13.2	0.93	1.972	93.87	97.53	2.601	93.28	98.12

4. $s = 49.25$

5. (a) $\bar{X} = [(38.2)(15) + (37.0)(10)]/25 = 37.72$
   (b) $s_{within} = \sqrt{[(5.3)^2(14) + (4.8)^2(9)]/23} = 5.11$
   (c) $df = 23$
   (d) 99% limits: $37.72 \pm (1.02)(2.807) = 34.86 - 40.58$

CHAPTER 7

1.

	Control	Experimental
$\bar{X}$	104.50	72.17
SS	2399.50	2680.83
$s_{\bar{x}_c - \bar{x}_e}$	12.99	
t	$32.33/12.99 = 2.49$	

2. (a) H_0: $\mu_c \geqslant \mu_e$
   (b) $t = 1.89$
   (c) The difference of 1.28 attitude units between the means is significant at the .05 level (one-tailed test), and therefore we conclude that seeing the movie brought about a more favorable attitude toward religion.

3.

	Strain A	*Strain B*
$\bar{X}$	12.50	16.33
SS	112.00	247.67
s^2	10.18	22.52
$s_{\bar{x}_a - \bar{x}_b}$	1.6507	
t	3.83/1.6507 = 2.32	

95% limits: $3.83 \pm (2.074)(1.6507) = 3.83 \pm 3.42 = 0.41 - 7.25$
Interpretation: The mean difference between the stains is significant at the .05 level, with Strain A learning more rapidly than Strain B.

4.

	Control	*Experimental*
ΣX	507	729
ΣX^2	31,697	58,501
n	10	10
$\bar{X}$	50.7	72.9
SS	5992.10	5356.90
s^2	665.79	595.21
s	25.80	24.40
$s_{\bar{x}_c - \bar{x}_e}$	11.23	
t	22.2/11.23 = 1.98	
95%	$22.2 \pm (2.101)(11.23)$	
confidence	$= 22.2 \pm 23.59$	
limits	$= -1.39$ to 45.79	

Because the 95% confidence limits contains zero as one possible value for the population mean difference, there is no evidence that the two groups differ from each other at the .05 level of significance.

5. $s_{\text{within}} = 25.11$
   expected mean difference $= 15$
   $d = 25.11/15 = .60$
   $\alpha = .01$(one-tailed)
   power $= .90$
   n per group $= 74$

6. If $\alpha_1 = .01$, power $= .56$
   If $\alpha_1 = .025$, power $= .70$
   If $\alpha_1 = .05$, power $= .80$

If $\alpha_1 = .10$, power $= .89$

A meaningful experiment can be conducted with $\alpha_1 = .05$ and power $= .80$.

7. (a) The t test for matched pairs $= 14.25/5.17 = 2.76$. With 7 df this difference is significant at the .05 level.

(b) The 95% confidence limits $= 14.25 \pm (2.365)(5.17) = 14.25 \pm 12.23 = 2.02 - 26.48$.

(c) The t test for independent groups $= 1.17$.

CHAPTER 8

1.

Source	df	SS	MS	F
Between	2	3,379.14	1,689.57	4.63*
Within	18	6,573.43	365.19	
Total	20	9,952.57		

*$p < .05$.

2. SS within Treatment 1 $= 4192.86$
SS within Treatment 2 $= 1028.86$
SS within Treatment 3 $= 1351.71$
Total $= 6573.43 = SS_{within}$

3. Critical difference $= (50.56)(3.61) = 182.5$. The difference between the total for Treatment 3 and Treatment 2 exceeds the critical difference and thus is significant at the .05 level.

4. Critical difference $= (71.50)(2.40) = 171.61$. Neither of the differences between the total for the control group and the two treatment groups exceeded the critical difference, and thus we conclude that there are no significant treatment effects relative to the control group.

5. (a)

Source	df	SS	MS	F
Between	4	704.17	176.04	1.65
Within	30	3,208.00	106.93	
Total	34	3,912.17		

(b) (i) $MS = 10.29$; $F < 1$
(ii) $MS = 7.14$; $F < 1$
(iii) $MS = 302.29$; $F = 2.87$
(iv) $MS = 624.86$; $F = 5.84$*
(v) $MS = 116.67$; $F = 1.09$

6. (a & b)

Source	df	SS	MS	F
[Between]	[4]	[4,019.28]		
Linear	1	132.25	132.25	
Quadratic	1	3,351.61	3,351.61	7.32**
Remainder	2	535.42	267.71	
Within	45	20,609.60	457.99	
Total	49	24,628.56		

(c) Her interpretation is that the data support her hypothesis that there is a significant quadratic relationship between amount of information (as manipulated by the number of angles of the geometric figures) and the amount of attention paid to the stimuli as measured by the total time the babies looked at the stimuli.

(b) The overall Between MS is not significant when tested against the Within MS. Thus, her conclusion would have been that there was no relationship between information content of the figures and the amount of time spent looking at the figures.

7. (a)

Source	df	SS	MS	F
[Between]	[6]	[5,461.37]		
Linear	1	5,187.04	5,187.04	119.10**
Quadratic	1	269.65	269.65	6.19*
Remainder	4	4.68	1.17	
Within	21	914.46	43.55	
Total	27	6,375.83		

(b) Best fitting linear equation:

$$X' = 53.2607 + (-6.8054)(k - 4)$$
$$= 80.48 - 6.81k$$

(c) Best fitting second degree equation:

$$X' = 53.2607 + (-6.8054)(k - 4) + (-.8958)[(k - 4)^2 - 4]$$
$$= -.90k^2 + .36k + 69.73$$

(d)

k = level of litter size	$\bar{X}$ = actual means of litter weight	X' = predicted weights based on linear equation*	X' = predicted weights based on quadratic equation**
1	69.30	73.68	69.19
2	67.10	66.87	66.85
3	62.05	60.07	62.71
4	56.98	53.26	56.77
5	49.15	46.46	49.03
6	40.30	39.65	39.49
7	27.95	32.84	28.15

$$*X' = 80.48 - 6.81k$$
$$**X' = -.90k^2 + .36k + 69.73$$

CHAPTER 9

1. (a)

Source	df	SS	MS
Between treatments	2	131.72	65.86
Classes within treatments	9	117.17	13.02
Subjects within classes within treatments	108	702.10	6.50
Total	119	950.99	

(b) $F = 13.02/6.50 = 2.00$. This value of F has probability of less than .25 associated with it. Thus Classes and Subjects cannot be pooled to yield a common error term.

(c) The appropriate F test = $65.86/13.02 = 5.06$ with 2/9 df. This is significant at the .05 level.

(d) Critical difference = $(22.82)(3.95) = 90.14$. The only significant difference is between the Aggressive set and the Caring set treatments ($397 - 298 = 99$, which is greater than the critical difference of 90.14).

(e) Critical difference = $(32.27)(2.61) = 84.23$. Neither of the two experimental instruction sets differs from the control or neutral set.

(f) Different hypotheses are tested in an exploratory experiment than in one comparing a control to several experimental means.

2.

Source	df	SS	MS	F
Between treatments	1	2,714.45	2,714.45	6.12*
Pooled subjects	18	7,978.50	443.25	
Litters within treatment	[8]	[4,109.00]	[513.62]	[1.32]
Subjects within litters within treatments	[10]	[3,869.50]	[386.95]	
Total	19	10,692.95		

3. (a)

Source	df	SS	MS	F
Trials	4	2,000.96	500.24	11.26**
Subjects	4	5,542.96	1,385.74	31.20**
Interaction	16	710.64	44.42	
Total	24	8,254.56		

(b)

Source	df	SS	MS	F
[Trials]	[4]	[2,000.96]		
Linear	1	1,824.08	1,824.08	41.06**
Quadratic	1	54.91	54.91	1.24
Cubic	1	8.82	8.82	<1
Quartic	1	113.15	113.15	2.55
Interaction	16	710.64	44.42	

4.

Source	df	SS	MS	F
Treatment	1	792.10	792.1	6.83*
Strain	9	8,272.90	919.21	7.92**
Treatment X Strain	9	889.90	98.88	<1
Within	20	2,321.00	116.05	
Total	39	11,959.90		

5.	Source	df	SS	MS	F
	Methods	2	12,907.72	6,453.86	5.87**
	Categories	3	6,426.22	2.142.07	1.95
	Methods × Categories	6	16,490.15	2,748.36	2.50*
	Within	108	118,731.90	1.099.37	
	Total	119	154,555.99		

There was a significant effect for Methods. Examination of the Methods means indicates that Method A was overall poorer than the other two methods. However, this generality is limited by the significant interaction which indicates that the effectiveness of each method is dependent on the psychiatric category with which it is associated.

CHAPTER 10

1. $r = .35$. With 14 df, this value of r does not differ from zero.

2. (a) $r = .50$. This is significant at the .05 level.
   (b) $Y' = 6.36 + .2527X$
   (c) $r^2 = .25$
   (d) $s_{y \cdot x} = 5.50 \sqrt{1 - .25} = 4.76$

3. (a) $r = .44$
   (b) $t = 2.58$ with 14 df
   (c) $t = 1.93$ with 28 df

4. $r = .9177$

5. $t = .38$. The judges show no evidence of systematic differences between themselves.

CHAPTER 11

1. $\chi^2 = 9.5757$. With 2 df this is significant beyond the .01 level.

2. If both candidates are equally preferred in the population, then the expected frequency for each in the sample is 87. When these data are evaluated with a chi-square test applying Yates' correction, χ^2 is found to be 3.59. With 1 df this is not significant at the .05 level. If we had set .05 to reject the null hypothesis, then we are not able to conclude that the voters prefer Candidate A over Candidate B.

3. The value of χ^2 corrected for continuity is 3.686. With 1 df this value falls between the .10 and .05 levels. If the researcher was making a prediction based on theory, she could use a one-tailed test. If so, she may conclude that the item is biased using the .05 level to reject H_0. Otherwise, she could not reject the null hypothesis.

4. $\phi = .1677$.

5. $\chi^2 = 51.836$ with $(3)(2) = 6$ df. This is significant beyond the .01 level and thus the researcher would conclude that there is a significant association between the mouse strain and the behavior patterns they exhibit in the exploratory apparatus.

6. $C = \sqrt{51.836/(51.836 + 120)} = .5492$.

7. $p = .0397$.

8. $\chi^2 = 4.5$. With 1 df this is significant at the .05 level.

CHAPTER 12

1. $z = 70/36.97 = 1.89$. This is less than .05 using a one-tailed test, and so the researcher may conclude that the population receiving testosterone is more aggressive than the population receiving saline in infancy.

2. $z = (9 - 22.5)/8.44 = 1.60$. This normal deviate does not reach the .05 level, and thus there is no evidence of a sex difference in susceptability to suggestion.

3. $H = 6.26$. With 2 df this is significant at the .05 level.

4. $\chi_r^2 = 18.84$. With 4 df this is significant beyond the .01 level.

5. $r_s = .34$. With 15 df this is not significantly different from zero.

Glossary

α *level* The same as the level of significance.

Alternative hypothesis The same as the experimental hypothesis.

A posteriori tests The same as *post hoc* comparisons.

Arithmetic series A sequence of numbers in which each succeeding number is increased by a constant amount over the preceding number.

β *error* The same as a Type 2 error.

Bimodal distribution A frequency distribution with two modes or major peaks.

Central limit theorem A theorem which states that, regardless of the distribution of raw scores in the population, the sampling distribution of means is approximately normal when the sample size is large.

Confidence limit An estimated range of values set around the sample mean with a given high probability (usually 95% or 99%) of including the population mean.

Confounding When two or more variables are combined in such a manner that it is not possible to evaluate their separate effects.

Contingency table A table containing R rows of nominal categories representing one classification and C columns of nominal categories representing a second classification. Within the table are listed the frequency of occurrence of the observed event.

Degrees of freedom Number of ways that a statistic can vary and still meet the restrictions placed on it. The degrees of freedom will always be some number less than n, the sample size.

Descriptive statistics The use of numbers (statistics) to describe, characterize, and summarize data obtained from the sample in one's study.

Distribution of differences A hypothetical population frequency distribution of the differences between μ_c and μ_e.

Error variance In the analysis of variance, this refers either to the error term in the sample used to evaluate treatment effects or to the variance in the population.

Expected value The average value of a statistic over an infinite set of samples.

Experimental hypothesis The hypothesis that the researcher wishes to evaluate. This is the alternative hypothesis to the null hypothesis and is generally designated as H_1.

Exploratory experiment An experimental design consisting of several qualitatively different treatment groups that are compared with each other. This is often called "I wonder what would happen if" kind of experiment.

Factor An independent variable in an experimental design.

Factorial experiment An experimental design in which two or more independent variables are systematically combined in all possible treatment combinations.

Fixed effects model An analysis of variance model in which the specific treatments or experimental procedures used are the only ones of relevance to the researcher. Generalizations cannot be made to other treatment conditions.

Fourfold table A contingency table in which both the row and column classifications contain two categories, thus resulting in four cells for the table.

Frequency distribution A systematic way of ordering a set of data from lowest to highest score showing the number of occurrences (frequency) at each score point.

Frequency polygon A graphic way of portraying a frequency distribution by plotting score values on the X axis, number of cases on the Y axis, and connecting the points to form a curve that is brought to the baseline (X axis) at both extremes.

Geometric series A sequence of numbers in which each succeeding number is a constant multiplier of the preceeding number. Taking logarithms of a geometric series converts it to an arithmetic series.

Hierarchical design The same as a nested design.

Histogram A graphic way of portraying a frequency distribution by plotting score values on the X axis and using bars to represent the number of cases on the Y axis.

Homogeneity of variance The assumption that the variances of the various groups in the experiment are homogeneous so that they can be pooled to yield one common value.

Honestly significant difference procedure After an overall significant F ratio has been found in an exploratory experiment, the honestly significant difference procedure is applied to determine which pairs of means differ significantly.

Independence Two or more events are said to be independent if the occurrence of one event in no way influences the probability of occurrence of the same or other events on subsequent trials.

Inferential statistics The logical procedures whereby one makes inferences about a larger group (the population) from information obtained from a sample.

Interaction The unique effect of combining two or more variables at the same time.

Interval scale A measurement scale in which numerical values that can be added and subtracted are assigned to the subject matter.

Level of significance The probability level set by the experimenter to reject the null hypothesis. This is also called the α level. The two most commonly used α levels are .05 and .01.

Main effect The overall effect of an experimental variable when evaluated separately.

Measurement A set of procedures or operations whereby a "score" (which can be quantitative or qualitative) is assigned to each element under study.

Negatively skewed distribution An asymmetrical and unimodal frequency distribution with the tail pointing toward the low score values. In such a distribution the median will be numerically greater than the mean.

Nested design An experimental design in which certain effects are restricted or nested within the treatment conditions.

Nesting When a particular effect is restricted only to one treatment condition.

Nominal scale A measurement scale that classifies the subject matter into two or more mutually exclusive and totally inclusive categories.

Nonparametric statistics Any statistical procedure based on nominal or ordinal measurement data and that does not make any assumptions about the nature of the population distribution.

Normal distribution A special kind of frequency distribution that is unimodal, symmetrical, and is defined by a particular mathematical formula relating the mean and standard deviation of the normal curve to the proportion of cases under the curve. (This curve had also been called a bell-shaped curve.)

Null hypothesis The hypothesis that assumes that there are no differences in the population between the control and experimental groups. This is also called the statistical hypothesis and is generally designated as H_0.

One-tailed hypothesis The situation in which a significant difference in only one direction is considered to be of importance to the researcher.

One-tailed probability values The proportion of cases remaining in the right-hand tail of a distribution beyond the limit set by a standard error score. This probability value must be doubled to give the equivalent two-tailed probability value. The normal curve table in the Appendix lists one-tailed probability values (Table 1).

Ordinal scale A measurement scale in which the subject matter is ranked in order on some dimension or attribute of interest to the researcher.

Orthogonal comparisons Statistical comparisons that are independent of each other.

Orthogonal polynomials A method for determining the linear, quadratic, cubic, etc., trends of a polynomial equation for the situation of equal n per point and equal spacing of the independent variable.

Parameter A numerical value that characterizes a population figure (e.g., the population mean μ).

Parametric statistics Any statistical procedure that assumes the numerical data have the property of additivity (i.e., are on an interval scale) and that also makes one or more assumptions about the nature of the population to which inferences are to be drawn (e.g., that the population is normally distributed).

Planned comparisons An experimental design in which specific hypotheses, which have been designated in advance, are tested.

Population A hypothetical unit of persons, organisms, or elements, quite large in size (often infinite) which we, as researchers, wish to characterize and draw inferences about.

Positively skewed distribution An asymmetrical and unimodal frequency distribution with the tail pointing toward the high score values. In such a distribution the median will be numerically less than the mean.

Post hoc *comparisons* Any statistical procedure used to determine which means differ from each other after an overall F test has been found to be significant. Also called postmortem tests and *a posteriori* tests.

Post mortem tests The same as *post hoc* comparisons.

Power The probability that a false null hypothesis will be rejected. This is defined as $1 - \beta$.

Precision The numerical value of the standard error of the mean indicates the precision with which the mean is estimated. The smaller the standard error, the more precise is our estimate of the population mean.

Principle of least squares A mathematical principle which states that a unique solution results when the sum of the squares of the deviations from the mean is at a minimum value.

Proportion A number ranging from 0 to 1.00.

Random effects model An analysis of variance model in which the treatments or experimental procedures used have been selected by a random procedure from a population of treatment conditions. With this model, generalizations can be made to the treatment population.

Random sample A sample obtained by following a procedure in which all elements of a population have an equal chance of being chosen to be in the sample.

Ratio scale A measurement scale that contains all the properties of an interval scale and also has an absolute zero.

Regression equations The best-fitting straight-line equations obtained from a correlational study to predict Y from a knowledge of X, or to predict X from a knowledge of Y.

Regression experiment An experimental design in which the independent variable is quantitative and is systematically varied by the researcher.

Robust statistic A statistic whose distribution and probability values remain much the same even though the assumptions underlying its derivation have been violated.

Sampling distribution A distribution of a statistic such as the mean or the variance.

Sampling error The difference in numerical value of a statistic from one sample to another.

Scatterplot A graphic way of depicting the distribution of pairs of numbers by plotting them on X-Y coordinates.

Single-classification analysis of variance An experimental design in which the various treatments are classified along a single dimension.

Standard error of a difference The standard deviation of the sampling distribution of differences between μ_c and μ_e.

Standard error of the mean The standard deviation of the sampling distribution of means.

Standard score A score based on standard deviation units. It is obtained by subtracting the mean of the distribution from a raw score and then dividing the resulting value by the standard deviation of the distribution. This is also called a z score.

Statistic A numerical value obtained from a sample (e.g., the sample mean $\bar{X}$).

Statistical hypothesis The same as the null hypothesis.

Tests for trend A statistical test of a regression experiment to determine if the linear, quadratic, cubic, etc., trends are significant.

Two-tailed hypothesis   The situation in which a significant difference in either direction is considered to be of importance to the researcher.

Two-tailed probability values   The proportion of cases remaining in both tails of the distribution beyond the limits set by a standard error score. The *t* table in the Appendix lists two-tailed probability values (Table II).

Type 1 error   Rejecting the null hypothesis when it is true. This is also called the α error.

Type 2 error   Not rejecting the null hypothesis when it is false. This is also called the β error.

Unbiased estimator   A statistic obtained from a sample whose value, in the long run, will be the same as the population parameter being estimated.

z score   Another name for a standard score.

References

Cochran, W. G., & Cox, G. *Experimental designs* (2nd ed.). New York: Wiley, 1957.

Cohen, J. *Statistical power analysis for the behavioral sciences.* New York: Academic, 1969.

Gaito, J. Unequal intervals and unequal *n* in trend analysis. *Psychological Bulletin*, 1965, *63*, 125–127.

Hays, W. L. *Statistics for psychologists.* New York: Holt, 1963.

Siegel, S. *Nonparametric statistics for the behavioral sciences.* New York: McGraw-Hill, 1956.

Snedecor, G. W., & Cochran, W. G. *Statistical methods* (6th ed.). Ames: Iowa State University Press, 1967.

Winer, B. J. *Statistical principles in experimental design* (2nd ed.). New York: McGraw-Hill, 1971.

Index

Absolute zero, 11, 14
Alternative hypothesis, 94, 120
Analysis of variance
   assumptions, 135–137
   factorial experiment, 210–218
   F test, 128–129
   nested designs, 187–201
   null hypothesis, 129–130
   randomized blocks design, 201–210
   single classification design, 134–182
Arbitrary zero, 11
Arithmetic mean, 26–30

Bimodal distribution, 24–25, 37

Central limit theorem, 68–70
Confidence limits, 70
   around a mean difference, 104
   and the normal distribution, 65–66
   and the t distribution, 80–83
Confounding, 190, 218
Contingency coefficient, 277
Contingency table, 262, 277
   $R \times C$, 262–265
   2×2, 265–266
Control group compared against several
   experimental groups, 144–145, 152–
   153, 159
Correction for continuity, 260–261
Correlated proportions, 274–276
Covariance, 225–226

Degrees of freedom
   for the analysis of variance, 130

for the t test, 74–76, 83
for χ^2 tests, 257–258, 263–265
Descriptive statistics, 2, 4

Error variance, 127, 182
Expected value, 127, 182
Experimental hypothesis, 120
   qualitative, 142–147
   quantitative, 148–150
Exploratory experiment, 142–144, 150–
   152, 159, 182

Factorial experiments, 210–218, 243–245
Fisher exact probability test, 269–274
Fixed effects model, 131–132, 182
Fourfold table, 265, 277
Frequency distributions
   bimodal, 24–25, 37
   negatively skewed, 23–24, 29, 37
   normal, 21–23, 37
   positively skewed, 23–24, 28, 37
Frequency polygon, 19–25, 37
Friedman test, 289–291
F test, 128–129

Goodness-of-fit test, 256–261

Hierarchical design, 188, 218
Histogram, 19–21, 37
Homogeneity of variance, 99, 120, 136
Honestly significant difference procedure,
   150–152, 182

Hypothesis
    alternative, 94, 120
    experimental, 89-94, 120
    null, 94, 120
    one-tailed, 91-93, 120
    statistical, 89-94, 120
    two-tailed, 90-93, 120

Inferential statistics, 2, 4
Interaction in factorial experiments, 212, 218
Interval scale, 8, 10-12, 14

Kruskal-Wallace analysis of variance, 288-289

Least squares, principle of, 33, 37, 230-231, 235
Levels of significance, 94-97, 120

Main effects, 212, 218
Mann-Whitney test, 283-286
Matched pairs
    principles involved in, 108-109
    and standard error of the difference
        between means, 107
    t test on, 105-109, 240-243
Mean, arithmetic, 26-30
Mean difference
    confidence limits around, 104
    t test of, 101-104
Mean square, 33-36
Median, 27-30
Missing values, 180-182

Nested designs, 187-201, 218
    choice of error term in, 189-190
    preliminary test on the model, 191-192
Nominal scale, 8-9, 12, 14
Nonparametric statistics, 12-14
Nonrandom sample, 42-44
Normal curve, 21-23, 37, 53-71
    and probability statements, 58-59
    table, how to use, 56-58
Null hypothesis, 94, 97-101, 120, 129-130

One-tailed hypothesis, 91-93, 120
One-tailed probability value, 78, 83

Ordinal scale, 8, 9, 12, 14
Orthogonal comparisons, 155, 182
Orthogonal polynomials, 162-177, 182

Parameter, 48-49, 52
Parametric statistics, 12-14
Phi coefficient, 276-277
Planned comparisons, 145-147, 153-159, 182
Population, 2, 4, 41
Post hoc test, 151-152, 182
Power of an experiment, 110-120
    how to increase, 113
    and matched pairs, 117-118
    and sample size, 113-118
    with single classification analysis of
        variance, 178-180
    and Type 2 error, 111
Preliminary test in nested designs, 191-192
Principle of least squares, 33, 37
Probability values
    one-tailed, 78, 83
    two-tailed, 78, 83
Proportion, 20, 22, 37

Random effects model, 132-133, 182
Randomization, 42, 44, 85-89
Randomized blocks design, 201-210
Random sample, 41-42, 52
Ratio scale, 8, 11-12, 14
Regression equations, 245-251
Regression experiment, 148-150, 183, 250-251
Relative frequency, 20, 22
Repeating an experiment, 13-14, 43-44

Sample, 2
    nonrandom, 42-44
    random, 41-42, 52
Sampling distribution, 50, 52
Sampling error, 50, 52
Scales of measurement
    interval, 8, 10-12, 14
    nominal, 8-9, 12, 14
    ordinal, 8, 9, 12, 14
    ratio, 8, 11-12, 14
Scatterplot, 226-228, 251
Significance levels, 94-97, 120

Skewed distributions
    positive skew, 23–24, 28, 37
    negative skew, 23–24, 29, 37
Spearman rank correlation, 291–293
Standard deviation, 32–36
Standard error
    of a difference, 99–101, 120
    of difference scores with matched pairs,
        107
    of estimate, 235–237
    and independent groups, 110
    and matched pairs, 109–110
    of a mean, 61–64, 70
    and the *t* distribution, 80–83
Standard score, 54, 70
Statistic, 48–49, 52
Statistical assumptions, validating, 13–14,
    43–44
Statistical hypothesis, 93–94, 120
Statistics
    descriptive, 2, 4
    inferential, 2, 4
    nonparametric, 12–14
    parametric, 12–14
Straight line, best fitting, 171–173
Sum of squares, 33–36

Tables
    chi-square, 319
    Dunnett's *t* statistic, 312–313
    F distribution, 304–307
        special, 308–309
    normal curve, 298–300
    orthogonal polynomials, 314
    power, 302–303, 315, 316, 317

q statistic, 310–311
    random numbers, 320–324
    significance of correlation coefficient,
        318
    t distribution, 301
Testing for association, 261–266
Test for trend, 159, 183
t distribution, 73–84
    confidence limits, 80–83
    standard errors, 80–83
    table, discussion of, 77–79
t test
    assumptions underlying, 104–105
    for difference between means, 101–104
    for matched pairs, 105–109
Two-tailed hypothesis, 90–93, 120
Two-tailed probability values, 78, 83
Type 1 error, 110, 120
Type 2 error, 110, 120

Validating statistical assumptions, 13–14,
    43–44
Variance, 33–36

Wilcoxin test, 286–287

Yates correction for continuity, 260–261

Zero
    absolute, 11, 14
    arbitrary, 11
z score, 57, 71